THE ARROW OF TIME

THE ARROW OF
TIME

*A voyage through science to solve
time's greatest mystery*

PETER COVENEY

AND

ROGER HIGHFIELD

FOREWORD BY
ILYA PRIGOGINE
NOBEL LAUREATE

Fawcett Columbine • New York

A Fawcett Columbine Book
Published by Ballantine Books
Copyright © 1990 by Peter Coveney and Roger Highfield

Originally published in Great Britain in 1990 by W.H. Allen

Library of Congress Cataloging-in-Publication Data
Coveney, Peter (Peter V.)
    The arrow of time : a voyage through science to solve time's
greatest mystery / Peter Coveney and Roger Highfield. — 1st
American ed.
       p.   cm.
    Includes bibliographical references and index.
    ISBN 0-449-90630-2
    1. Time.   I. Highfield, Roger.   II. Title.
QB209.C64   1991   90-85770
529—dc20   CIP

Manufactured in the United States of America

First American Edition: June 1991

10 9 8 7 6 5 4 3 2 1

TO
SAMIA, PAT AND JIM,
AND
JULIA, DORIS AND RONALD

Time glides by with constant movement, not
    unlike a stream.
For neither can a stream stay its course, nor can
    the fleeting hour.

Ovid
Metamorphoses XV, 180

Contents

Colour plates
between pages 160–1

Scanning tunnelling microscope image of atoms (*IBM*)
Overall view and closeup of the hexagonal convection pattern
in a layer of silicone oil 1 millimetre deep (*M. Velarde*)
A strange attractor (*M. Markus and B. Hess*)
Graphical representation of the logistic equation, a simple
model in population dynamics, being put through its
paces (*M. Markus and B. Hess*)

Illustrations

between pages 288–9

Foreword

BY

ILYA PRIGOGINE
WINNER OF THE 1977 NOBEL PRIZE FOR CHEMISTRY

IT is a great pleasure for me to write a foreword for this book by Peter Coveney and Roger Highfield.

Is there an arrow of time? This question has fascinated Western philosophers, scientists and artists since the Presocratics. However, at the end of this century, we may ask this question in a new context. For a physicist, the scientific history of our century can be divided into three parts. First, we had the breakthrough associated with two new conceptual schemes, relativity and quantum mechanics. Secondly, the disclosure of unexpected findings, including the instability of 'elementary' particles, evolutionary cosmology and non-equilibrium structures, which include a variety of phenomena such as chemical clocks and deterministic chaos. The third – and present – period confronts us with the necessity of rethinking physics, taking into account these new developments.

A remarkable point is that all this emphasises the role of time. To be sure, in the nineteenth century the importance of time was already recognised in fields such as biology and the social sciences. But it was widely accepted that the fundamental level of physical description could be expressed in terms of deterministic, time-reversible laws. The arrow of time would then correspond only to a phenomenological level of description. However, this view is difficult to maintain today.

We now know that the arrow of time plays a critical role in the formation of non-equilibrium structures. As has been shown in recent years, the evolution of these structures can be simulated on computers programmed with dynamical laws; this makes clear that self-organising processes cannot be the effect of phenomenological assumptions, and must be inherent in some classes of dynamical systems.

We are today in a position to understand better the message of entropy, a quantity which always increases according to the Second Law of Thermodynamics and therefore gives an arrow of time. Entropy is basically a property of highly unstable dynamical systems. These are dealt with in Chapters Six and Eight of this book. Much remains to be done and numerous problems are still open. It is therefore not surprising

15

that I do not necessarily agree with everything that is stated in this book. But I do agree with the general description which the authors advocate: the arrow of time is an exact property of important classes of dynamical systems.

These questions are so important that I warmly welcome this book which is written on a high scientific level while being accessible to a wide public. Peter Coveney was uniquely qualified to write this text, as he has himself made important contributions to the subject. Roger Highfield has brought in an element of stylistic presentation which makes the book very attractive.

In October 1989, the Nobel Conference of the Gustavus Adolphus College (St Peter, Minnesota) was devoted to a challenging theme: 'The End of Science'. The organisers wrote: 'there is an increasing feeling that . . . science, as a unified, universal, objective endeavour, is over'. They go on to say that 'if science does not speak of extra-historical, universal laws, but is instead social, temporal and local, then there is no way of speaking of something real beyond science, that science merely reflects'. This statement opposes extra-historical laws to temporal knowledge. Indeed, science *is* rediscovering time, and in a sense this marks an end to the classical conception of science; will it mark an end to science proper?

Indeed, as I have already mentioned, the research programme of classical science was focused on a description in terms of deterministic, time-reversible laws. Actually, this programme was never completed as, in addition to laws, we need also events, which introduce an arrow of time into our description of nature. Often the goal of classical science seemed to be near to completion; but always something went wrong. This gives to the history of science an element of dramatic tension. For example, Einstein's goal was to formulate physics as a geometry of nature. But general relativity paved the way to modern cosmology, only to meet the most striking of all events: the birth of the universe.

The law—event duality is at the heart of the conflicts which run through the history of ideas in the Western world, starting with the pre-Socratic speculations and continuing right up to our own time through quantum mechanics and relativity. Laws were associated to a continuous unfolding, to intelligibility, to deterministic predictions and ultimately to the very negation of time. Events imply an element of arbitrariness as they involve discontinuities, probabilities and irreversible evolution. We have to face the fact that we live in a dual universe, whose description involves both laws and events, certitudes and probabilities. Obviously the most decisive events we know are related to the birth of our universe and to the emergence of life.

16

'Will we be able some day to overcome the Second Law of Thermodynamics?' is the question that the civilisation of Asimov's *Last Question* keep asking to a giant computer. The computer answers, 'The data are insufficient.' Billions of years pass by, stars and galaxies die, while the computer, directly connected to spacetime, continues to collect data. Finally there is no information left to be gathered any longer, nothing 'exists' any more; but the computer goes on computing and discovering correlations. Eventually it reaches the answer. There is no longer anyone there to learn, but the computer now knows how to overcome the Second Law. *And there was light*. . . For Asimov, the emergence of life or the birth of the universe is an anti-entropic, anti-natural event.

The new frame of thinking, to which this book is an excellent introduction, leads to a new physics which includes both laws and events, and brings us closer to a better understanding of the universe in which we are embedded.

Acknowledgements

FOUR years ago we first thought of writing *The Arrow of Time*, after a dinner at Keble College, Oxford. The idea emerged during a late-night conversation about an aspect of time which seems blindingly obvious to all but is elusive in science. Since then, Peter Coveney has outlined some of his ideas about time in a review for the journal *Nature*. Our book attempts to deliver this latest thinking on the arrow of time to a general readership, within the broad spectrum of scientific thought.

In addition to discussion of the scientific ideas, the book contains descriptions of some of the personalities involved as well as the historical development of some of the major fields of science. There is no mathematics involved. Technical points aimed at the scientific reader are safely tucked away at the back of the book in the numbered notes, which also contain detailed references to the research literature. Anyone whose interest is kindled will be able to pursue ideas further by means of the bibliography, and readers may also find the glossary of terms useful.

We owe a very deep debt of gratitude to a large number of people who helped us during the writing of this book. Many thanks are due to Max Hastings, editor of the *Daily Telegraph*, for allowing Roger Highfield to take up a Leverhulme Fellowship at Queen Elizabeth House, Oxford University, during Trinity Term (summer), 1989. During that term he received a great deal of support for his research from the programme organiser, Neville Maxwell, and from his 'tutor', Julian Lewis of the Imperial Cancer Research Fund. Thanks are also due to the University of Wales, Bangor, for supporting this project and Peter Coveney's research in general.

Many scientific friends and colleagues have given generously of their own time to provide information and explanations. We are particularly indebted to Peter Landsberg, Alastair Rae, Gerald Whitrow and Art Winfree for providing invaluable assistance, criticism and advice. Thanks to Mario Markus and Benno Hess of the Max Planck Institute for Nutritional Physiology in Dortmund for providing the inspiring graphics that grace the cover of the book. We also wish to express gratitude to Ilya Prigogine for his encouragement, for his comments on the manuscript and for writing the Foreword.

In terms of constructive criticism on draft chapters, we are especially grateful for comments made by Josephine Arendt, Eric Barton, Kenneth

Denbigh, Dieter Flamm, Claude George, Leon Glass, Jeremy Kilburn, Dilip Kondepudi, Charalambos Kyriacou, Cedric Lacey, Mario Markus, Michael Mackey, Robert May, John Mulvey, Jim Murray, Samia Néhmé, Gregoire Nicolis, Denis Noble, Paul Nurse, Stephen Parker, Oliver Penrose, Roger Penrose, Antoine Schlijper and Rick Welch.

We would also like to thank the following for providing valuable information and material for the book: John Barrow, Walter Bodmer, Richard Dawkins, Matthew Freeman, Norman Gilinsky, Bill Hamilton, Stephen Hawking, Benno Hess, Vic Norris, Peter Schuster, John Ross, David Ruelle, Sarah Smith, Isabelle Stengers, Manuel Velarde, Christian Vidal, Thomas Wehr and Christopher Zeeman.

Errors that remain are, of course, our responsibility alone. None of the above necessarily agrees with the opinions we express in this book.

Many friends read and commented on parts of the book and helped to test whether we had made the science clear. They include Julia Brookes, Jon Dagley, Richard Daly, Evan Davis, David Johnson, Charles Parsons, Steve Mangham and Yasmin Carter. Adrian Berry also made some helpful suggestions.

By far our toughest and most constructive general critic was Paul Carter. His penetrating insight and the awkward points he all too frequently raised were priceless. To him we owe a very special debt, particularly for his comprehensive assistance in editing and organising the entire book.

We also wish to thank Gulshan Chunara of the *Daily Telegraph* for her unfailing secretarial help; Alan Gilliland, Glenn Swann, Roy Castle and Debbie Obeid of the *Telegraph*'s graphics team, who produced the illustrations; Susanne McDadd, who did so much to promote the book and who first pushed us to write it; and the staff of W. H. Allen for all their efforts. Most important of all, we relied on Samia and Julia for their love and moral support and, indeed, for putting up with us during what turned out to be a drawn-out and exhausting project.

Peter Coveney, University of Wales, Bangor
Roger Highfield, *Daily Telegraph*

January 1990

Prologue

LUDWIG BOLTZMANN was on a seaside holiday in an Adriatic village. It was meant to be a relaxing break from his studies in Vienna to help him overcome a period of illness and depression. But Boltzmann was agitated.[1]

A professor since his twenties, he had battled for years to understand the sole piece of scientific evidence for one of man's most fundamental assumptions – that the passage of time is irreversible. In this grand quest he had failed. His work on entropy, a measure of change that always increases with time, was brilliant but still inconclusive. The enigma of time's direction remained a flaw at the centre of science. And for Ludwig Boltzmann, time had run out.

In spite of his appearance – a bulky man sporting a formidable beard – he was a soft and vulnerable character. He was overworked and plagued by ill-health. Now 62, he had almost completely lost his sight and he suffered agonising headaches. Wildly fluctuating moods had taken him to the brink of despair and led to a stay in an asylum near Munich. Even the smallest irritation could cause him deep distress – such as his wife's insistence today on delaying his return to Vienna by taking his suit to be cleaned.

Frau Boltzmann took the suit with her as she and her daughter set off for a swim in the Bay of Sistiana. It was then that her husband committed the ultimate irreversible act. He tied a short cord to the crossbars of a window casement and made a noose round his neck. Then, alone in his rented apartment, he killed himself.[2] His daughter Elsa returned to find him hanged.[3]

Boltzmann's suicide is one of the most vivid examples of the way time mocks and defeats those who seek to unravel its mysteries. His loss was deeply felt. George Jaffé, one of his pupils in Leipzig, wrote: 'Boltzmann's death is one of the tragic events in the history of science, like the decapitation of Lavoisier, the commitment of R. J. Mayer to a lunatic asylum and the crushing of Pierre Curie under the wheels of a truck. It is all the more tragic as it happened on the very eve of the final victory of his ideas.'[4]

These ideas concerned the existence of atoms. Some commentators have portrayed Boltzmann as the victim of an intellectual 'thirty-year war' against those who did not accept the theory of atomism. His opposing army included a range of prominent nineteenth-century thinkers among whom were the Frenchmen Pierre Duhem, Auguste Comte and Henri Poincaré, the Germans Wilhelm Ostwald and Georg Helm, and others in the United States and England such as William Rankine and John Stallo. The battles between Boltzmann and his greatest adversary, fellow countryman Ernst Mach, pushed him into intellectual isolation. He once confessed to a colleague that absolutely nobody understood his most supreme theories.[5]

Eventually, Boltzmann's beliefs about atoms and molecules held sway. However, he had hoped to go even further and use them to explain the direction of time, a feature of nature which he thought about more than any other scientist.[6] It was in this daring ambition that Boltzmann was defeated by the manic depression which pushed him to commit suicide. As we shall see, he had succeeded in making a crucial connection between the two ideas. But his great dream was still unrealised at his death.

Boltzmann was not the last person to die in sad circumstances while attempting to express the arrow of time and other features of the world we inhabit in the language of atoms and molecules. As David Goodstein of the California Institute of Technology wrote in the opening lines of his book, *States of Matter*, 'Ludwig Boltzmann, who spent much of his life studying statistical mechanics, died in 1906, by his own hand. Paul Ehrenfest, carrying on the work, died similarly in 1933. Now it is our turn . . . perhaps it will be wise to approach the subject cautiously.'[7]

Images of time

Ruine hath taught me thus to ruminate:
That Time will come and take my love away.
This thought is as a death which cannot choose
But weepe to have that which it feares to loose.

William Shakespeare
Sonnet 64, 11–14

TIME is one of the greatest sources of mystery to mankind. Throughout history, human beings have restlessly puzzled over time's profound yet inscrutable nature. It is a subject which has captivated poets, writers and philosophers of every generation. But not, so it seems, the modern scientist. Contemporary science – in particular, physics – has sought to suppress if not to eliminate the role of time in the order of things. Time has been described as the forgotten dimension.[1]

We are all aware of the irreversible flow of time which seems to dominate our existence, where the past is fixed and the future open. We may yearn to turn back the clock, to undo mistakes or to relive a wonderful moment. But alas, common sense is against us: time and tide wait for no man. Time cannot run backwards.

Or can it? Disturbingly, there is little support for the common-sense view of time in many scientific theories, where time's direction makes little difference. The great edifices of modern science – Newton's mechanics, Einstein's relativity and the quantum mechanics of Heisenberg and Schrödinger – would all appear to work equally well with time running in reverse. For these theories, events recorded on a film would be perfectly plausible no matter which way the film was run through the projector. Uni-directional time, in fact, comes to appear as simply an illusion created in our minds. Frequently scientists who investigate this problem refer to our everyday sense of the flow of time, rather sneeringly, as 'psychological time' or 'subjective time'.

Could it be that somewhere in the universe the direction of time may flow against the time with which we are familiar, in a world where people rise from the grave to lose their wrinkles and eventually return to the womb? It would be a world where perfume mysteriously condenses into bottles; where ripples of water in ponds converge to eject stones; where the air in rooms spontaneously separates into its components;

23

where wrinkled pieces of rubber expand and seal themselves into balloons; where light would shine out of astronomers' eyes to be absorbed by stars. Perhaps the possibilities do not end there. Could it be that if this line of thinking is correct, time might be thrown into reverse here on Earth? Could we all be sucked back into the past?

This contradicts all the evidence we have that time flows in a single direction. For example, compare time with space. Space surrounds us, yet time is experienced bit by bit. The distinction between right and left is trivial compared with that between past and future. We can shuffle around freely in space yet by our actions we can only affect the future, not the past. We have memory, not precognition (clairvoyants apart). Materials generally seem to decay rather than to assemble spontaneously. So it seems that although space has no preferred directional characteristics, time does.[2] It points from the past to the future, like an arrow. The evocative term 'the arrow of time' was first coined by the astrophysicist Arthur Eddington in 1927.[3]

In this book we shall investigate the role of time in present-day scientific theories, weigh the consequences and show how it is indeed possible to achieve a unified vision of time: a vision which is consistent rather than in conflict with time as we directly experience it. The arrow of time may even point towards the need for a deeper and more fundamental theoretical framework to describe nature than any currently in use.

Time in literature

The common-sense view of time finds its most eloquent expression in some of the great works of literature.[4] Uni-directional time gives us the idea of transience, captured in the title of Proust's autobiographical novel *A la recherche du temps perdu*. Uppermost in such authors' minds is the knowledge that we have only a finite – and short – amount of time to live and that there can be no going back. Moments must be snatched as time continues its ineluctable progress, each moment appreciated with poignant intensity. The mystery of life is made all the more wonderful owing to its very ephemerality, while our sense of time's irreversibility is heightened by death. It is no coincidence that the symbolic figure of Father Time shares his attributes of a scythe and an hour-glass with death's skeletal Grim Reaper, who will mow us all down when our time is done.

The flow of time is described in literature and poetry again and again. One of the most striking meditations upon it can be found in the writings of the Persian philosopher-poet Omar Khayyám (d. 1123), immortalised by Edward Fitzgerald's free-ranging translation:

> The Moving Finger writes; and, having writ,
> Moves on: nor all your Piety nor Wit
>   Shall lure it back to cancel half a Line,
>   Nor all your Tears wash out a Word of it.[5]

Here irreversibility is revealed as the ultimate source of the pathos of human life. Unspoken, but implicit, is the final triumph of death. And here we have a link with science, for the fact that every living creature dies is the most tangible evidence for the flux of time. It is a crucial issue if we are to make sense of the world about us. In the words of Arthur Eddington: 'In any attempt to bridge the domains of experience belonging to the spiritual and physical sides of our nature, time occupies the key position.'[6]

Cultural time

The idea of directional time has not always been with us. The tides, solstices, seasons and the cyclic movements of the heavenly bodies led many primitive societies to regard time in terms of organic rhythms, as essentially cyclic in nature. They thought that since time was inseparable from the circular movement of the heavens, time itself was circular. Day follows night, new moon follows old, summer follows winter, so why not history? The Maya of Central America believed that history would repeat itself every 260 years, a period of time called the *lamat*, or fundamental element, of their calendar. They also believed in cyclic catastrophes: when a group of invading Spaniards landed in 1698 members of one tribe, the Itza, fled because they believed the cycle had turned full circle and calamity had come.[7] They were right, but not by prediction or even coincidence: the Spanish knew what to expect because their missionaries had learnt of the Mayas' belief in cyclic time eighty years earlier.

The cyclic pattern of time was a common feature in Greek cosmological thought. Aristotle observed in his *Physics* that 'there is a circle in all other things that have a natural movement and coming into being and passing away. This is because all other things are discriminated by time and end and begin as though conforming to a cycle; for even time itself is thought to be a circle.'[8] The Stoics believed that when the planets returned to the same relative positions as at the beginning of time the cosmos would be renewed again and again. Nemesius, Bishop of Emesa in the fourth century AD remarked: 'Socrates and Plato and each individual man will live again, with the same friends and fellow citizens. They will go through the same experiences and the same activities. Every city, village and field will be restored, just as it was. And this restoration of the universe takes place not once, but over and over again – indeed to all eternity without end.'[9] It was as though historical events were decked around a great celestial wheel. This notion of eternal return reappeared in modern mathematical form as 'the Poincaré recurrence', named after Henri Poincaré, one of the world's foremost mathematicians, who was active at the turn of the twentieth century.

Time's arrow aroused deep fear – even terror – because it implied instability, flux and change. It also pointed towards the end of the world rather than to rebirth and renewal. In his work on time's arrows and cycles, *The Myth of the Eternal Return*, the Romanian anthropologist and historian of religion Mircea Eliade maintained that most people throughout mankind's existence have clung to the comfort of time's cycle, where the past is the future, there is no real 'history' and mankind is resigned to rebirth and renewal. Significantly, he wrote: 'The life of archaic man . . . although it takes place in time, does not record time's irreversibility; in other words, [it] completely ignores what is especially characteristic and decisive in a consciousness of time.'[10]

It was the Judaeo-Christian tradition which had established 'linear' (irreversible) time once and for all in Western culture. 'Christian thought tended to transcend, once and for all, the old themes of eternal repetition,' wrote Eliade.[11] Through the Christian belief in the birth and death of Christ and the Crucifixion as unique events, unrepeatable, Western civilisation came to regard time as a linear path that stretches between past and future. Before the advent of Christianity only the Hebrews and the Zoroastrian Persians preferred this progressive view of time.[12]

Irreversible time profoundly influenced Western thought. It prepared the human mind for the idea of progress, for the concept of 'deep time', the shocking discovery by geologists that human evolution is only a late and brief episode in the Earth's history. It paved the way for Darwin's theory of evolution, our union through time with more primitive creatures. In short, the emergence of the idea of linear time and the intellectual evolution which it entailed have underpinned modern science and its promise of improvement of life on Earth.

Aspects of time in biology are analogous to both cyclical and linear cultural experience. Cyclical time appears in cell division and the

26

orchestra of different rhythms in our bodies, ranging from high-frequency nerve impulses to leisurely cycles of cell turnover. And the notion of irreversible time is manifested by ageing in the passage from birth to death. Ordinary clocks also express both these facets of time. They compound a succession of pendulum swings or crystal oscillations to reveal 'the time', which on Earth is expressed as a 12- or 24-hour cycle. The flow of time is manifested indirectly by dissipation: the running down of batteries, slackening of the mainspring or the falling of weights.

Time in philosophy

Time has been the subject of repeated speculative investigation by philosophers. The mathematician Gerald Whitrow, in his influential work, *The Natural Philosophy of Time*,[13] highlights how the ideas of Archimedes and Aristotle represent two extreme views of time: Aristotle regarded time as intrinsic and fundamental to the universe: Archimedes did not. Their debate has continued in one form or another through the centuries.

In Plato's cosmological work *Timaeus*, time was born when a divine worksmith imposed form and order on primeval chaos. *Timaeus* begins with the distinction between Being and Becoming, two concepts which reappear in various guises in modern scientific theories. For Plato, the world of Being is a fundamental world 'apprehensible by intelligence with the aid of reasoning, being eternally the same', while that of Becoming (the realm of time) 'is the object of opinion and irrational sensation, coming to be and ceasing to be, but never fully real'.[14] He was making the same distinction as between a journey (becoming) and its destination (being), claiming only the latter was real. This distinction, in which the physical world, including time, has only a secondary reality, dominated Plato's entire philosophy.

In this view Plato was preceded by Parmenides who believed that reality was both indivisible and timeless. His pupil, Zeno of Elea in southern Italy, teased us with his famous paradoxes aimed at undermining our whole concept of time. One of the best known is usually referred to as that of Achilles and the tortoise,[15] claiming to show that motion is impossible if time can be infinitely subdivided. Achilles is pictured chasing a tortoise: during the time it takes Achilles to reach the point from which the tortoise started out, the latter has advanced a (small) distance; in the time Achilles takes to cover that distance, the tortoise has again moved on; and so on *ad infinitum*.[16]

Opinions differ on the significance of this and the other of Zeno's paradoxes. In the 24 centuries since their formulation, they have been either written off as absurd or treated as most profound in the massive literature they have generated. In his careful analysis, Whitrow concludes that there are but two ways in which the paradoxes may be resolved. Either one can seek to deny the notion of 'becoming', in which case time assumes essentially space-like properties; or one must reject the assumption that time, like space, is infinitely divisible into ever smaller portions.[17]

Just as the colour red can induce different subjective impressions on different observers but may nevertheless seem an essential part of what we see, the philosopher Immanuel Kant maintained that while time is an essential part of our experience, it is devoid of objective reality: 'Time is not something objective. It is neither substance nor accident nor relation, but a subjective condition, necessary owing to the nature of the human mind.'[18] Kant's 'subjectivist' viewpoint finds close parallels in the way some scientists attempt to explain time in present-day science. One very simple and obvious way out, and one which has been popular with idealists in all ages – Parmenides, Plato, Spinoza, Hegel, Bradley and M'Taggart – is to say that time is riddled through and through with contradictions, and hence cannot be real. A withering remark on this kind of metaphysical evasion came from the logician M. Cleugh: 'Merely to say that because time is self-contradictory it must be appearance only, is, so far from solving the problems, not even an answer to them.'[19]

Boltzmann dubbed metaphysics a 'migraine of the human mind'.[20] 'The most ordinary things are to philosophy a source of insoluble puzzles,' he remarked. 'With infinite ingenuity it constructs a concept of space or time and then finds it absolutely impossible that there be objects in this space or that processes occur during this time . . . To call this logic seems to me as if somebody for the purpose of a mountain hike were to put on a garment with so many long folds that his feet become constantly entangled in them and he would fall as soon as he took his first steps in the plains. The source of this kind of logic lies in excessive confidence in the so-called laws of thought.'[21] Boltzmann criticised several philosophers virulently, singling out Hegel, Schopenhauer and Kant: 'To go straight to the deepest depth, I went for Hegel; what unclear thoughtless flow of words I was to find there! My unlucky star led me from Hegel to Schopenhauer . . . Even in Kant there were many things that I could grasp so little that given his general acuity of mind I almost suspected that he was pulling the reader's leg or was even an imposter.'[22]

Time: Newton and Einstein

But what of time in science? The invention of the first successful pendulum clock in the middle of the seventeenth century by Christiaan Huygens[23], and the progressive increase in the precision of 'timekeeping' that followed, fostered the image of a mechanical and predictable side to nature. The technological development of clocks disentangled time from human events and helped to create belief in an independent world of science.[24] The 'classical' science that emerged in the seventeenth and eighteenth centuries portrayed a universe in which free will and capricious chance were redundant, a universe that to all intents and purposes was a cosmic machine.

We can trace the birth of a truly scientific time back to Sir Isaac Newton, who discovered mathematical expressions for the movement of bodies. His achievement was breathtaking: mathematical description could describe the motion of objects ranging from apples to moons, fusing celestial and terrestrial mechanics. The dazzling ability of his expressions to describe the movement of the heavens using only a few assumptions, and their aesthetic appeal, rapidly brought about acceptance of his ideas.[25] Thus Newton laid the foundations of modern physics.

Newton was no doubt influenced by the mathematician Isaac Barrow who, on resigning as the Lucasian professor at Cambridge in 1669, saw to it that Newton succeeded him. Barrow had remarked that 'because mathematicians frequently make use of time, they ought to have a distinct idea of the meaning of that word, otherwise they are quacks'.[26] Yet in spite of the grandeur of Newton's scientific achievement, time was only incorporated in his equations as a primitive, undefined quantity. It was, like space, absolute. That is to say, all events could be regarded as having a distinct and definite position in space and occur at a particular moment in time. Everywhere, from the Greenwich Observatory to the tip of a distant spiral galaxy, was connected by the same moment of 'now'. As Newton said in his *Principia*, 'Absolute, true, and mathematical time of itself and from its own nature . . . flows equably without relation to anything external.'[27]

Newton's mechanics promises vast predictive power, allowing one instant to provide all possible information about the past and future of the universe. Take the positions and speeds of all the stars in our universe at any instant and plug these values into a cosmic computer that solves Newton's equations. Frozen in that instant is the past and the future: the computer could calculate the positions and speeds of the stars

29

at all times. But what his equations fail to do is to decide which direction of time constitutes the actual past and future of our universe. Instead they strip time of its sense of direction, leaving no room for its relentless march onward. We could highlight this symmetrical time with a film of planetary motions taken by, say, the *Voyager 2* space probe, which was launched to explore the outer solar system in 1977. Such motions were the first to be reduced to mathematical law by Newton. Yet the film would be consistent with his laws of celestial mechanics whether it was run forwards or backwards. This belief in a deterministic world, where time has no direction and the past and future are preordained, has played a pre-eminent role in the development of physics. Its power is shown in a remarkable statement made by Einstein when he learnt of the death of his lifelong friend and confidant Michelangelo Besso. In a letter written on 21 March 1955 Einstein seized upon this unshakable conviction in the 'timelessness' of the laws of physics to offer some comfort for Besso's family. Death was not so final, he suggested: 'For we convinced physicists, the distinction between past, present and future is only an illusion, however persistent. . .'[28] Perhaps the letter was also designed to comfort Einstein himself, for he added that Besso 'has preceded me briefly in bidding farewell to this strange world'. Einstein died a month later.

Newton's theory of motion is now known to fail when applied to bodies moving with speeds close to that of light, to vast masses, including black holes, when gravitational forces become enormous, and to the smallest of length scales involving atomic and sub-atomic particles. But the two great revolutions of twentieth-century theoretical physics that rule in these regimes – Einstein's relativity and quantum mechanics – are also built on the same directionless notion of time. They too remain unable to bridge the gap between the irreversible time of history and literature, and the symmetrical time of Newton's laws.

That is not to say that they did not throw up many fascinating new ideas about time. Einstein's theories of relativity shattered Newton's common-sense concept of absolute time – that any event in the universe should be considered to take place at a particular point in space and at a given instant in time which is the same everywhere. Instead, Einstein put forward the idea of a four-dimensional existence in spacetime (three dimensions of space plus one of time) rather than the evolution of a three-dimensional existence in time.[29] Our perception of time can be warped by illness or by drugs. But Einstein's theory of relativity shows that it also depends on one's point of view – the faster a clock travels, the slower it ticks. In the wake of relativity, even the possibility of time travel was to achieve a certain level of scientific respectability, through the work of Kurt Gödel, one of the greatest ever logicians.

Nevertheless, Einstein's remarkable relativity theories are silent on the one-sided nature of time. As with Newton, the structure of his equations makes it possible to know the past and the future of any system – say the rotation of a star round a black hole or the evolution of the universe itself – if one has a precise knowledge of it at any instant. But there is still no clue as to which is the past and which the future. Fundamental doubts are also raised by the embarrassing presence in the mathematics of 'singularities', where the description of space, time and matter breaks down. The best known singularity is the so-called Big Bang, the super-dense fireball of creation widely believed to have spawned the universe. At this singularity, where vast energies are condensed into a single point, observable quantities in the theory blow up into infinities and hence become meaningless. As the cosmologist Dennis Sciama remarked: 'General relativity contains within itself the seeds of its own destruction.'[30]

Quantum time

In our hunt for a scientific basis for the direction of time, the quantum theory governing the atomic and molecular world looks more promising. It gives a highly successful (although quite baffling) description of the vagaries of atoms and molecules. It can explain the behaviour of lasers, sub-atomic particles in nuclear reactors, electrons in computers and much more. Perhaps, upon a quantum description of the vast agglomerations of atoms and molecules which make up the world, one could construct a description of the arrow of time so keenly felt by our senses. This idea follows an honourable tradition. Ever since the Golden Age of Greek civilisation, the philosophical legacy of describing the world in terms of its component atoms and molecules – atomistic reductionism – has been paramount in the development of scientific thought.

A glimpse of a quantum arrow of time does emerge from two tantalisingly elusive elements we will encounter in Chapter Four – the strange case of a sub-atomic particle called the long-lived kaon and the mystery which surrounds interpreting the very act of measurement in quantum theory. Nevertheless, the core of quantum theory follows other 'fundamental' theories in making no distinction between the two directions of time. Like Einstein's relativity, quantum theory also has deep intrinsic difficulties – it can explode into unpleasant infinities when put to work on real problems, such as the way light is absorbed and emitted by atoms. Although physicists have learned ingenious tricks to sidestep these problems, one has the feeling that they provide further evidence that something is badly amiss.

Thus quantum mechanics and Einstein's theory of relativity sit uneasily side by side. In the long run, some scientists, such as Roger Penrose of Oxford University, believe that a proper unification of the two would produce a quantum theory of gravity (or some entirely new theory) in which an arrow of time would finally be made explicit. Such a development seems some way off and would quite possibly still be unsatisfactory. For there is a serious danger in basing our scientific world-view too firmly on the time-reversible, microscopic level of atoms, molecules, particles and fields – a world which is never directly observed. As the Nobel laureate Ilya Prigogine remarked: 'In spite of the fact that it was told to me by the greatest physicists, I could never understand how, out of reversibility, you could ever find the evolutionary patterns of our universe, culture and life.'[31]

Time and thermodynamics

A second kind of description, at what scientists call the macroscopic level, deals with phenomena on a scale that we can see, taste, feel and touch. It is at this level that the discipline known as thermodynamics applies. Developed by Black, Carnot, Clausius, Boltzmann, Gibbs and others with the advent of steam power during the nineteenth century, it was originally concerned with the performance of heat engines. Within a formal theoretical framework it set out the relationship between heat and work, spelling out how heat can be converted into or exchanged with other forms of energy.

Classical mechanics, relativity and quantum mechanics confounded our idea of time's flow, but thermodynamics rides to the rescue. Just as we are aware of a direction to time, so is the Second Law of Thermodynamics, the law which says that heat can only flow from a hotter body to a cooler one, that snowmen melt and that statues crumble. The link between the Second Law and our sense of time could be illustrated with a film of a bull in a china shop. We would expect, if time were running in the right direction, that the film would show fine bone china being sent flying and crockery trampled under hoof. But if the bull strolled backwards into a wrecked china shop and emerged with every last teacup neatly stacked, we would know the film was running in the wrong direction. The Second Law thwarts

32

such events and manufacturers of perpetual motion machines by showing that in any process energy is wasted as heat: in this case, the energy of cracking china is turned into heat and sound which can never be recovered. This irreversible loss is connected with our sense of the passage of time: through the Second Law, we discover that a quantity called entropy (a measure of the capacity to change) is intimately linked with time. Increasing entropy is a signpost indicating the direction of time. The impeccable credentials of the Second Law were emphasised by Arthur Eddington in the warning: '. . . if your theory is shown to be against the Second Law of Thermodynamics, I can give you no hope: there is nothing for it but to collapse in deepest humiliation.'[32]

In Newton's mechanics, as we have seen, any one moment in the past, present or future is like any other. In this sense, mechanics is 'timeless' and evolution has only a trivial meaning. But in thermodynamics, moments are distinguished by entropy in a universe that is truly evolving.

Just as ancient man regarded irreversible time with fear, and some philosophers have dismissed it as an illusion, so many scientists have attempted to bury the implications of the Second Law. They reject it as something that is more to do with the way we interpret time in our minds than with the objective passage of time. So great has been the apparent success of theories that are reversible in time that these scientists have sought to describe the Second Law's arrow of time as nothing more than an illusion. But we shall show that this view, if true, would render almost everything, including the rhythms and processes of life itself, the result of our own limitations and approximations; for the arrow of time is a fundamental part of the mathematical apparatus explaining living processes.

One popular approach to the problem focuses on how mechanics is used in practice. After all, in order to apply a law of physics one needs not only the law itself but also to feed numbers into it – the initial (or 'boundary') conditions, such as the positions and speeds of all the particles in the universe. Many physicists contend that the thermodynamic arrow of time is somehow associated with these initial conditions alone, without being inherent in the laws of physics themselves.

According to this line of thought, the explanation of time's arrow lies in considering the grandmother of all initial conditions, the birth of the universe. It rests on the claim that the universe started off small and highly compact, in a highly organised state with a very low entropy. Then, the argument goes, the passage of time inevitably corresponds to increasing entropy or disorder as expansion proceeds and the energy of the universe is dissipated into a disperse soup of waste heat, a process known as the Heat Death of the universe.[33] However, the idea runs into difficulties if at some future point the universe begins to contract.[34] One might conclude that at that instant entropy starts to decrease and time then goes backwards. We will see later on that such arguments involving an appeal to boundary conditions are still at best subjective and at worst irrelevant. It is clear that the irreversibility of time cannot be explained merely by initial conditions.

Others, like the mathematical physicist Stephen Hawking, wish to explain the initial conditions as a result of the cosmological theory itself. They stretch speculative time-symmetric theories of the universe to the limits and then maintaining that the boundary condition is that there is no boundary condition. There would be no lawless singularities and no 'edges' of time and space – the universe would be self-contained like the surface of a sphere.

Against all these kinds of cosmological arguments, Peter Landsberg of Southampton University wrote: 'It seems an odd procedure to "explain" everyday occurrences, such as the diffusion of milk into coffee, by means of theories of the universe which are themselves less firmly established than the phenomena to be explained. Most people believe in explaining one set of things in terms of others about which they are more certain and the explanation of normal irreversible phenomena in terms of cosmological expansion is not in this category.'[35]

Rather than pursuing this sterile approach, a more rewarding route is to return directly to the Second Law. For supporters of reversible ('unreal') time, the basic argument against thermodynamics is that its approach to the world is too superficial – unlike relativity and Newton's mechanics, which can deal with the 'fundamental' and invisible microscopic world. The way to meet this argument in the case of, for example, the melting of ice in a tumbler of gin and tonic, is to reconcile the Second Law with the dynamical properties of molecules of alcohol, quinine and other ingredients as well as the large-scale quantities used to describe the gin and tonic such as entropy, volume and temperature. Here Boltzmann's contribution stands out, although his attempt to rediscover the arrow of time in terms of atomic and molecular behaviour did not convince many of his contemporaries. He shocked physicists of the time by making a link between entropy and probability, thus becoming the first person to give a fundamental law of physics a statistical interpretation.[36] His pioneering work was not in vain. It can be used to estimate, for example, the probability of water molecules staying together at room temperature to make an ice cube, a state of low entropy, compared with the much better odds of them making a little puddle of water, a more random state of high entropy. The precise mathematical relationship expressing this, called Boltzmann's principle by Einstein, is now part of the accepted toolkit of physical scientists. It adorns Boltzmann's tomb in the *Zentralfriedhof* in Vienna.[37] And although it does not answer our problem, it points to the direction in which we should go.

At first sight, the Second Law seems to be contradicted by another discovery which shook the nineteenth-century world – Darwin's theory of evolution. While classical mechanics portrayed the universe as a perfect machine, thermodynamics appears to imply that the machine is running down to complete disorganisation. Darwin's work, on the other hand, shows that life has become more – not less – organised through time, as simple creatures evolved into more complex ones. The evolution of the wealth of living things that fly, swim and walk on Earth might seem irreconcilable with a theory preaching inexorable decline. In fact there is no contradiction. For there is something magical hidden within the Second Law of Thermodynamics which enables creative (rather than purely destructive) evolution to occur. Boltzmann may have glimpsed this as long ago as 1878,[38] but the development of this idea had to wait until a more recent reassessment of the Second Law showed that it does not imply a monotonous decline into disorder: instead, the universe can harness thermodynamics to create, evolve and unfold. This confers a new degree of sophistication – and even greater credibility – on the Second Law's arrow of time.

Creative time

By means of a twentieth-century brand of thermodynamics pioneered and developed largely by the group led by Ilya Prigogine at the Free University of Brussels, one can grasp how order can emerge from disorder in terms of a new scientific paradigm, that of 'self-organisation'. This argues, contrary to received wisdom, that the Second Law is *not* synonymous with an inexorable collapse into disorder. While this might be the final state of matter and points towards a corrupted and decayed universe at the end of time, the Second Law most certainly does not claim that this trend occurs uniformly throughout space and time.

First we need to make a distinction between equilibrium thermo-dynamics, where all potential for change is spent, and non-equilibrium thermodynamics. The former applies to a cup of coffee that has reached room temperature and can cool no further, the latter to the moment when you first add milk to the coffee and it still has the potential to mix and cool. Self-organisation arises naturally from the Second Law if the law is applied to the real world where things evolve and change rather than fester in the dead state of thermodynamic equilibrium. The contrast between equilibrium and non-equilibrium thermodynamics is as stark as that between being and becoming or the words of this sentence and the full stop that ends it. During the putative run-down of the universe to heat death we can find spectacular examples of how order can emerge spontaneously. On adding milk to coffee, the end state is the usual mud-coloured liquid. But *en route*, swirling patterns and structures of white milk in black coffee enjoy a fleeting existence.

One laboratory example of self-organisation is found in a 'chemical clock', a particular kind of chemical reaction which changes colour at regular intervals and which can also display beautiful spiralling structures. To sustain these patterns such chemical reactions must be constantly replenished. They also have special ingredients: an interlocking series of chemical reactions – involving feedback loops – in which the products formed also participate in the same chemical reaction or even catalyse their own manufacture. Astonishingly, countless trillions of molecules in the chemical clock all seem to know exactly what each other is doing – they can 'communicate' with one another.

These ideas have far-reaching implications in biology, where the end-state of change – equilibrium – is death. Thermodynamics provides a natural language through which biological processes can be described, and these are processes that must be maintained away from equilibrium if change is to occur. We exist by grace of a complex web of intricately synchronised rhythms. The biochemical reactions that turn within these rhythms are generically similar to the chemical clock. Participating in them are the very threads of life, the long chain-like genetic molecules DNA and RNA, which indirectly catalyse their own production.[39] Thus non-equilibrium thermodynamics bridges the divide between classical thermodynamics and Darwin's theory of evolution. It shows in general terms how creatures like man, containing structures of exquisite sophistication, might yet emerge in a universe where entropy is increasing.

Even more fascinating are the implications for time. Self-organisation – in particular, the example of the chemical clock – shows how the Second Law of Thermodynamics not only furnishes an arrow of time but also has within it the seeds of the temporal cycles and patterns which we discern in the world around us. Both aspects of time are important. Time's arrow represents progress: each instant is branded with an individual marque. But the metaphor of time's cycle is vitally important in seeking patterns within natural phenomena which are ruled by the same laws – in the same way as beat and rhythm distinguish music from mere noise. The Second Law provides a foundation for both of our most important images of time.

But if the Second Law's arrow of time is too sophisticated and well grounded to be dismissed as an illusion, our main problem still remains. How can this irreversible time be reconciled with the microscopic world described by 'timeless' mechanics? Boltzmann got only part of the way to finding an answer to this riddle. In Chapter Eight we will see that the answer may lie in the burgeoning field of dynamical chaos, the sister subject of self-organisation.

Chaos and the arrow

In the context of irreversible processes, chaos does not mean pure mayhem but, rather, a bizarre form of order. It turns out that the equations which describe the behaviour of the chemical clock provide a luxuriant spectrum of possibilities, not only for describing self-organisation but also for 'deterministic chaos', a paradoxical predictable randomness. In the chemical clock, chaos is seen as a random sequence of colour changes. It is called deterministic because chaologists have unravelled this behaviour to reveal a subtle form of underlying organisation. Chaos is believed to underlie the weather, where forecasts may work in the short term but are worthless over longer intervals of time. Now scientists worldwide are rushing to find chaos in the waxing and waning of gypsy moth populations, in epilepsy and in a host of other phenomena, from politics to economics.

Chaos also exists inside Newton's time-symmetric equations of motion, a quite surprising discovery with profound implications. Physicist Joseph Ford, a self-proclaimed 'evangelist of chaos', was quoted as saying: 'We are in the beginning of a major revolution. The whole way we see nature will be changed.'[40] Research has shown that chaos can emerge in the simplest of situations – even if only three particles interact with each other. This demolishes the centuries-old myth of predictability and time-symmetric determinism,[41] and with it any idea of a clockwork universe. For dynamical chaos is the rule, not the exception, within our world. While the past is fixed, the future remains open and we rediscover the arrow of time.

Chaos takes us from a deterministic description of the world based on

37

predictable behaviour to one that is couched in terms of probabilities. Using this description we can find a quantity that behaves like entropy in that it increases with time. In this way the divide between time-symmetric laws and the Second Law of Thermodynamics can be bridged to yield the elusive arrow.

We are at last beginning to grasp how the notion of an open future fits within the most basic of the sciences as surely as it does in the more complex. For chaos seems to be an integral part of how the arrow of time fits within Newton's mechanics and quantum theory, allowing for the very possibility of creative evolution. This idea is beginning to lay the basis for the solution to a problem which has beset science since the time of Boltzmann.

The rise of Newtonian physics: time loses its direction

Science! true daughter of Old Time thou art!

Edgar Allan Poe
'To Science'
from *Al Aaraaf, Tamerlane and Minor Poems*, 1829.

ISAAC NEWTON was born in a small manor house in Woolsthorpe, Lincolnshire, on Christmas Day 1642, so premature that he was given little chance of survival; his mother Hannah said that he could have fitted into a quart pot.[1] After he died, at the age of 84,[2] Newton's life was celebrated with pomp, pageantry, poems and statues. 'He was buried like a king who had done well by his subjects,' wrote Voltaire.[3]

Newton made sense of motion, from the orbits of the planets round the Sun to the paths of arrows on their way to a target. While the Great Plague raged around him, he developed the first major mathematical description of the universe involving time, by fusing the astronomical ideas of Copernicus and Kepler with Galileo's new theories of motion. Many have called 1666 his *annus mirabilis* because of the giant strides he made then in mathematics, optics and celestial dynamics. But it is clear that 1666 was no more miraculous than 1665. In an account he gave fifty years later, Newton describes his remarkable list of achievements and concludes: 'All this was in the two plague years of 1665–1666. For in those days I was in the prime of my age for invention & minded Mathematicks and Philosophy more than at any time since.'[4]

Without doubt, the contributions of Newton to mathematics and physics were of unequalled importance, opening the way to a completely new approach to the analysis of the physical world. He revealed Nature's unity through laws that ruled the heavens and the Earth. His life's achievement is neatly summarised by the greatest of the Augustan poets, Alexander Pope, who proposed the celebrated epitaph for Newton's tomb in Westminster Abbey:

Nature, and Nature's Laws lay hid in Night.
God said, Let Newton be! and All was Light.[5]

The presentation of Part One of Newton's *Principia Mathematica* to the Royal Society on 28 April 1686 was the turning point in the science of physics. Some rate the *Principia* as the greatest scientific book of all time, a jewel in the crown of scientific literature.[6] Others have compared it to a great edifice soaring above ramshackle and temporary constructions around it. Newton regarded it as his greatest achievement in print. At the start of Part Three he proudly declared: 'I now demonstrate the frame of the System of the World.'[7]

Time was to appear in this first scientific description of motion just as surely as the earliest clocks harnessed motion to convert the passing of time into an easily measurable quantity involving the traversing of space, like the swing of a pendulum. Within his *Principia*, Newton put forward three laws of motion that rewrote the science of moving bodies.[8] He showed that bodies on Earth and in the heavens were governed by one and the same force – gravity – even though planets are maintained in an orbit while objects like apples fall to the ground. By uniting terrestrial and celestial mechanics, he also solved a problem that had obsessed Man since his most primitive origins – the movement of planets in space – and which provided the key to timekeeping and navigation.

To use the heavens as an accurate clock or calendar needed not only information on the motion of the Sun and the stars but a means to weave that data into a theory. Newton derived mathematical expressions for the movement of the heavenly bodies with a precision never before obtained. Ironically, although this fascination with the heavens had its roots in theology, Newton's ideas became the anvil on which Man could sever this link. It was the very regularities of the motions of heavenly bodies, their orbits and repetitive cycles, which eventually led people to shift the emphasis of understanding away from magic, witchcraft and gods to scientific and mathematical principles. As Bertrand Russell wrote: 'Almost everything that distinguishes the modern world from earlier centuries is attributable to science, which achieved its most spectacular triumphs in the seventeenth century.'[9]

The way that the concept of time emerged from the pursuit of astronomy highlights the evolution of scientific ideas. The ancients recognised the importance of marking the passage of time in order to anticipate when there might be floods, the start of winter or the first day of spring. Just as we use the hands of a clock sweeping through the 24-hour cycle to measure the passage of time, so they relied on the rising and setting of the Sun, the monthly movement of the Moon or the nightly turning of the sky as it carried celestial bodies from one horizon

40

to the other. It is therefore not surprising that the rudiments of astronomy exist in almost all cultures.

Life is organised around the vital energy that constantly streams down onto the Earth from the Sun. The celestial timepiece has ticks and rhythms that are so fundamental to life on Earth that they are reflected in biological rhythms that beat in almost every living creature (to which we shall return in Chapter Seven and the Appendix). Energy is pumped into each region of the Earth from the time the Sun rises in the east to when it sets in the west. This is an apparent motion caused by the rotation of the Earth on its axis, and is one that can be harnessed for timekeeping – Newton was said to be able to tell the time of day by where shadows fell.[10] A second apparent motion of the Sun provides us with a seasonal unit of time – the year – which is nearly 365.25 days. It is because the period of the Earth's orbit around the Sun does not add up to a whole number of days that an extra day is added every fourth year, or leap year, to prevent an error accumulating. Seasonal changes of the Sun's position in the sky and the seasons themselves arise from the Earth's orientation in space, its axis of rotation being tilted with respect to the plane of its elliptical orbit. Tribes in Borneo used the Sun to watch the passing seasons by observing changes in its height. On Salisbury Plain stands Stonehenge, a megalithic temple used to measure the seasons by the alignment of the Sun with its stone pillars. Other celestial timepieces were used. Most important of all in ancient times was the lunar cycle, which lasts around 29.5 days between each time the crescent of a new Moon appears in the western evening sky – a motion that roughly approximates to our month. The constellations are also linked to the seasonal cycle, with some groups of stars dominating in the summer and others in the winter.

Astronomy originated earlier in human history than the other natural sciences, in prehistoric times from which no records survive. In the earliest civilisations, the divine or occult was used to explain the movement of the stars and Sun: the Aztecs, for example, believed that the Sun had to be nourished with a sacrifice of blood and a beating human heart, or else it would vanish. In those dark days there was no conflict between science and religion: a priest, magician or shaman would jealously guard scientific knowledge of the seasons and calendar. Knowledge, regarded as a sign of divine work in the world, conferred immense status on priests within their community, because they were able to foretell the future with some success. Astronomy meant power over the people. It guided Man through the seasons, showing when to plough, to harvest or to move herds. Religious and sacrificial acts also

41

had to be performed on specific occasions, for example to coincide with the phases of the Moon or solar solstices. Astronomy also helped to guide the traveller. No wonder the three men who followed the star to Bethlehem in the Bible were said to be wise.

With the rise of the calendar, human activities could be linked more accurately with the seasons and thus better coordinated. Three of the periods used in calendars – days, months and years – are based on astronomical cycles that have the greatest bearing on human life. The very earliest calendars relied on the Moon, which not only rose and set but changed its shape over a period of around a month that is convenient for describing the seasons. Later calendars evolved to take account of the Sun's annual cycles. The ancient Egyptians are credited with one of the most advanced solar calendars of ancient times. Astronomy was their key to predicting the great annual event, the inundation of the Nile which coincided with the pre-dawn appearance on the eastern horizon of Sothis (Sirius), the brightest star in the sky. To the ancient Egyptians the event was so important that they dubbed the rising of Sirius the 'Opener of the Year', and the calendar was arranged around this event.

Many ancient calendars had 12 months based on the average period of around 30 days between new moons, making a year that fell short of the solar year and had to be extended: the original Egyptian calendar had 12 months and, being linked to the Moon's 30-day cycle, produced a 360-day year. Later it had an additional five days added at the end of the year to keep the lunar month and the solar-based seasons coordinated and thus the calendar in step with the Opener of the Year.[11]

We can thank the Egyptians for the 24-hour day, running from sunrise to sunrise. A 12-hour cycle fitted the movement of the stars across the sky, with an hour marking out the time between each star, or group of stars, rising on the eastern horizon. In this way, the 12-hour night evolved and, probably for the sake of symmetry, a 12-hour day. The Egyptians measured the passing daylight hours with water clocks, where the flow of time was measured by water trickling through an orifice in a stone vessel, and with sundials and shadow clocks, where a sweeping shadow showed the passing hours.[12] The last two furnished merely 'temporal hours'; they were not equal and varied according to the seasons.[13] In Japan, temporal hours were still in use in the nineteenth century and mechanical clocks adapted accordingly, but in Europe, the day was carved into 24 equal hours in the fourteenth century, when towns introduced mechanical clocks.

Our own calendar is derived from that employed by the Romans. They used a lunar month, squeezing in a month every so often to make up a solar year. By the time Julius Caesar ruled, this process had fallen into such disarray that the winter months fell in the autumn. It had been abused by pontiffs, state administrators who determined the intercalary month for political reasons to prolong a term of office or to bring an election forward. By 47 BC the calendar was out of step with the solar year by three months. The following year, under the guidance of the Greek astronomer Sosigenes, Caesar not only made the usual intercalation of 23 days but inserted two additional months, amounting to a total of 445 days. It became known as the 'year of confusion'.[14] From that time, each of the twelve months has had its present duration.

Unfortunately, the Julian calculation of 365.25 days to a year was too long by 11 minutes and 14 seconds. As the centuries advanced, the shifting of the calendar dates of the seasons meant that the equinox which occurred on 25 March in the time of Caesar fell on 11 March in 1582. That year, Pope Gregory XIII introduced a new and more accurate calendar, directing that the day after 4 October be designated 15 October. Protestants, however, were reluctant to fall into line with a Catholic innovation. In England it was not until 1752 that the Gregorian calendar replaced the Julian. Then a belated 11-day correction was necessary, provoking riots in the streets of London and Bristol, where several people died.[15] Workers demanded to be paid for the days they had lost; many thought they had lost part of their lives. The change also had implications for Newton's birth, which the modern Gregorian calendar now puts at 4 January 1643. However, it was not until 1924 that Orthodox Churches moved to the Gregorian calendar, though some fairs and local festivities still adhere to the Julian. Muslims have a lunar calendar, which explains why their holy month of Ramadan moves forward appreciably each Gregorian year.

The week does not have a basis in the motions of the heavens. As Michael Young, Director of the Institute of Community Studies in London, put it: 'The Sun has not been the only master. Humans can create their own cycles without having to rely on the ready-made ones. No other creature has demonstrated so much independence from astronomy. No other creature has the week.'[16] The week probably arose from the practical need for societies to have a time unit smaller than a month but longer than a day.[17] Communities run more smoothly when there are regular opportunities for laundry, worship and holidays. Ancient Colombia used to have a three-day week.[18] The ancient Greeks favoured ten-day weeks while some primitive tribes today prefer a week of only four days. The seven-day week derived from the Babylonians who in turn

THE ARROW OF TIME

influenced the Jews (though the former ended it with an 'evil day' rather than a Sabbath, when taboos were enforced to appease the gods – perhaps the origin of the restrictions on Sunday activities[19]). Its popularity has defeated a number of attempts at change. The French tried to decimalise it after the Revolution but their ten-day week was scrapped by Napoleon.[20] In 1929 the Soviets attempted to introduce a five-day week and in 1932 extended it to six, but by 1940 the seven-day week had returned.

Just as the week ignores astronomy, so does the modern technology of timekeeping. As modern societies have developed, time has been counted in smaller and smaller subdivisions. In fact, the division of hours into minutes and seconds, like the week, owes its origins to the earliest scientific astronomers, the Babylonians, who carried out all their reckonings in multiples of 60 when they compiled their star catalogues around 1800 BC. But it is only since the Industrial Revolution, with its train timetables and other detailed working schedules, that the minute has grown to its present importance. These trends to subdivide time have continued with the rise of science, because of the need to cope with extremely rapid processes. For example, the pulse of a laser can be used to capture events like the movement of atoms within a decaying molecule which occur over timescales of a few femto-seconds, a millionth of a billionth of a second.

The Greeks, ancient astronomy and science

So far as Newton's work is concerned, the most important astronomical legacy came from the ancient Greeks. Not only did they collect information on the world about them, as had the very earliest of civilisations, but they also attempted to rationalise the workings of the universe without having to resort to gods, magic or superstition.

For his ideas to be used to explain the heavens, Newton not only needed his mechanics but a realistic model of the heavens. Thales of Miletus near Ephesus (circa 625–547 BC) is often called the first philosopher. He believed that there was only one basic substance, water, and he also thought that the Earth floated on water in a universe that was spherical, a symmetrical model that was to appear again and again. Further influential views followed in the sixth century BC with Pythagoras (born around 560 BC), who became the leader of an intriguing group of classical hippies, the Pythagorean Order. They avoided alcohol, preferred not to wear animal products like wool, and were vegetarians. They believed that the soul could leave the body.

44

They also had a great faith in mathematics, possessing a love of symmetry and believing that 'numbers were the ultimate essence of reality'.[21] Based on their knowledge of the mathematics of sound and the orbital times of the planets, they also believed there was a musical aspect to the heavens, the so-called music of the spheres. This way of looking at things had a profound effect on the development of astronomy. It has even been said that Newton was a seventeenth-century Pythagorean, 'his life being devoted to the study of universal harmony'.[22]

The symmetry of circular motion, manifested for example in the guise of a spinning wheel, was central to ancient models of the heavens. Just as the circle has influenced the development of tools for half a million years, so it has influenced theoretical models, which are tools of thought.[23] Being the most perfect of curves, it was the most aesthetically appealing model of how the planets revolved round the Earth. From this love of circular symmetry, it followed for the ancients that the heavens and Earth must be spherical.

During the Golden Age of Greek thought, around 400 BC, Plato's beliefs were pre-eminent. Like the Pythagoreans, Plato (428–347 BC) attached great significance to mathematics. Unfortunately, his theory of knowledge rested on the notion that the world we observe does not resemble the real world. He believed a good model of the universe should not so much describe what we observe as display divine perfection. His model owed much to the Pythagoreans (although not to Philolaos, one of Pythagoras's disciples, who believed that the Earth was in orbit like other planets). To Plato, observations of the real world and experiments were irrelevant in the search for knowledge: true reality could only be contemplated with the mind. In his great cosmological work, *Timaeus*, Plato's universe emerges as an ordered place, the Earth being at the centre, with all other heavenly bodies moving in spheres of varying radii. There can be no doubt that, in spite of his mathematical bent, or perhaps because of it, Plato's dislike of the experimental method severely inhibited the development of science.

The birth of Aristotle in 384 BC heralded a new age in Greek science. A pupil of Plato, he too saw the Earth lying at the centre of the universe. The other planets turned on spheres while all the stars bedecked the outermost one. Aristotle valued observations more highly than his predecessors, laying the foundations on which modern scientific work could be done. Thus he advocated mutual interplay between observations of the way the world works and theories explaining its behaviour.

45

But, with the benefit of hindsight, we can see that Aristotle set back the development of science even more than Plato. This father of Western philosophy preferred a teleological chain of explanation to a causal one. *Teleology* is concerned with the search for purpose in the phenomena of the world. Whereas we would explain the existence of the humpback whale by invoking a causal argument – Darwin's theory of evolution – a teleological argument would ascribe it to the action of a beneficent creator (God) for the benefit of Mankind. Perhaps the most famous example of teleology is the 'argument from design' put forward by many theologians as evidence for God's existence – notably by William Paley in his *Natural Theology* of 1802. In his treatise, Paley passionately argued that the machinery of life is so intricate that it must have had a designer – God.

Today, this line of argument is regarded as the scientific apotheosis of putting the cart before the horse. Paley's methodology is beautifully described and pulled to pieces by a modern advocate of the Darwinian theory of evolution, Richard Dawkins, in his book *The Blind Watchmaker*. Intriguingly, however, the argument from design has regained credibility among some present-day astronomers and cosmologists in the guise of an extreme version of what is called the anthropic principle, which we shall meet in Chapter Three. In short, it argues that we see the universe as we do because we would not be here to observe it if it were different.

Aristarchos of Samos (thought to have lived between 310 and 230 BC) was the first to propound a celestial model in tune with modern understanding, a so-called heliocentric model. He maintained that the Earth moved in a circle around a fixed Sun rather than being located at the centre of the universe. Yet his ideas, lacking empirical support, were rejected under the influence of Aristotelianism and lay dormant for almost two thousand years. Thanks to the support of Thomas Aquinas (1225–1274), who worked with manuscripts preserved by the Arabs, Aristotle's model came to be cherished by the Catholic Church because it made Man the centre of the universe. The decline of the notion of divine intervention and the concomitant rise of scientific ideas about the universe occurred during the Scientific Revolution, which began in the fifteenth century and carried on until the end of the sixteenth. As Galileo, who died the year Newton was born, said: 'The Bible shows the way to go to Heaven, not the way the heavens go.'[24]

46

The Scientific Revolution

The Scientific Revolution represented the simultaneous flowering of several lines of thought which can be traced back to the emergence in ancient Greece of mathematics as an independent discipline.[25] Several early celestial models, like that of Aristarchos, contained the germ of the 'modern' view that the Earth orbits the Sun. Yet owing to the influence of the Aristotelian perspective, no significant intellectual figure took this seriously until the Polish canon Copernicus (1473–1543). He lived during the Renaissance period when a proper mathematical notion of perspective emerged, giving an extra dimension to art. In a similar way, Copernicus explained the complicated paths of the planets by changing his perspective of the universe, imagining it from the viewpoint of the Sun rather than that of the Earth.

With Copernicus, the Sun was brought to a standstill and the Earth launched into the heavens. He placed the Sun at the centre of the planetary system, as he described in his book *De Revolutionibus Orbium Cælestium*, published in 1543. A century later, for all acceptable models, the Sun had replaced the Earth as the centre of planetary motions.[26] Sir Richard Blackmore, who died in 1729, wrote:

> Copernicus, who rightly did condemn
> This eldest system, form'd a wiser scheme;
> In which he leaves the sun at rest, and rolls
> The orb terrestrial on its proper poles.[27]

For a long time, however, the Copernican model of the universe was not generally accepted. The prevailing view was based on a literal interpretation of the scriptures in which Mankind occupied centre-stage. Even before the publication of *De Revolutionibus*, which contained the new world system, Luther had complained in 1539 that 'This fool wishes to reverse the entire science of astronomy; but sacred Scripture tells us (Joshua 10:13) that Joshua commanded the Sun to stand still, and not the Earth.'[28] Strangely, there was no outraged rejection of Copernicanism in Roman Catholic countries until Giordano Bruno (1548–1600) actively promulgated it, as an enthusiastic apostle of the new doctrine.[29] Then the ideas of Copernicus became muddled up with Bruno's advocacy of magic rather than of science. Bruno was imprisoned by the Inquisition and eventually burned at the stake and Copernican theory then became a target of the Roman Catholic Church.

The next great and independent contributions that led eventually to a mathematical framework incorporating time fell to Kepler (1571–

1630) and Galileo (1564–1642). Voltaire wrote: 'Before Kepler, all men were blind. Kepler had one eye, Newton had two.'[30] Johannes Kepler settled in Prague a year before the death of his mentor, Tycho Brahe (1546–1601), the greatest of the astronomers who made do without telescopes.[31] On his deathbed, Brahe asked Kepler to construct a table of planetary motions intended as support for Brahe's own theory (as opposed to that of Copernicus) wherein the Earth was still rooted firmly at the centre of everything. The Moon and Sun were in orbit around it, although he conceded that the planets could orbit the Sun.

Instead, Kepler's work showed that neither Brahe nor Copernicus was correct. In his book *The New Astronomy* (1609), Kepler broke with the tradition which had originated with the Greeks by showing unequivocally that not only did Mars orbit the Sun, but that its speed varied and it had an elliptical orbit.[32] The departure from circular planetary motion – with the implied loss of godly perfection – represented another nail in the coffin of Aristotelian pseudo-science. Kepler enunciated a set of three laws of planetary motion: in his first he described the elliptical shape of an orbit; the second accounted for the varying speed of a planet in its orbit; and the third linked the size of an orbit with its orbital period. Yet his view of the way the world worked was still saddled with theological and mystical baggage. Thus in his *Harmonics of the World* (1619) he related the greatest and least speeds of the planets to musical harmony, the 'music of the spheres' dating back to the Pythagoreans.[33]

Into our story steps Galileo, the creator of the modern scientific method. Galileo Galilei, the son of a composer and musicologist, was born in Italy in 1564, the same year as William Shakespeare. It is impossible to overestimate Galileo's contribution to the modern science of dynamics. Crucially, however, he formulated the concept of *acceleration* which underlies Newton's second law of motion.[34] The notion of *velocity* as a quantity measuring the change of position of an object with time was already commonplace and widely understood in general terms. Acceleration builds in more time: the acceleration of a body in motion requires us to find the change in the velocity – in direction or speed – within a fixed time interval. Galileo maintained that a moving body would only experience a change of velocity, and thus accelerate, if acted upon by some force. Anyone who has tried to walk down the aisle of a moving train while balancing a full cup of coffee has had first-hand experience of how changes in force cause changes in velocity.

Galileo put the Copernican revolution on a firm experimental foundation through his pioneering scientific use of the telescope. With the telescope, whose inventor is still a matter of scholarly debate[35], he could focus on the imperfections of the real universe rather than rely on the ethical preconceptions of the ancient Greeks. In 1609 Galileo first learnt of the work by Hans Lippershey in Holland on the perspicillum, as it was then called.[36] Within a few days Galileo had made his own instrument. The next year, in *Sidereus nuncius* (Starry Messenger), he became the first to publish the results of astronomical observations with a telescope: 'Oh! When will there be an end put to the new observations and discoveries of this admirable instrument?'[37] He had seen clear evidence from the revolutions of Jupiter's moons, the phases of Venus and the rotation of the Sun, that Copernicus was substantially right. Already a Copernican, he turned each discovery into ammunition for the heliocentric model: for instance, he used the observation of Jupiter being ringed by moons as a kind of metaphor for the solar system.

But his work brought him directly into conflict with the Church. In 1952 Einstein wrote in a foreword to Galileo's *Dialogue Concerning the Two Chief World Systems* that Galileo possessed 'the passionate will, the intelligence, and the courage to stand up as the representative of rational thinking' in the face of priestly superstition.[38] The *leitmotif* of Galileo's work, he said, was 'the passionate fight against any kind of dogma based on authority'.

Galileo was hauled before the Inquisition for spreading ideas which contradicted the teachings of the Catholic Church; in 1633 he was threatened with torture twice, though he recanted before he came to any harm. When summoned, Galileo was told he was 'vehemently suspected of heresy' but that he would be absolved provided that 'with a sincere heart' he 'abjure, curse and detest the aforesaid errors and heresies' and be imprisoned.[39] Under house arrest in Arcetri near Florence, the 69-year-old Galileo continued to work on his ideas on mechanics. He also played another role in the scientific formulation of time: towards the end of his life he applied himself to the idea of timekeeping and the use of the pendulum to regulate clockwork. His experiments were later put into practice by the Dutch scientist Christiaan Huygens in 1656, heralding a new era of precision timekeeping.[40]

By that time Galileo's book on the 'world systems' had been hailed all over Europe, while Protestantism had established itself in several countries in opposition to the Catholic Church. In these European states, governments were basically secular: no matter how much Protestant clerics raged against Copernicus, they lacked any authority to suppress his ideas.

Newton and mechanics

The ancient Greeks understood static entities such as geometric shapes but they lacked a clear idea of motion; of how, for example, an arrow changes its position from one moment to the next during flight. It is no wonder, for to study the way things move in nature one needs a good clock and they lacked this essential technology. Upon the foundations that Galileo had laid, through showing that a pendulum was a reliable timekeeper, 'Newton was able to erect a cathedral of superb grandeur,' as Roger Penrose put it.[41]

Newton's *Principia Mathematica* was written between 1684 and 1687. Some see Newton as the great synthesiser, uniting the astronomical revolution of Copernicus and Kepler with the theories of motion put forward by Galileo and Descartes. Others believe there are enough novel features to argue that Newton's ideas were truly revolutionary. In spite of their revolutionary nature, Newton claimed that they grew from the writings of the ancients and 'principally of Thales'.[42] Indeed, he believed that his inverse square law of *gravitation* lay locked inside the Pythagorean music of the spheres, an idea which, as we have already seen, inspired Kepler. It has been argued that by referring to these classical ideas, Newton was trying to lend greater respectability to his own.[43]

His *Principia* consists of an introduction and three books (or parts). The introduction contains the laws of motion.[44] They read as follows:

1. Every body continues in its state of rest or uniform motion in a straight line unless it is compelled to change that state by forces impressed on it.[45]
2. The change of motion is proportional to the motive force impressed; and is made in the direction of the right line in which that force is impressed.
3. To every action there is always opposed an equal reaction: or, the mutual actions of two bodies upon each other are always equal, and directed to opposite parts.

From the mathematical framework he erected in his first two books, he derived in his third the motions of the planets, the comets, the Moon and the sea. Within the Newtonian framework, Kepler's earlier laws concerning planetary orbits emerged naturally as a consequence of supposing that the acceleration of a planet is inversely proportional to the square of the distance separating it from the Sun.

A planet moving in an orbit around the Sun must be subjected to a force to keep it on its path, for although it travels at a constant speed, the direction of motion is constantly changing. Newton regarded the 'force' between planet and Sun as the cause of the change in velocity: specifically, he stated that acceleration is proportional to the applied force. This led him to formulate his celebrated law of gravitation which says that the force between two bodies like the Sun and the Moon is directly proportional to the product of their respective masses and inversely proportional to the square of the distance separating them.[46] In other words, if one mass is doubled, so is the force, but if the distance is doubled, the force is cut to one quarter of its former value.

Voltaire said that 'Sir Isaac Newton walking in his garden had the first thought of his System of Gravitation, upon seeing an apple falling down from the Tree.'[47] According to *The Newton Handbook*, it does seem that Newton at least used the idea to illustrate gravity, although it would be an exaggeration to say that he was hit by the apple in question or that it explains the origins of his ideas on universal attraction. One thing that can be said for certain is that the apple tree in question was at Woolsthorpe and that it produced a variety of cooking apple known as the Flower of Kent, one that is flavourless, pear-shaped and red with streaks of yellow and green. Though the tree was blown over in 1820, it lives on thanks to scions grafted to trees belonging to Lord Brownlow at Belton.

The need to produce some firm predictions from his laws on the motion of planets and apples – not to mention mice sliding down poles and other dynamical phenomena – led Newton to invent the branch of mathematics known as calculus which to this day remains a cornerstone of advanced mathematics and theoretical physics. It was to be the subject of one of the most notorious and bitter quarrels in the history of science, for Newton and the German philosopher-mathematician, Gottfried Wilhelm Leibniz (1646–1716), each claimed to have invented it. The heat of the argument was no doubt fuelled by the fact that Leibniz had for many years made it plain that he found Newton's science unsound.[48] In 1711 Leibniz complained to the Royal Society of London; two years later the Society gave an 'impartial' adjudication in Newton's favour. Many years later it emerged that Newton himself had, as President of the Society, been an author of the adjudication report, the *Commercium epistolicum*. His biography by Richard Westfall, *Never at Rest*, suggests that this is typical of Newton's character: ruthless, self-seeking, arrogant and deceitful.[49] 'Isaac Newton was not a pleasant man,' admitted the cosmologist Stephen Hawking, who is the present incumbent of the famous Lucasian chair at Cambridge University which Newton once occupied.[50]

Rather than enter into the mathematics of the calculus itself, which provides an elegant shorthand for describing much of the world, we will try to portray in words how time appears in Newton's equations. As we have seen, the motion of objects ranging from planets to fleas concerns the change in position of the body with time. Newton regarded time as absolute, a quantity which he did not set out to describe so much as to use. First there was a problem for Newton to overcome in seeking to describe accurately the motion of a body. If a coach and horses travels 200 feet in 100 seconds, then its average speed is 2 feet per second. But suppose he asked how fast it was moving after 50 seconds had elapsed. At that instant in time – which is of zero duration – the coach clearly does not move at all. Asking for the instantaneous velocity of the coach seems to be the same as asking how far it moves during a time interval so short that there is no motion.

Newton overcame this by studying the way an object moves within a series of short intervals. This would give a mathematician, or a highwayman, the detailed performance of the movement of a coach and horses, how fast it took bends, where it slowed down and so on, so that he could estimate when it would pass a certain point. With a clock, our highwayman could measure the velocity of the coach over successive periods of a few seconds. His measurements could in principle be pushed to the limit, by recording the coach's velocity over a series of time intervals of ever-decreasing duration. The instantaneous velocity emerges by extrapolating to the limit in which the time interval shrinks to zero.[51] In this infinitesimal interval of time, the change in the position of the coach will also become infinitesimal. The instantaneous velocity of the coach is still a finite quantity: it is given by dividing the tiny change in its position by the equally tiny interval of time. In calculus, this quantity is known as a first derivative. The idea of using such a limiting procedure to tease out a precise description of the speed of a stage coach at any one instant underlies the whole theory of calculus: a similar limit can be used to produce a value for the coach's instantaneous acceleration.

Newton applied this new mathematics to astronomy. He recognised that gravity curves the paths of the planets around the Sun into ellipses or ovals. Thus, a planet in motion around the Sun is constantly falling. Its orbit arises because it is being subjected to a gravitational force which changes its velocity at every instant in time. To calculate the planet's orbit properly by calculus, we must know the precise acceleration at every instant, not a crude average value. For this reason, it is the instantaneous acceleration which arises in the mathematical formulation of Newton's laws of motion, which were then put to work to give an ever more detailed description of the workings of the solar system. Thus, in the words of the Nobel laureate Stephen Weinberg: 'Newton broke down the barrier between the celestial and terrestrial styles in physics, and in so doing he not only demystified the heavens, he opened up the possibility of an understanding which would embrace the heavens and the earth, all in one synthesis.'[52]

Newton's description of motion transformed Mankind's perception of the structure of the universe, and thus of time, beyond recognition. Bertrand Russell went so far as to maintain that owing to the rise of Newtonian science, 'In 1700 the mental outlook of educated men was completely modern; in 1600, except among a very few, it was still largely medieval.'[53] The role of a deity in the order of things was made remote: God, if he existed – and Pierre Simon de Laplace (1749–1827), among others, questioned the necessity even for that hypothesis – may have set everything in motion at the beginning of time but, on the face of it, no further divine intervention was required.[54] Indeed, it was by no means clear that there had been a beginning of time. The motion of heavenly bodies was now predictable. Though Mankind's place in the cosmos became obscure and unimportant, people began nevertheless to admire the power of human intellect and ingenuity. Inevitably, this led to a theological reassessment of God's relationship with Man. Church dogmatism was almost eclipsed.

Newton, however, was aware of the limitations of what he had done. He wrote: 'I seem to have been only like a boy playing on the sea shore, and diverting myself in now and then finding a smoother pebble or a prettier shell than ordinary, whilst the great ocean of truth lay all undiscovered before me.'[55] (This particular analogy was probably second-hand; Newton had never been to the sea or even walked on a beach.) Indeed, Newton incorporated God in his view of nature because of his strong religious convictions.[56] He insisted that the solar system would require attention from God from time to time if it were not to suffer instability. For Newton, God was the creator and maintainer of the universe, leading to the accusation by Leibniz that Newton's God resembled a second-rate watchmaker, incapable of making a clock which did not need his regular attention each time it broke down.[57]

In checking the predictions of his new theory, Newton drew on observations of the Moon's motion made by the first Astronomer Royal at Greenwich, John Flamsteed (1646–1719), another scientist with whom Newton later quarrelled over intellectual property rights.[58] The Greenwich Observatory had an important role to play, outside the realm of pure research, in making celestial observations which enabled sailors and mariners to find longitude at sea (latitude being found by measuring the angle of elevation a fixed star makes with the horizon), but even with the work of Flamsteed and his successors, it was 1767 before this was achieved. It was at about this time that John Harrison, a carpenter from Barrow on Humber, developed the first chronometer which could provide the 'home time' at sea, but it was another century, in 1884, before it was decided that this home time should be that at Greenwich; hence Greenwich Mean Time was established to standardise timekeeping throughout the world.[59] By comparing GMT with the local time, the longitude could be obtained. For instance, if sunrise occurred at sea four hours later than it did in Greenwich, then a captain knew he had travelled one-sixth of a rotation round the Earth.

The anatomy of Newton's equations

Newton's equations of motion link the acceleration of a body directly to the applied force, and that produces a curious result for time, which appears twice in the instantaneous value of the acceleration: acceleration is the rate of change of the velocity with time, itself the rate of change of position with time. Thus acceleration is the rate of change in time of the rate of change of position in time. In calculus, such a quantity is known as a second derivative, here of position with respect to time. This has a very important consequence in Newton's equations of motion, where time is raised to the power of two, or squared. If we replace time running forwards ('positive time') by time running backwards ('negative time') the equations are unaltered, for the product of two negative numbers, like that of two positive numbers, is always positive: negative time squared equals positive time squared. Thus Newtonian mechanics cannot distinguish between the two different directions of time. By themselves, Newton's equations have nothing to say about whether we are growing older or younger, the most important aspect of the human condition.

For every solution of Newton's equations describing a ball bouncing off a bat, or the way Mercury orbits the Sun, an equally admissible one may be obtained by simply reversing the direction of time. It is tantamount to imagining time running backwards.

One is immediately struck by the very special nature of this *time-symmetry*: there are only a few phenomena of which we have direct experience about which the same could be said. Consider the case of two planets revolving around the Sun shown in Figure 1.

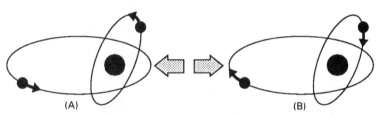

Figure 1. The reversibility of Newton's mechanics. There is no distinction between time running forwards and backwards. Consider two planets orbiting the Sun as in A. If we film them and run the film in reverse, we obtain B. But how can we alone distinguish which is really going forward in time?
[Adapted from P. V. Coveney. *La Recherche* <u>20</u>, 190 (1989).]

Both (a) and (b), which are distinct states of motion, are equally permissible solutions of Newton's equations. Thus, a ciné film of the situation (a) shown backwards would produce state (b); but whichever way the film is run, there is no way of telling which represents the true direction of time and which shows time running backwards. We can say that Newton's equations of motion describe a perfectly reversible world.

But is this the world as we know it? Consider, for example, the filming of a bull in a china shop (*see* Figure 2). The comical time reverse – in which the broken china is miraculously reassembled and the bull runs backwards out of the shop – is never seen in the natural world. It is self-evident which of the two processes represents the correct direction in which time is passing.

There are countless other examples: one never witnesses a cup of tea heating up spontaneously – it always cools down. We only observe the seasons themselves in the same order, spring, summer, autumn, winter and never, for example, autumn immediately before summer. Probably our most direct contact with one-way processes is ageing. There has never been a single recorded example of a dead organism coming back to life, growing younger and finally ending up being 'born' in reverse. One-way processes like life are said to be irreversible.

Yet Newton's laws clearly predict that such impossible time-reversed processes could all occur. Bulls can stroll into wrecked china shops to reassemble and stack the china. Are we then to conclude that Newton's theory of motion is false, since we have found so many processes which apparently conflict with its predictions?

A popular way to avoid this conclusion is to resort to an argument based on the initial conditions required to carry out the time reverse of the bull in a china shop. Reassembling of broken china by a bull is in

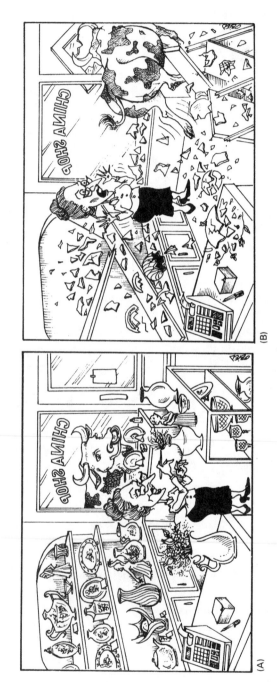

Figure 2. In the real world one never expects to encounter perfectly reversible systems. Everyday processes are irreversible, the archetypal case being a bull entering a china shop (A before, B after his entry). One *never* sees the time-reverse process with B occurring before A. [Adapted from P. V. Coveney. *La Recherche* 20, 190 (1989).]

principle possible, but it is highly unlikely. There are many ways to smash a plate but only one way to put the pieces back together again. So, the argument goes, the arrow of time emerges because the initial conditions required to start off Newton's equations to reassemble crushed crockery are so improbable that we do not see reformist bulls wreak construction on demolished china shops.

The argument is flawed. The arrow of time, it is said, is not an intrinsic feature of nature but arises owing to the special orderly initial state of the china shop. It is analogous to saying that a ball will always roll one way − downwards − if placed on top of a hill. However, Newton's time-symmetric equations still mock this argument: without presupposing a direction of time, a set of initial conditions can still be the starting point for events going backwards as well as forwards in time. Similarly, the selection of the special starting condition does not say why a ball at the bottom of a valley does not spontaneously roll up to the brow of a hill.

Perhaps the arrow emerges from dissipation, a key property distinguishing reversible from irreversible processes, which concerns the redistribution of energy in or loss of energy from the system under study. When crockery collides, 'friction' or other forces operate which can dissipate energy by transferring it between the pieces of china and their surroundings. We know that the relative movement of the china is slowed by friction, not speeded up, as the energy of movement is transferred into heat. This 'damping' effect of friction can be included in Newton's equations simply by accepting its existence and adding an extra term to represent (but not explain) it. This term breaks the time-symmetry of the original equations, leading to irreversible behaviour.

The fundamental difficulty is not resolved, however, because if we take the modern atomic point of view, friction has still to be accounted for in terms of the motions of atoms and molecules, which are themselves governed by Newton's reversible laws (or the equally 'timeless' laws of quantum mechanics, to be discussed in Chapter Four). In short, the problem has been swept under the rug. So long as Newton's equations are unconcerned by the arrow of time, it is possible for a broken teapot to gather the necessary energy to reassemble spontaneously by sucking in sound waves, relying on billions of molecules in the scattered pieces of china to move in concert.

Perhaps Newtonian mechanics is not universally applicable, and we have sought to push it too far to explain the arrow of time. However, this arrow *is* an intrinsic feature of many processes, including life, which have no special dependence on 'initial conditions'. We shall begin to discuss these processes in Chapter Five.

Electromagnetic time

Newton formulated his laws in order to study the effects of gravitation on massive bodies. But there are other forces in nature, such as the electrostatic force – the one that makes clean hair stand on end when it is combed. The laws discovered to explain electrostatics eventually culminated in the electromagnetic theory, which formed the second major theoretical structure of physics. And here, too, time presents a knotty problem.

An intriguing aspect of Newton's theory of gravity is that it describes an instantaneous interaction between two massive bodies, such as the Sun and the Moon, despite the fact that they have no direct contact. This phenomenon came to be known as action at a distance. It troubled scientists and philosophers of the day, because there was no obvious mechanism for it. In his *Principia*, Newton stated: 'I wish we could derive the rest of the phenomena of Nature by the same kind of reasoning from mechanical principles, for I am induced by many reasons to suspect that they may all depend upon certain forces by which the particles of bodies, by some causes hitherto unknown, are either mutually impelled toward one another and cohere in regular figures, or are repelled and recede from one another. These forces being unknown, philosophers have hitherto attempted the search of Nature in vain.'[60]

Physicists and philosophers were content with the idea of the force of impact, like that given by a punch or a slap, but regarded other forces of attraction or repulsion – like Newton's gravity – as occult. Leibniz, Newton's primary scientific opponent, commented on Newton's work that 'Gravity (and, by implication, any of his active principles) must be a scholastic occult quality or the effect of a miracle.'[61] Newton circumvented this problem by conceiving of a gravitational field emanating from each gravitating mass, which permeates instantaneously the whole of space, and whose strength diminishes with the inverse square power of the distance from the centre of mass of the body; thus the strength of the gravitational field is diminished by a factor of four each time the distance is doubled.

Electrostatic forces – for example, between a charged comb and hair – act over space in the same manner. And for this effect to occur at a distance, an electric field was postulated, just like Newton's gravitational field. It was in 1785 that the Frenchman Charles Coulomb achieved the necessary accuracy in the laboratory to provide the basis for a theory of the electrostatic forces. He had been led, on the basis of laboratory investigations, to a law quantifying the interaction between electrically charged bodies. Using a torsion balance, a device which measured the electrical forces between pairs of charged balls, Coulomb showed that electrical charges of the same sign repel each other while those of opposite signs attract each other, in both cases with a force that varies precisely according to the inverse square of the distance between them (and is proportional to the product of the two charges involved).

The similarity of Coulomb's law to Newton's law of gravitation is striking: both employ the field concept and an inverse square law to explain action at a distance. To be sure, there are important differences. Electrical charge comes in two varieties, positive or negative. Charges of the same sign repel, opposite charges attract. Gravity, on the other hand, has only one kind of 'charge' – mass – which is always attractive: the Sun, Moon and stars all exert a tug on each other.

The related subject of stationary magnetic fields has a similar history to that of electrostatics, displaying many formal similarities. It became clear through the ingenious work on electricity and magnetism by Michael Faraday in the 1820s, when director of the Royal Institution in London, that moving or dynamic magnetic effects were closely related to static electrical effects and *vice versa*. Moving charges generate magnetic fields, while magnetic fields in motion produce electric current flow through conducting materials. (In Chapter Three, we shall discuss the deeper reasons for this symmetry.) Faraday's seminal contributions were followed through by the powerful theoretical reasoning of the Scotsman James Clerk Maxwell (1831–79). In 1864, as a professor at King's College, London University, he succeeded in demonstrating that electrical and magnetic effects were distinct manifestations of a single electromagnetic force.[62] The resulting mathematical equations were so elegant that they prompted Boltzmann to quote Goethe: 'Was it God who wrote these lines . . .?'[63] By casting Faraday's laws of electromagnetism into a mathematical form, now known as the Maxwell equations, Maxwell predicted as a necessary consequence that an electromagnetic signal should move in a vacuum at a constant speed – the same as the speed of light.

The compelling inference was that light itself is a form of electromagnetic effect. It was not long before other forms of *electromagnetic radiation* were discovered and, from then on, visible light was regarded as occupying just one region of an entire electromagnetic spectrum spanning the whole gamut from radio waves to X-rays and beyond. The spectrum of electromagnetic radiation with which we are familiar, from red to violet in colour, is simply that part which excites the retina of the human eye.

However, just as we found with Newton's equations, Maxwell's make no distinction between the past and the future. They are invariant (i.e., unaltered) whether a positive or a negative value for time is used, and thus they contain no intrinsic distinction between past and future. According to Maxwell's equations, a charged massive particle, such as an electron moving in the presence of a combination of electric and magnetic fields, will, due to the combined action of the fields, experience a force dubbed the Lorentz force after the Dutch physicist Hendrik Lorentz. Thus the electron's motion is described by Newton's equation of motion, in which the acceleration of the particle depends on the product of the Lorentz force and the particle's mass.

Once again we have lost the arrow of time. As was true of the theoretical framework governing motion under gravity, we are once more left with a reversible mechanical description for electrodynamics. Experiments on charged particles in electric fields, magnetic fields or both verify that the solution of these time-symmetric equations of motion do indeed provide a correct account of the dynamics. Yet it is again clear that many electromagnetic phenomena have a temporal direction too. One never sees light waves converging from a brightened room onto the filament of a lamp where they are all absorbed, nor is light emitted from our eyes and absorbed by the Sun or another conventional source. For this reason, some people like to talk about an electromagnetic arrow of time which rules out such 'inverse' processes on the grounds that the initial conditions necessary to realise them are so unlikely. The argument is very similar to one we dismissed earlier, about bulls and china shops.

Electricity and electromagnetic radiation have had a wider technological significance for timekeeping. They heralded the end of local time, which depended on the accuracy of local timekeeping, and the beginning of national time, a nationwide sense of 'now'. Radio waves could be used to synchronise the time given by many clocks spread across a nation. When the first electric telegraph was introduced in Britain in 1838, it was realised that a similar arrangement could be used to transmit a signal from a master clock.[64] Electricity contributed to the ever-increasing precision of clocks. In the United States' Bell Laboratories, the quartz crystal clock was born in the latter half of the 1920s with the help of electrical circuitry. Inside these clocks, a quartz crystal vibrates like a tuning fork at a constant and very precise frequency. This frequency is a property of the quartz crystal and, unlike mechanical clocks, is essentially independent of engineering design.

In 1948 the US National Bureau of Standards in Washington succeeded in associating timekeeping with a molecular vibration, paving the way to atomic clocks, whose ticking frequencies are totally independent of the vicissitudes of engineering design. The vibration in question was that of ammonia, a pyramidally shaped molecule consisting of three atoms of hydrogen and one of nitrogen. The 'tick' of this clock occurs as the nitrogen atom hops back and forth through the hoop of hydrogen atoms. Such atomic clocks threw into sharp relief the shortcomings of our most ancient timekeeper – the rate of rotation of our planet. The freezing and melting of the ice caps, the friction of the tides and other effects deep in the Earth's interior mean that the length of the day fluctuates by a millisecond or so during the year, making it quite inadequate for modern ultra-precise timekeeping.[65]

Of fields, æthers, space and time

The electromagnetic waves predicted by Maxwell were conceived as being disturbances of an electromagnetic field which 'lies in the space surrounding the electrified and magnetic bodies', as he put it.[66] These waves are characterised by their *wavelength* – the distance between identical points on consecutive waves (*see* Figure 3). Radio waves have wavelengths of several metres or more, while the gamma rays emitted from radioactive atoms are more than a million million times smaller – about one-hundredth of the diameter of a hydrogen atom. The visible

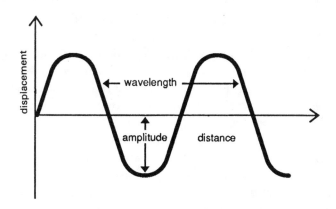

Figure 3. The meaning of wavelength.

light capable of being detected by the retina of our eyes has wavelengths several thousand times the diameter of an atom, and thus lies in between these two extremes.

But what is the medium in which these electromagnetic waves are embedded? To us, the most familiar kind of waves are not the electro-magnetic ones raining down from the Sun onto the head of a sunbather, but those from the sea breaking on a sandy shore. Surely Maxwell's electromagnetic waves must ripple through something? Maxwell postu-lated a medium for electromagnetic waves known as the *æther*. This belief in the æther, which survived until it was swept away by Einstein, had an interesting genesis in the concept of absolute space.

Absolute time and absolute space owe their origins to Greek thought and to Aristotle in particular. Newton and many others thought of something akin to an enormous grid stretching across the universe, known as a frame of reference, or state of absolute rest, to which the motion of all the objects in their experiments and theories could be compared. For example, the velocity of a passenger walking to the buffet carriage of a train could vary from a few to thousands of miles per hour depending on whether the motion is measured relative to the train, to the track or to Newton's hypothetical state of absolute rest. As we have already mentioned, Newton also envisaged an *absolute time*, independent of space, which flowed at the same rate everywhere: 'Gone were the attempts to relate time to the motion of the stars (as proposed by Plato), to the "number of motion" (Aristotle), to the mind (Augustine), to the world and mankind (Averroës) or to life and feeling. Time became a type of universal order that existed by and in itself, regardless of what happened in time.'[67]

Theological models of the cosmos in Newton's time had no difficulty in defining absolute space and time as attributes of the Creator. But it was far less clear in Newtonian physics where such an absolute space and time were to be found. Newton himself decided that absolute space (and time) coincided with the centre of mass of the solar system; this was later refined by others to be the so-called 'frame of the fixed stars' which took as its reference the most distant stars seen from Earth, which, on account of their distance, were apparently stationary.

In point of fact Newtonian mechanics, while upholding the notion of absolute time, could no longer logically support absolute space. There is no absolute state of rest, lacking motion – it all depends on one's personal point of view or, more technically, on one's frame of reference. Consider the simplest example, a universe cleared of everything except two balls. Imagine sitting on one. If the distance between the balls is steadily increasing it is impossible to say whether one, the other or both are moving. It is impossible to assign a privileged status to any single such frame of reference. This book may be lying 'stationary' on a table in a house. But, though stationary with respect to the Earth, it is certainly not with respect to the Sun. Nor can we assume that the Sun is fixed or indeed that there is any place in the universe that is absolutely at rest. The fact that Newton's theory actually denies the existence of such a notion was pointed out by many scientists and philosophers, including such people as his contemporaries Bishop Berkeley, Leibniz and, latterly, Poincaré, Mach and Einstein.

Precisely because there is no privileged cosmic grid stretching across the universe or a point in the cosmos that can be uniquely identified, like the origin of a graph, Newton's laws describe the same physics in all such frames of reference. It is an idea that we all take for granted: an air steward expects that the laws of fluid flow are the same regardless of whether he pours a cup of coffee at rest on the runway or flying with constant velocity at an altitude of 8,000 metres. All that has to be done is to take into account the effects of relative motion: in the case of a train travelling at 50 mph past a railway guard sitting on a platform, an apple core thrown by a passenger in the direction of motion at 30 mph would, relative to the guard, travel at 80 mph. This can be expressed mathematically by means of a device called a Galileo transformation which relates measurements made by two or more observers in uniform relative motion, that is, travelling with respect to one another at constant velocity.

Though logically untenable, the idea of absolute space became so deeply ingrained that it was quite natural for nineteenth-century physicists to assume that electromagnetic waves were disturbances within the nebulous æther which pervaded absolute space. Any measurements of distance could be made relative to this æther, a property of the universe, like references on a map. But to those physicists the æther was more than just a cosmic grid reference. By virtue of its motion with respect to absolute space, the Earth was believed to be rushing through a continuous 'æther wind'. If it was to be anything other than a fairytale, psychologically induced by the festering corpse of Aristotelian thought, it had to be demonstrated that the æther led to some tangible scientific results. The search for the æther was the motivation for the famous Michelson–Morley experiment, the ramifications of which were far-reaching indeed, and which we shall examine in the next chapter.

The power of prediction in Newtonian physics

Newton's equations have another startling property with respect to time: they are *deterministic*. We can see what this means by thinking about any system, whether it be colliding billiard balls or the orbit of Mars around the Sun. Newton's equations of motion are such that, no matter what the positions and velocities at an initial time of observation – the *initial conditions* – the behaviour of the system is determined for all future and past times. Whether the bodies are interacting through electromagnetic forces, like electrons buzzing around an atom, or through gravitation, like the planets orbiting the Sun, Newtonian mechanics in principle enables us to determine their entire past and future behaviour from a knowledge of their velocities and positions at any single instant in time. We may think the future is uncertain but, according to Newton's equations, it is fixed down to the very last detail. This 'determinism' is a direct consequence of the mathematical structure of Newton's equations. Determinism is closely related to *causality*, the notion that every event has a cause of which it is the effect – in the present case, the first cause can be taken to be the initial conditions, since we do not ask how these are brought about.

We are apparently obliged to conclude that everything was decided by the putative initial conditions of the universe – presumably fixed by God when he lit the blue touch paper. All the arguments in favour of free will adduced by philosophers, who so often wish to 'prove' what their particular prejudices have led them to believe, are blown sky-high by Newtonian determinism. It also belittles the role of God and Mankind in the evolving universe. Newton's clockwork universe would seem to be a mortal blow struck at the heart of Christian theology; it is no small wonder that the Church has had an unhappy relationship with science.

Einstein's remark, quoted in the last chapter, that the difference between past, present and future is merely an illusion, no doubt arose from the deterministic, causal structure of Newtonian mechanics, which also underpins the theory of relativity. In Newtonian mechanics, which is a time-symmetric, deterministic theory, there can be no distinction between past, present and future – all three are trivially related to one another. Take the known coordinates and velocities of the planets at any one instant: the description of the solar system at all 'later' and 'earlier' times is completely determined by inserting those data into Newton's equations. Therefore, the Newtonian description of a mechanical system at any instant in time contains at one and the same

64

moment both its past and its future. Indeed, our very sense of cause and effect is undermined, for the time-reversal symmetry implies that effect can become cause and cause effect. As far as Newton's equations are concerned, a time-reversed cricket match where every ball ends up being caught by the bowler is quite acceptable. The 'cause' is the heat soaked up by the ball as it begins to roll and bounce on the grass. The 'effect' is the ball leaping against the bat, sucking in a sound wave and bouncing into the outstretched hand of the bowler as he runs backwards.

Poincaré's return

Newton's mechanical equations have no intrinsic arrow of time, so there is no reason to choose one direction in time in preference to the other. But things are worse still. There is a theorem due to Henri Poincaré which purports to show that, given a long enough interval of time, any isolated system (for example, the universe itself) will return to its initial state; in fact, in an unlimited amount of time, it will do so infinitely often.[68] This is the cyclic time of the Stoics, mentioned in Chapter One.

Poincaré's theorem arises because isolated systems have a limited size: a billiard ball limited in its movement by the surface of a frictionless snooker table will eventually return to its original position with its original velocity. This eternal return is easy to envisage in the case of the snooker table. However, for many systems of interest to us, there is such a large number of particles (for example, atoms and molecules) present[69] that the 'recurrence time', as it is called, is many times the present age of the universe (about 10^{10} years or one with ten zeros after it). Nonetheless, this endless cyclic recurrence undermines the essential notion of time's arrow, and negates the concept of evolution in anything other than the most trifling sense.[70] *Poincaré's recurrence* (or Poincaré's return), in spite of its limitations,[71] has proved to be one of the most potent soporifics in the minds of theoretical physicists, whose principal response, by and large, has been to insist on a subjective explanation of irreversible time.

The scope of Newtonian physics

Newtonian physics admits the notion of absolute time but not of absolute space. As we have remarked above, in the example of tossing an apple core out of a train, the Newtonian description is unchanged under a Galileo transformation, the mathematical device relating a single phenomenon recorded by two observers whose frames of reference differ by virtue of their travelling at different uniform velocities. Such an invariance principle is necessary if the physics is not to depend on the frame of reference employed to describe it. No doubt astronauts would have been surprised, to say the least, had they found the laws of physics changed when they landed on the Moon. It should therefore come as no surprise to learn that invariance principles play a central role in modern physics. However, according to Newton's equations, while the positions of two observers differ by virtue of their relative motion, both have an identical perception of time, which does not depend on the frame of reference. It remains Newton's absolute time or indeed anyone else's. If synchronised at some earlier time, the watches of both observers in the passing train would show the same time, according to Newton's view of the world. This sounds very reasonable and fits in with our common-sense view of time.

Unfortunately, as will become evident later, common sense is often a poor guide to the real world. A hint that trouble was in store for absolute time came when it was found that the Maxwell equations of electromagnetism are not invariant under the Galileo transformation. In other words, electromagnetic phenomena vary according to the velocity or Galilean frame of reference. Yet the Galileo transformation exists because of the very lack of a state of absolute rest in Newtonian mechanics. This misbehaviour of electromagnetism hinted at a preferred frame of reference and the resurrection of the spectre of absolute space. Peculiar effects seemed to be possible: for instance, the laws of light propagation should depend on the relative speed of the observer. Then Hendrik Lorentz discovered a new transformation – later to be known as the *Lorentz transformation* – which left the Maxwell equations invariant. This transformation was quite different from its Galilean analogue; in particular, it appeared to mix up the space and time coordinates used by two observers in uniform relative motion so that the laws of light propagation would be independent of speed. Lorentz's curious transformation was overlooked as being of marginal interest until the advent of Einstein's special theory of relativity in 1905, which forms part of the subject of Chapter Three.

Apart from this minor embarrassment, Newton's physics worked well in describing the dynamical behaviour of bodies interacting via gravitational and electromagnetic forces on a large scale, whether apples falling off trees or the orbits of the planets around the Sun. What, though, of the invisible microscopic world of atoms and molecules, the building blocks of matter?

At the stage we have now reached in our story – towards the end of the nineteenth century – an atomic theory of matter was by no means generally accepted. But evidence in its favour steadily accrued, built up by the work of such people as Ludwig Boltzmann, who provided a molecular basis for the kinetic theory of gases; by Maxwell, whose efforts linking molecular behaviour to viscosity were likened by Boltzmann to 'a great symphonic poem'[72]; and subsequently by Einstein, whose work on the molecular theory of Brownian motion (1905) we will encounter again in Chapter Four. All used Newtonian mechanics to describe molecular motion.

Opposition to the atomic theory of matter was largely inspired by the forerunners of what later came to be known as the Vienna Circle of 'logical positivists'. The proponents of this philosophical doctrine, among whom were numbered such influential figures as Ernst Mach and the German chemist Wilhelm Ostwald[73], maintained that there could be no meaning in statements concerning a putative atomic theory since we have no means of directly verifying the existence of atoms and molecules. Boltzmann, convinced as he was of the scientific necessity for an atomic theory, was constantly at loggerheads with such obstinate opponents, who gradually lost the intellectual battle.

By the early years of the twentieth century, when the atomic theory had become firmly established, most physicists were of the opinion that essentially everything was now fully understood. There may have been a few small problems in need of an explanation – such as the exact way in which certain materials absorbed heat and some uninterpreted spectra (patterns of lines of light) emitted by atomic vapours – but the physicists contended that just a few more years of work would put paid to these uncertainties. To take just one example, Michelson commented confidently in 1903: 'The more important fundamental laws and facts of physical science have all been discovered, and these are now so firmly established that the possibility of their ever being supplemented in consequence of new discoveries is exceedingly remote.' As Weinberg wrote, 'physicists have laughed at this ever since'.[74] Yet hostages are still, from time to time, given to scientific fortune. In 1928, Nobel-laureate-to-be Max Born commented that 'physics as we know it will be over in six months'. The discovery of nuclear forces sent that thought quickly into the dustbin. But even in 1988, Stephen Hawking in his best-selling book, *A Brief History of Time*, was saying that he believed 'there are grounds for cautious optimism that we may now be near the end of the search for the ultimate laws of nature'.[75] Time will tell.

In any event, the rise of atomic theory sounded the death knell for

Newtonian physics in some regimes. The next two chapters describe the two revolutions that broke upon the world of Newtonian physics in the first quarter of the twentieth century, in the wake of mounting evidence demonstrating its failure adequately to describe the very fast, the very massive and the very small. Quantum theory transformed our understanding of the 'fundamental' microscopic world beyond all recognition. And Einstein's theory of relativity swept away the notion of absolute time once and for all. As Sir John Squire wrote, in reference to Pope's heroic couplet: 'It did not last: the devil howling Ho, Let Einstein be, restored the status quo.'[76]

Time loses its direction

The nature of time, a fundamental feature of our everyday existence, remains obscure in Newtonian physics. Time was introduced in mathematical terms by Newton in order to formulate the concept of motion, which is defined as change of position in time. Time remains, nevertheless, a primitive quantity, which is not itself defined in Newtonian physics: motion is explained by time, not time by motion. Whereas we can place a body at will in space, we have no control over its location in time.

In a Newtonian universe, clocks in uniform relative motion agree on the rate of flow of time regardless of their position or speed. The mythical man on the Clapham omnibus is thought to be most comfortable with this view of time, yet it is contradicted by Einstein's theory of relativity, which will be investigated in Chapter Three.

There is a paradox in Newtonian time. Human experience reveals that time marches ever onward. The passage of time is what enables us to observe motion, but the reason for time's arrow remains unexplained. In spite of their power, Newton's equations of motion produce results that are counter-intuitive: their time-symmetry makes them oblivious to the direction of time. Through Poincaré's return, they ensure that history will repeat endlessly, and their determinism makes it possible, with enough information on a system, to predict all future and past events. No wonder the Romantic poets hated Newton's view of nature. Keats, in his *Lamia* (1819), wrote:

> Philosophy will clip an angel's wings,
> Conquer all mysteries by rule and line,
> Empty the haunted air, and gnomed mine –
> Unweave a rainbow.[77]

If Newtonian mechanics is universally valid, one has to conclude that all processes that can ever occur can be expressed in terms of the motions of the constituent atoms and molecules. Because his mechanics is deterministic, the future and past behaviour can be predicted from knowledge about those movements at any one moment. Since our brains are made of atoms and molecules, then there can be no such thing as free will. The French philosopher Henri Bergson, among many others, was worried by this improbable picture of the world, for he felt that 'everything is given once and for all in classical physics: change is nothing but the denial of becoming and time is only a parameter'.[78] The same difficulty also led another French philosopher of science, Alexandre Koyré, to consider motion described by Newton's mechanics as 'a motion unrelated to time or, more strangely, a motion which proceeds in an intemporal time − a notion as paradoxical as that of change without change'.[79]

Newton's mechanics provides a vital body of theory which is still being put to work in all manner of ways, from the motion of billiard balls to galaxy formation and the technology of space exploration. Trajectories of planets, missiles, rockets, satellites and space probes such as *Voyager* are all worked out in advance on the basis of the 300-year-old theory. As Werner Israel and Stephen Hawking stated: 'It does its job with unbelievable accuracy − better than one part in a hundred million for the motion of the earth around the sun − and it remains in daily use.'[80] In describing the trajectories of planets and other 'simple' systems consisting of a very small number of bodies in motion, it is often the case that small uncertainties in the knowledge of their positions do not matter because Newton's equations predict a very similar future from a very similar present. But such determinism is often irrelevant in the real world where it is impossible to possess perfect knowledge of the way a complex system comprised of many bodies behaves. In the overwhelming majority of cases, such as the vast weather systems which the first man in space gazed down upon, the slightest uncertainties in the description of the present conditions will lead to vastly different futures, putting paid once and for all to the deterministic dream of Laplace, according to which everything would be predicted if only we know the positions and velocities of all the particles in the universe at any single instant of time. And with its demise comes, at last, the possibility of rediscovering a more unified and consistent view of time.

Relativity:
how time defeated Einstein

Relativity has taught us to be wary of time
Wolfgang Rindler
Essential Relativity

AT the age of 26, Albert Einstein demolished the 300-year-old idea of absolute time. He overthrew the entire foundations of classical Newtonian physics and replaced them with a revolutionary reassessment of reality in which time and space took on a new meaning. This was *relativity*. From it blossomed many completely novel effects, including *time dilation* (when time slows down for one person relative to another), time travel through *wormholes* in space, and the bizarre prospect of auto-infanticide, in which time travellers go back to their birth and murder themselves. Einstein smashed almost everything that common sense had to say about time. However, as we shall see, one essential aspect of time eluded him. He failed to account for time's arrow.

It would be wrong to think that everything changed overnight with the publication of Einstein's relativity papers. The seeds of the destruction of absolute time had been sown as early as the seventeenth century, by a Dane called Ole Roemer. Indeed, according to Eddington, time, as we now understand it, was discovered by Roemer.[1] He had been the first to assign a speed to light signals in 1675 while investigating the irregularities of Jupiter's moons at the Paris Observatory. He told members of the Académie des Sciences that because light travelled 'gradually' rather than instantaneously – as had been thought hitherto – the forthcoming eclipse of the innermost moon would lag ten minutes behind the time calculated from previous observations.

Once people began to think about light taking time to travel from a candle or light bulb into their eyes, rather than propagating instantaneously, it became clear that when we gaze at the heavens we see distant stars and galaxies as they were long ago. It was not until 1728, when Roemer's claim was confirmed by the English astronomer James Bradley, that a finite speed for the propagation of light was widely accepted. (The modern value is approximately 300,000,000 metres per second.) But Roemer's work marked the beginning of the end of the most naïve conception of 'absolute' time, the assumption that events perceived at the far-flung corners of the universe were labelled with the same instantaneous time as that on Earth. Einstein went on to show that, even taking into account the speed of light that transmits information about these events, absolute time could not be upheld. Time measurements made by observers in different states of motion do not agree with one another: relative to a stationary observer, time measured by a clock depends on its speed and, if gravity is taken into account, even on its location in space.

The traditional image of Einstein is of a white-haired, genial old eccentric. But it was a dapper young man, with dark curly hair and a moustache, who shook the scientific world in the years before the First World War. One biographer wrote of the young Einstein: 'It is noticeable that he appears to have been particularly happy in the company of women. The feelings were often mutual. The well set-up young man with his shock of jet black wavy hair, his huge luminous eyes and his casual air, was distinctly attractive.'[2] This was a young man with a mission, too. Einstein's dream was to create a consistent description of the world detached from all petty human considerations, a world in which objectivity reigned supreme. His driving ambition was to rid the laws of physics from their dependence on the role of the observer, and to do this he insisted that all observers, no matter what their location or state of motion, must be treated equally. There could be no preference given to any particular observer or *frame of reference* – the laws of physics must be independent of such trivia. The first stage in Einstein's odyssey was the special theory of relativity, based on the new principle that the velocity of light has the same value for all observers regardless of their velocity; later this theory broadened in scope to explain gravity and became known as general relativity.

The Michelson–Morley experiment

Before the theory of relativity could take the stage, the idea of the æther had to leave it. The æther, it will be recalled, was inspired by everyday experience. Sound consists of waves in air and ripples consist of waves in water. So nineteenth-century physicists naturally thought that light also had to consist of vibrations in something, which they dubbed the æther. As Abraham Pais put it, the æther was 'a quaint hypothetical medium which was introduced for the purpose of explaining the transmission of light waves'.[3] This æther provided a benchmark for measuring absolute space. It could be thought of as a veil cast across the universe which provided a reference for the measurement of distances, like a grid on a map.

The success of Maxwell's theory of electromagnetism, which we encountered in the previous chapter, inspired experiments attempting to locate the speed of the Earth through the æther as it orbits the Sun – the 'æther drift'. In fact, it was one of those experiments – in 1887 at the Chase School of Applied Mechanics in Cleveland – which helped lay the ghost of the æther to rest.[4] This was the most famous of the æther-quest experiments, performed by a physics professor, Albert Michelson,[5] with his colleague Edward Morley, a chemistry professor. Essentially, their experiment was a repetition of an earlier one by Michelson which was riddled with errors and ambiguities. Two beams from a single light source were shone at two equidistant mirrors and reflected back, one beam being aligned in the direction of the supposed motion of the Earth with respect to the æther, the other being aligned at right angles to this direction.

Michelson and Morley were looking for differences in the speed of light travelling in two directions at right angles to each other, which might be brought about by motion of the Earth through the æther. For instance, a light ray shining in the direction of the Earth's orbital motion might be travelling against the æther wind. It would thus move more slowly than a ray shining in the direction of the breeze. Analysis of the motion of light showed that light travelling perpendicular to the æther flow should return before light which has made the journey along the direction of the æther wind and back. The time difference in the arrival of the two beams back at the source could be calculated and compared with the experiment.[6] But Michelson and Morley found no time difference at all. They even repeated the experiment at various times in the year in case the direction of the æther wind varied as the Earth rotated round the Sun. But no matter what they tried, no time lag was observable. There was no æther.

It was a fantastic result, one which at the time the Nobel prize winner Robert Millikan thought to be an 'unreasonable, apparently inexplicable experimental fact'.[7] It seemed that the velocity of light did not change, whether it was travelling with or against the flow of the æther or the motion of the Earth. If the æther was untenable, perhaps the very structure of mechanics was in need of alteration. The 'æther at rest' had provided the absolute frame of reference which had been such a psychological necessity for Newton, even though Newtonian mechanics showed that the æther was redundant: there is no way to detect absolute motion – if you drop a ball in a train it will fall vertically whether the train is at rest or travelling at a constant speed. All one can observe is the movement of two bodies relative to each other, not some absolute frame of reference.

There was still, however, an apparently logical need for the æther because electromagnetism did not satisfy this same principle of relativity.[8] Electromagnetic effects varied under changes of perspective and did seem to be relative to some absolute frame: at the time different explanations were used to describe how to transform motion into electricity with a dynamo and how to transform electricity into motion with a motor. Because the idea of the æther was no longer tenable, Einstein demanded a new theory to provide a single perspective for all natural phenomena. He was to reformulate physics in such a way that the null result of the Michelson–Morley experiment became a natural consequence of the new principles.

It is undoubtedly an oversimplification to claim, as many authors have done, that the Michelson–Morley experiment, together with some earlier and later experiments, was the last word on the æther. A number of well-known physicists valiantly attempted to reconcile the Michelson–Morley result with the æther hypothesis. Of all these efforts, the most notable came from the Dutchman Hendrik Lorentz and the Irishman George Francis Fitzgerald. They tried to explain the Michelson–Morley result in terms of a physical contraction of objects moving through the æther, using the Lorentz transformation formulae mentioned in the last chapter, an approach that saved the æther hypothesis at the expense of an unexplained deformation of moving bodies. As we shall see, this length contraction is close to the kind of effect we have to get used to in the world revealed by Einstein.

Lorentz did actually come close to a formulation of special relativity but he could not rid himself of the 'classical' notion of absolute time formulated by Newton, and clung resolutely to the æther theory. The French mathematician and physicist Henri Poincaré saw clearly the problems posed by Newtonian mechanics. He asked: 'What is the æther, how are its molecules arrayed, do they attract or repel each other?' And he also closely anticipated the radical solution proposed by Einstein. He remarked: 'Perhaps we must construct a new mechanics, of which we can only catch a glimpse, . . . in which the velocity of light would become an unsurpassable limit'.[9] In 1904 Poincaré even codified a 'principle of relativity'. However, by Einstein's own account, it appears that he went to his death never having understood the physical content of special relativity.[10]

73

Einstein himself was largely ignorant of all these efforts until much later, and devised his theories quite independently. He was not familiar with the contents of up-to-date research papers published in physics journals. Indeed, he knew nothing of Lorentz's work after 1895; in particular, as we shall see, he had not heard of the Lorentz transformations which were to reappear in his own work. It is not even clear that Einstein regarded the Michelson–Morley experiment as being in any way crucial to his subsequent development of special relativity, though in 1916 he did explicitly state during an interview in Berlin with his friend the psychologist Max Wertheimer that he had been influenced by it.[11] However, in a letter he wrote in 1954 he maintained: 'In my own development, Michelson's result has not had a considerable influence. I even do not remember if I knew of it at all when I wrote my first paper on the subject [1905] . . . in my personal struggle Michelson's experiment played no role, or at least no decisive role.'[12]

The path to special relativity

Thus we come to Einstein himself. His appearance at birth at 11.30 a.m. on 14 March 1879 rather shocked his mother, Pauline. The back of his head was large and angular and she feared she had given birth to a deformed child.[13] He developed slowly, his ability for language being so poor that those around him thought he might never learn to speak. After entering the Luitpold Gymnasium (secondary school) in Munich at the age of eight, the young Einstein was told by his Greek tutor that he would never amount to very much. His family moved to Italy in 1894 and Einstein was left to stew at a school he disliked because of its strict regime, with German military service looming at the age of sixteen. He hardly endeared himself to the school, being 'precocious, half-cocksure, almost insolent'.[14] The Greek teacher went so far as to suggest that Einstein should leave school, and indeed within six months of his parents' departure he had followed them across the Alps. Einstein wrote later: 'I was summoned by my home-room teacher who expressed the wish that I leave the school without the diploma that would ensure entry to university. To my remark that I had done nothing amiss he replied only: "Your mere presence spoils the respect of the class for me." I myself, to be sure, wanted to leave school and follow my parents to Italy. But the main reason for me was the dull, mechanised method of teaching.'[15] Once expelled, Einstein gratefully seized on the opportunity to hike around northern Italy before rejoining his parents.

In 1895 it was decided that Einstein should try to enter the Swiss

74

Federal Polytechnic School at Zurich (now known as the Eidgenössische Technische Hochschule or ETH for short) with a view to becoming an electrical engineer. But he failed the entrance examination. It was only after further preparation as a student at the Swiss cantonal school at Aarau that he passed. During his student years at ETH he was taught by Russian-born Hermann Minkowski, who once described him as a 'lazy dog' who 'never bothered about maths at all'.[16] Certainly Einstein relied on his friend and fellow student Marcel Grossmann for notes on those lectures he did not attend.[17] Minkowski, as it turned out, later played a pivotal role in the development of Einstein's ideas.

In 1900 Einstein graduated and occupied himself in Zurich by tutoring and part-time teaching. By 1902 he had been accepted in the Swiss Patent Office in Berne, helped by a recommendation from Grossmann's father. However, although Einstein had applied to be a Technical Expert (second class) he was offered a post as Technical Expert (third class). It was while in this somewhat improbable job that he produced a colossal scientific theory on a par with Newton's.[18] At the Patent Office he met Michelangelo Besso, an engineer who became a lifelong friend and who was the only person acknowledged by Einstein in his first relativity paper.

The year of 1905 was Einstein's *annus mirabilis*. He was then an unorthodox individual with a reasonable academic record to his credit. The qualities that set him apart from his contemporaries were those of the true genius able to liberate himself from the yoke of long-accepted ideas. Einstein turned the puzzles surrounding Newtonian mechanics and electrodynamics on their heads by formulating radically new postulates for the whole of physics, a truly revolutionary step. The consequences which flowed from these postulates indicated that 'common sense' born of our limited experience of time and space can be deceptive.

The famous photographer Philippe Halsman once linked Einstein's disdain for socks with this approach. He asked the great man about his eccentricity and Einstein's secretary, Helen Dukas, chipped in, 'The professor never wears socks. Even when he was invited by Mr Roosevelt to the White House, he did not wear any socks.'[19] Einstein gave the prosaic explanation that 'I found out that the big toe always ends up by making a hole in the sock. So I stopped wearing socks.' Perhaps it also had something to do with what Einstein's 1901 military service book, under a category headed 'diseases or defect', listed as *Pes Planus and Hyperidrosis ped* (flat and sweaty feet).[20] But Halsman took the more romantic view that 'this detail seemed symbolic of Einstein's absolute and total independence of thought'.

75

The abolition of absolute time

Scientists are faced with a choice when confronted with perplexing results – they may either bend, twist and bludgeon the existing theory to fit (which is relatively easy but may not work) or create their own new one (which is more difficult, and may even be too hard). The daring Einstein chose the latter course, abandoning imagery based on everyday experience. This trend away from the pictorial to the abstract, which has continued to be highly successful in modern physics, is fully in line with Einstein's famous remark that 'Nature is subtle, but she is not malicious.'[21] Although ripples travel in water and sound waves travel in air, Einstein decided that electromagnetic waves were ultimate, irreducible realities which did not have to travel through the æther to exist. By using abstract mathematical representations, he enforced a line of approach begun by Maxwell's electromagnetic theory. He was not interested in naïve models which provided a comforting pictorial reality – typified by the then popular image of the atom as a plum duff, a spherical, positively charged 'pudding' dotted with negative 'currants'. Einstein was interested in the truth – however strange or surprising it might prove.

For example, Einstein thought that there were unacceptable anomalies within the descriptions of electromagnetic phenomena; that is, effects involving both electrical and magnetic features. Recall that at the time, different explanations were put forward to account for the behaviour of motors and dynamos. He was convinced that a motor which converts electricity into motion is ruled by the same physics as a dynamo converting motion into electricity. That the customary view differed according to whether the conductor or the magnetic field was in motion struck Einstein as inconsistent, and implied that one should be able to detect absolute motion.[22] Instead of agonising over how to incorporate these 'problems' within the existing framework of physical laws on the basis of 'special effects', Einstein unified the descriptions in terms of the relative motion of motor and dynamo.

In 1905 Einstein put forward two fundamentally new postulates of physics. They appeared in his first scientific paper on relativity, published in the premier German physics journal *Annalen der Physik* under the title 'On the Electrodynamics of Moving Bodies'. These foundations of special relativity, which refer to observers moving at constant velocities, are:

1. A 'relativity principle': the laws of physics must be the same everywhere in the universe, no matter what the velocity of the observer.
2. The speed of light is constant and independent of the motion of the light source.

Einstein's second principle, the constancy of the speed of light, sounds outrageous. Imagine measuring the speed of a rifle bullet and discovering it to be the same, whether the bullet is fired at you by a motionless foot soldier or from an aeroplane cruising at supersonic speeds. Since the speed is not the same for bullets, why should it hold for the speed of light? Einstein dealt a body blow to common sense by showing that no matter how fast one observer is travelling with respect to another, *both* frames record the *same* value for the speed of light.

Einstein turned prevailing arguments upside down by asserting the constancy of the speed of light, and that the relativity principle was universally valid for all physical phenomena. He thus went beyond the ideas of Newton, which related to purely mechanical phenomena, and relativised the whole of physics – a bold step indeed.

Newtonian mechanics, as enshrined within Newton's three laws of motion, is relativistic too: there is no privileged frame of reference in the universe to which other observations can be related in absolute terms. As we found in the last chapter, if there are only two objects in the universe and the distance between them is increasing it is impossible to say whether one or other or both are moving. This idea about relative motion went back as far as Galileo, who considered the motion of gnats, flies, small winged creatures and 'certain fishes' in a ship at rest and in motion at constant velocity. He wrote: 'When you have observed all these things carefully . . . have the ship proceed with any speed you like, so long as the motion is uniform and not fluctuating this way and that. You will discover not the least change in all the effects named, nor could you tell from any of them whether the ship was moving or standing still.'[23] But Newton (who nevertheless liked to believe in a state of absolute rest) also implicitly used a second postulate, namely that of an absolute time which was the same throughout the universe. In Newtonian physics, time passed at the same rate whatever the speed or position of the observer.

Because the laws of Newtonian mechanics hold true for all observers moving at constant relative velocities – said to define 'inertial frames of reference' – time and space coordinates in different frames are related by the Galilean transformation. Use of the transformation could translate one 'point of view' of motion into another, say from that of a gnat in Galileo's ship to that of a sailor standing on the shore. Significantly, this transformation could not be applied to the laws of electromagnetism that governed the behaviour of light. If we use instead Einstein's second postulate and ensure that the speed of light, and also the laws of physics, remain unchanged no matter what the velocity of the observer, then the mathematical formulation for the change of perspective – which ensures that Maxwell's equations describing light and other electromagnetic phenomena remain unchanged whatever the observer – is called a Lorentz transformation.

This mathematical transformation had been previously developed by Sir Joseph Larmor and Hendrik Lorentz.[24] But, in a startling demonstration of the power of his ideas, Einstein deduced the Lorentz transformation from 'first principles': that is, solely on the basis of the foundations of his own theory – without reference to their work. In fact, the June 1905 paper contains not a single reference to previous work. Moreover, Einstein dismissed the two-centuries-old idea of the æther in a single sentence and simply abolished the notion of absolute time flowing at a constant rate throughout the universe.[25]

The abandonment of absolute time is a profound result, so profound that it was important to check the second postulate – the constancy of the speed of light – from which it arose. In addition to the result of the Michelson–Morley experiment, arguments in favour of the constant velocity of light, whatever the speed of the source, were adduced by the Dutch astronomer Willem de Sitter in 1913 from analysis of light from double stars rotating about a common centre. However, direct laboratory verification only emerged as late as 1963.[26] This does not mean that people had questioned the validity of special relativity until then: the remarkable success of the theory with regard to a myriad of other experimental observations and predictions was enough to establish its ascendancy over Newtonian concepts. The confrontation of theory with experiment is, of course, what distinguishes science from philosophy.

New worlds at the speed of light

Despite Einstein's reassessment of time, for the most part Newton's ideas stand up remarkably well after 300 years of experiment. It is no wonder that one astronaut remarked on his return from the first circumlunar voyage in 1968, 'I think Isaac Newton is doing most of the driving right now,' emphasising how the Apollo mission had relied on Newton's laws to calculate the trajectory of the spacecraft.[27] It is only when we have to deal with objects travelling at speeds approaching that of light itself that Newtonian mechanics breaks down. Such situations are usually quite alien to our everyday experience, except where light and electromagnetic effects are concerned (which is why Maxwell's theory did not sit happily within the Newtonian scheme). Indeed, in the limit where the velocities of moving objects are small with respect to the speed of light, such as a car on a motorway, the Lorentz transformations of special relativity can be shown to be equivalent to the Galileo transformations of classical physics. In other words, special relativity reduces to Newtonian or classical physics in these situations. Accordingly, many of the strange phenomena arising from relativity that we shall describe in this chapter are only significant if the relative motion involved approaches the speed of light itself.

Length contraction is a good example. Einstein showed that objects moving near the speed of light appear to a stationary observer to be squashed along the direction of their motion. It is purely a relativistic effect: the body does not actually shrink in any way, it merely appears so to an observer. To illustrate what happens, imagine a high-speed train, the Relativistic Express. Consisting of a single coach, it rattles along a railway line at a constant speed relative to observers sitting on a railway platform. At very high speeds it will appear to the observers to contract. But from the point of view of the train's passengers it is the platform which is in motion and so it – rather than the carriage – is perceived to shorten. The extent of the contraction is dependent on the actual speed of the moving body: as the relative velocity reaches that of light, the length shrinks to zero. The appearance of objects travelling at relativistic speeds is an interesting subject in its own right; rods become curved, bicycle wheels look like boomerangs, and so on.[28]

If the people on the platform could measure it, they would find that the mass of the Relativistic Express had also changed at high speeds.[29] And at the same time, passengers could detect a change in mass of the station. This is because relativity predicts an increase in mass for moving bodies. The 'proper mass' of a body is that measured in the reference frame in which it is at rest. But seen by another observer travelling with uniform relative velocity, the mass increases with the speed of the body, actually by the same 'Lorentz factor' by which bodies are contracted. This relativistic mass increase has been observed experimentally for the minuscule particles that abound in particle accelerators all over the world, and has been found to conform quantitatively with Einstein's predictions. As the speed of the body approaches that of light, its mass becomes infinite: thus an infinite force would be required to accelerate it to exactly the speed of light. In this way, one can understand why it is impossible for a massive body to reach the speed of light, let alone exceed it. Only particles of zero rest mass can travel at the speed of light: an example is the photon – the particle of light which in quantum theory is associated with the electromagnetic field. The photon can *only* travel at the speed of light.[30]

79

Simultaneity and time dilation

From the 'common-sense' point of view, the most remarkable properties of special relativity arise from the relativisation of time. The very concept of simultaneity – events taking place at the same moment – depends on the relative speed of an observer and is not, as Newton would have it, an absolute concept. As we said earlier, even in the Newtonian world with a finite speed of light, one never sees the world right *now* because light does not enter our eyes instantaneously, but only as fast as the speed of light. When you glance at a watch, what you see is slightly 'out of date', owing to the time it takes for light to get to the retina of your eye from the watch face and then for the signal to be transmitted by nerve impulses into the brain. According to Newtonian physics, by making a suitable correction one could still reconstruct an absolute time for events as they occur, which all observers would agree on. But Einstein's relativity prohibits this.

To illustrate the difficulty with 'now' – the fact that simultaneity is relative – let us return to the Relativistic Express. Viewed from within the express, a flash of light from a lamp situated in the centre of the only carriage reaches the passengers at either end simultaneously. However, observers on the platform see light from the flash shining on passengers at the rear end of the carriage *before* it reaches those at the front, because the motion of the train means that the former are moving into the light beam (which thus has less distance to travel), whereas the latter are being carried away from it (so the beam has further to travel to reach them). Seen from the platform the arrival of the light at the two ends of the carriage is clearly not simultaneous. If the passengers at either end of the carriage are equipped with clocks, these will show the same time when the light strikes each end of the coach. However, that time will be different from the time recorded on the platform clock.

Equally startling is the phenomenon in special relativity called time dilation. Moving clocks tick more slowly than those at rest. Suppose that passengers in the train and those on the platform are supplied with identical clocks. When the train is at rest relative to the platform, both sets of clocks tick at the same rate, say one tick of the second hands each second. (This is known as the 'proper time', meaning the time recorded by a clock at rest with respect to an observer, or equivalently, the time recorded by a clock in its own rest frame.) According to observers on the platform, the interval between ticks of clocks on the moving train is longer than one second, by an amount that depends on the speed of the train (again given by the Lorentz factor). Indeed, as the express's speed approaches that of light, the interval between ticks of the moving clocks increases without limit. Like length contraction, this is a purely relativistic effect. From the point of view of passengers travelling in the carriage, the ticking of clocks on the platform is slowed down by the same amount.

Time dilation has been experimentally verified many times. Certain elementary particles called *muons*, which are created in the Earth's atmosphere at altitudes of 10 kilometres by the impact of exceedingly fast particles known as cosmic rays, undergo radioactive decay so rapidly (in their own rest frame) that most would never reach us were it not for the fact that their decay time is dilated in our own frame of reference. If their clocks ticked as if they were at rest, the muons would only travel about 600 metres before decay: they do not live longer from their own perspective yet they survive up to nine times as long according to ours.[31] Many similar experiments performed by physicists with the aid of particle accelerators also demonstrate that the lifetimes of muons can be extended by accelerating them to very high velocities.[32]

The twin paradox

Further oddities about time in special relativity are brought out in the twin paradox, the oldest of the so-called paradoxes of special relativity. Imagine two identical twins, Tweedle-dum and Tweedle-dee. While Dee stays at home on Earth, Dum sets out on a round-trip, relativistic (high-speed) space journey. Take the outward part of the journey and assume that each twin has a clock in the form of a beacon which sends out a pulse every five minutes. As Dum's speed increases, the pulses Dee receives from him on Earth gradually spread apart; from Dee's point of view, Dum's clock ticks more slowly, so at the end of his voyage Dum returns younger than Dee (though both have aged). Seen from the point of view of Dum, however, it might appear that the converse should follow – namely that Dee should be younger than Dum when they meet again. From Dum's point of view it will be Dee's pulses which spread apart, and Dee's clock which slows down, as Dum sets out on his journey.

Evidently, both results cannot be correct: both twins cannot be younger than each other when they meet again. The paradox is resolved as soon as it is appreciated that Tweedle-dum and Tweedle-dee have not been through equivalent experiences. Dum, unlike Dee, must have accelerated initially on leaving Earth, decelerated and then accelerated back. Finally, he would have decelerated to get back into Dee's frame of reference. Since Dum has not travelled at constant velocity (or, equivalently, he has not remained in an inertial frame), we cannot apply the analysis of special relativity to describe how he sees things and, in particular, to his experience of the passage of time. Although this disposes of the 'paradox', it does suggest that something very strange is happening. In a sense we can say that, thanks to relativity, Dum has been able to perform time travel into Dee's future.[33] One can actually observe this effect by flying an accurate clock on a commercial airliner and, after the round trip, comparing its time with an atomic clock left at the airport. However, as Stephen Hawking remarked, 'it would take an awful lot of frequent flyer miles to prolong one's life by a day this way'.[34]

Spacetime

Underlying these weird effects in special relativity is a whole new way of thinking about time. Relativists like to speak of something called *spacetime*, a concept which makes the mathematics of relativity simpler. It arose from the mathematical properties of Lorentz transformations, which imply that space and time should not be treated individually but rather as an inseparable whole. This fusion was first noticed by Hermann Minkowski who, inspired by his student's special theory of relativity, remarked in Cologne in September 1908: 'Henceforth space by itself, and time by itself, are doomed to fade away into mere shadows, and only a kind of union of the two will preserve an independent reality.'[35]

In special relativity, the nature of spacetime is spelled out by its metric structure, an abstract yet fundamental notion associated with the geometry of the universe. This metric structure is intrinsic and quite independent of any observers, a property which satisfies the quest of relativity to ensure that the laws of physics hold true regardless of speed or position. In relativity, the 'geometrical' properties – such as the paths described by light rays – are defined in terms of space and time as if these were inextricably bound together. Consequently, it is not a great step to claim that space and time are really just aspects of a single spacetime. In a sense, although Einstein eliminated the ideas of absolute space and absolute time, he introduced absolute spacetime. However, he regarded spacetime merely as a relational quantity, representing the totality of events. Thus a human body would appear as a four-dimensional 'worm' on the spacetime landscape, each three-dimensional slice of which corresponds to that body at a particular moment in time.

In relativity we can treat space and time as a four-dimensional entity: space is three-dimensional and time one-dimensional. But physically speaking, they are quite distinct, a point that we should never lose sight of. Most important of all, the time we experience – and as this book aims to show – is uni-directional, whereas no such restriction applies to space itself. It was Eddington who made the telling remark: 'The great thing about time is that it goes on. But this is an aspect of it which the physicist sometimes seems inclined to neglect . . .'[36] Once again we are back to time's arrow – and face to face with a serious flaw in special relativity.

Notice that in the twin paradox the twins are both supposed to have aged. But the notion of ageing as a one-way process in time is not itself explained in special relativity. This is because relativity, like classical mechanics before it, admits no distinction between the two directions in which time might possibly flow, forwards or backwards. It conceives only of one-dimensional, not uni-directional time. So far as the time-symmetric theoretical structure of relativity is concerned, one could have concluded that Dee, who stayed on Earth, grew younger than Dum. But we would have to dismiss this as absurd, since we know as a matter of fact that living beings age in time, rather than the reverse. Relativity itself, however, provides no explanation for why this should be.

Acceleration and absolute space

There is another major flaw in special relativity. It too is associated with the role of time, being caused by acceleration (the change in speed of a body over time). As we have repeatedly said, owing to the relativity of uniform motion in Newtonian mechanics, absolute space had no meaning. Nevertheless acceleration, caused by the presence of a force like that of gravity, is absolute in Newton's theory. To put it another way, acceleration is the same regardless of the observer's state of motion. A physicist on the back of a horse can dispute whether it is the horse or the ground beneath which is in motion, but he will be left in no doubt if his mount is brought up short and throws him flying from the saddle. This force and the resulting acceleration are the same, from whichever reference frame they are observed – horse, ground or moving train. Accelerating observers seem to have a different status from ones travelling at constant velocity. This implies privileged frames of reference even though the view that the Earth is accelerating towards the physicist is just as tenable as the one in which the physicist accelerates towards the Earth. A similar problem is present in special relativity, which is also only concerned with observers in uniform relative motion. Nothing in special relativity explains why these observers should be endowed with such a privileged position: they are simply built into the postulates.[37]

It was clear to Einstein that gravitation broke his intuitively appealing criterion – that the same physical laws should hold good, no matter what the observer's state of motion. In other words, the mathematical operation used to translate the view of one observer into that of another – the Lorentz transformation – failed to apply to gravity within special relativity's description of the universe. The root of the problem was that, contrary to what is implied by special relativity, acceleration is not absolute but relative. This can be illustrated with an example that also serves to show why a theory incorporating acceleration describes gravity. Suppose an astronaut is placed with some bathroom scales in a spaceship on a launch pad. If he stands on the scales, they will show his weight. When the launch button is pressed and the rocket accelerates away from Earth, he will feel much heavier and the scales will show that he has gained weight. Suppose that the take-off is aborted and the spaceship plummets back to Earth. The ill-fated occupant will float freely in the rocket during the few seconds until the moment of impact; if he takes his keys out of his pocket and drops them during the descent, they will not fall to the floor of the spacecraft. He will also be weightless: within his frame of reference, there is no force of gravity. Yet to an observer watching the spacecraft fall, the force of gravity is only too apparent. Thus the effects of acceleration *are* relative. Einstein was also inspired by free fall – not of an astronaut in a rocket but of a man falling off a roof in Berlin. The man, who survived without injury, told Einstein that he had not felt the effects of gravity.[38]

Einstein was in any case well aware of the limitations of special relativity, and on æsthetic grounds wished to rid physics of the lingering presence of privileged frames of reference. He set out, in developing his general theory, to attack the much more difficult problem of formulating a broader version of physics which would be valid for all observers, regardless of their relative states of motion. It is no wonder that this can only be achieved by a more complex relationship between time and space than that in special relativity. Indeed, the development of the general theory of relativity required the use of unfamiliar mathematics – the tensor calculus – and it was not until 1915, after many years of intellectual struggle, that Einstein was ready to publish it. As he remarked during this effort: 'Every step is devilishly difficult.'[39]

When it finally appeared, general relativity also provided a beautiful and rather complete theory of gravitation. If we again consider the astronaut in the spacecraft it can be seen why: it is impossible for him to tell whether any force he is experiencing is due to gravity or acceleration. He could not decide whether it was the effect of gravity showing on his weighing scales or his own inertia – the reluctance of his body to be budged – as the rocket accelerated through space. Recognition of this had led Einstein in 1907 to formulate a new fundamental principle – the *equivalence principle* – which he insisted be valid for the whole of physics. In effect it asserts that gravity and acceleration are equivalent.

There are at least two versions of the equivalence principle. One, the 'weak' equivalence principle, dates back to Galileo and his experiments, performed, as legend would have it, at the Tower of Pisa.[40] Galileo found that all bodies fall to Earth with the same acceleration (neglecting air resistance). The equivalence principle shows why: taking a relativistic point of view, one could consider that it is the ground which is accelerating upwards while the objects remain at rest, in which case it is clear that all the objects must be approached at the same rate of acceleration. This had been rather a mystery until Einstein's paper on the relationship between accelerated motion and gravity was published in 1914, stating that a uniform gravitational field is entirely equivalent to an appropriate acceleration as far as experiments within any given laboratory are concerned. The equivalence principle also demonstrated that special relativity is a purely local theory: there are few real-life observers who are not accelerating, because gravity dominates our universe, which is scattered with matter in the form of stars, planets and so on. The 'strong' equivalence principle, which Einstein advocated, declares that all the laws of physics are the same for all observers everywhere and at all times in the universe, regardless of the effects of motion or gravity. For Einstein, the moment had come to move on from special relativity to construct a theory of the universe, one that must transcend the local descriptions based on special relativity.[41]

The road to general relativity

The construction of general relativity was to take Einstein eight years of arduous and single-minded endeavour, peppered with flashes of insight as well as many excursions up blind alleys, before the new theory was finally laid down in its full intellectual splendour. In early July 1909, Einstein resigned from the Patent Office in Berne to take up a faculty appointment at the University of Zürich, the first of several professorial appointments he held in the following five years.

From the end of 1907 until mid-1911, Einstein remained silent on the subject of gravitation. Although he was still thinking about the matter a great deal, the newly born quantum theory (which we shall discuss in the next chapter) had by now also come onto the stage. Einstein had already made major contributions to this nascent theory; it continued to preoccupy his thoughts during this period and was to become a major source of worry to him throughout the rest of his life. Among the moves Einstein made during this time, his return to Zürich from the Karl-Ferdinand University in Prague in August 1912 appears to have been crucial to the mathematical development of the general theory of relativity.

He left Prague firm in the conviction that time and the paths of light rays are warped by gravity. But this idea had to be put on a sound basis. No sooner had he arrived in Zürich than he turned for help to his old friend and former fellow student Marcel Grossmann, then professor of geometry and dean of the mathematics-physics section at ETH. He told Grossmann: 'You must help me or else I'll go crazy!'[42]

Flat and curved spaces

To understand how Einstein solved the problem of gravitation, we must first consider the geometry of the world as we perceive it and as it was enunciated by the Greek mathematician Euclid who lived in Alexandria between 320 and 260 BC. For Einstein discovered that the laws of Euclidean geometry are valid only in restricted regions of space. These geometric properties – described by the metric structure – are very useful on Earth but break down if applied to the large-scale structure of the universe.

It is easiest to think of spacetime as if it were space alone and use the speed of light as a measuring stick (remember that the speed of light is absolute). An interval of time can be converted into a spatial length simply by measuring it in terms of the distance travelled by light during that interval. Astronomers cope with huge galactic and intergalactic distances by using the resulting light year, which is around 10 million million kilometres, and the parsec, which is 3.26 light years; these units avoid the use of too many zeros in their work.[43] For example, the Sun is at a distance of only 8 light minutes, while Sirius is 2.7 parsecs away; and the Gemini galaxy cluster is at 350 million parsecs.

In special relativity, the metric properties imply that spacetime is geometrically flat, like the green baize of a snooker table. But in general relativity, we have to get used to the idea of spacetime being curved.

86

Intuitively, everyone knows what is meant by a flat surface, a space of two dimensions. A pristine piece of paper sitting on a desk top describes a flat space (it has no curvature). On the other hand, the surface of a sphere is curved.

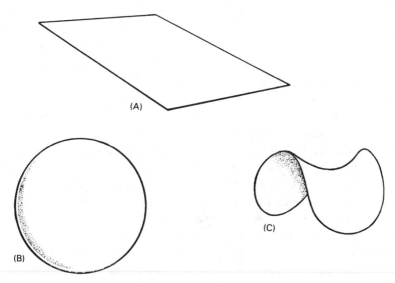

Figure 4. Flat (A) and curved (B and C) spaces. The sphere (B) has positive curvature while the saddle (C) has negative curvature.

These two-dimensional spaces or surfaces (called manifolds) are easily illustrated because they are embedded in the three-dimensional space with which we are all familiar. With the possible exception of mystical experiences, it is not possible for us to visualise the geometries of dimensions higher than three. Yet it is important to recognise that the property of flatness or curviness is intrinsic to the space. No reference is needed to some putative higher-dimensional space.

Of fundamental significance is the fact that the geometry on flat surfaces differs from the geometry on curved surfaces. Children are taught the geometry of flatness, worked out by Euclid over two thousand years ago. Every secondary school pupil knows that the sum of the three angles in a triangle is 180°, and that the circumference of a circle of radius R is $2\pi R$. Einstein referred to the 'noble building of Euclid's geometry . . . that magnificent structure, on the lofty staircase of which you were chased about for uncounted hours by conscientious teachers'.[44] But in fact its results are true only for flat spaces. Triangles drawn on the surface of a sphere have angles totalling more and circles have circumferences measuring less than those drawn on a flat piece of paper, by an amount which depends on the curvature of the sphere.

Although we are unable to envisage a curved three-dimensional space, its existence nevertheless can be inferred in the same way. Consider the world of Flatland, first described by the Victorian schoolmaster Edwin Abbott in 1884.[45] Abbott related the adventures of A Square, a two-dimensional creature with no sense of up or down, that moves only on a surface. For our purposes, let us imagine A Square on the surface of a sphere. Though oblivious to a third dimension, Square would soon discover that he lived in a curved space. All the creature would need to do is set out in a straight line and at some point he would find that he had returned to his point of departure. Actually this feature is, properly speaking, an aspect of the global topology, or large-scale shape, of the world Square inhabits, rather than a local one. Nevertheless, beings like A Square or ourselves – who live in three-dimensional space – have only to measure such (local) geometrical properties as the circumferences of circles to test whether they obey the laws of Euclidean geometry (in which case we live in a locally flat world) or not (in which case our space is curved). The great nineteenth-century German mathematician and astronomer Carl Friedrich Gauss (1777–1855) recognised this and performed experiments to try to determine the extent to which our three-dimensional space departs from flatness. Neither he, nor anyone else subsequently, has been able to detect any curvature of space in terrestrially based experiments. This comes as no great surprise, since Euclidean geometry is known to hold pretty well for us, otherwise it would not be taught in schools.

However, pure mathematicians are not generally concerned with the physical world. In the nineteenth century, they started to contrive abstract spaces of arbitrary dimension and curvature, and to describe their geometrical properties in exquisite detail. This work was pioneered by Gauss, developed by his student Georg Friedrich Bernhard Riemann (1826–66) and later elaborated principally by Bruno Christoffel, Gregorio Ricci-Curastro and Tullio Levi-Civita. These brilliant mathematicians demonstrated that the metric structure tells us about a space, and in particular whether it is flat (Euclidean) or curved (non-Euclidean).[46]

At the time these discoveries were made, they were only of academic interest to a small circle of mathematicians. It was not until Einstein's work that their intellectual contributions were widely recognised to be endowed with profound physical significance.[47] Moreover, it was only due to Einstein and his followers that time also came to be included within geometry. As we mentioned earlier, Minkowski's work on special relativity had shown that, for the purposes of mathematical physics, one could treat time as if it were analogous to another spatial dimension. In this way, it is possible to talk not only of flat and curved space, but of flat and curved spacetime.

Arriving in Zürich, Einstein was blissfully unaware of the work of Riemann and its significance for his own thinking. But when he discussed the gravitation problem with Marcel Grossmann, the latter told him that what he had been seeking was a spacetime possessing so-called Riemannian geometry as opposed to the flat Euclidean properties of special relativity.

Spacetime's key feature is that, while curved on a grand scale, it appears flat on smaller scales, just as the Earth appears flat to someone standing on a cricket pitch. It follows that for the description of events confined to local regions of spacetime, special relativity and the Lorentz transformations remain valid. But such a picture breaks down for extended regions over which the curvature of spacetime becomes noticeable. In an analogous way, a cricket pitch looks flat to a cricket team but the continent on which it resides looks curved when viewed by an astronaut. The larger the radius of a sphere, the smaller its curvature and the greater the region surrounding any point which appears to be locally flat.

The transition to Riemannian geometry was the crucial insight that led Einstein, initially with Grossmann's collaboration, to his ultimate formulation of post-Newtonian gravity. The general theory of relativity can be regarded as having been finally completed in Berlin, where Einstein had moved in 1914, with the presentation of his paper 'The Field Equations of Gravitation' to the Prussian Academy of Sciences on 25 November 1915.

General relativity

The extent to which spacetime is curved is determined by the distribution of matter in the universe: the greater the density of matter in a region, the higher the curvature of spacetime. Thus spacetime is warped more around the Sun than the Earth because of the greater mass of the Sun. The view of the universe taken in general relativity means that gravity no longer exists as such; it is transformed into the geometry (curvature) of spacetime. In Einstein's new vision, one might say that gravity is born in the transition from the flat spacetime of special relativity to the curved spacetime of general relativity.

This radically transforms our view of everyday events such as, for example, an apple falling to Earth. Rather than thinking of gravity as some kind of mysterious force acting at a distance through space, one envisages that a massive body like the Earth distorts space and time as well. The simplest way to gain an intuitive grasp of this statement is to imagine spacetime as a sheet of rubber stretched flat. Massive objects distort the sheet by stretching it locally, the amount of distortion depending on the mass of the object. The Sun, being by far the most massive body in our solar system, stretches spacetime the most. The trajectories of the planets can be represented by the paths of ball bearings of differing masses rolling over the rubber sheet and trapped in the deep well surrounding the Sun. In the case of the apple, it is not drawn to the Earth by a force but is merely rolling into the local spacetime 'well' created by the Earth.

The laws of motion of bodies in curved spacetime are generally quite different from those in flat spacetime. Instead of a free body moving through three-dimensional space at constant speed in a straight line, the new law of motion incorporating gravity states that bodies follow *geodesics*. Basically, geodesics are the lines of minimal 'length' connecting any two points within curved or flat spacetime, provided these are sufficiently close together[48] (*see* Figure 5). At very low velocities and very low densities of matter,[49] geodesic motion reduces to the kind described by Newton. It is clear that this 'reduction' of general relativity must occur, since the predictions of Newton's physics are so successful within its limited range of validity. However, Einstein was able to put

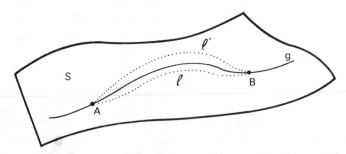

Figure 5. Geodesics define the paths of motion in the curved spacetime (S) of general relativity. If A and B are two points sufficiently close together on a geodesic g in S, then all neighbouring lines joining A and B (l and l') have greater length than that part of the geodesic between A and B. In general relativity, the Earth's elliptical orbit about the Sun is interpreted as being due to geodesic motion in spacetime curved by the Sun's mass.

[Adapted from W. Rindler. *Essential Relativity*, p. 106.]

geodesic motion to work in explaining problems that Newton could not answer.

The first example concerned a small but important detail in the orbit of Mercury, the planet closest to the Sun. Although this was far from Einstein's mind when he developed relativity, it provided a splendid test for his new theory. A single planet orbiting the Sun should, according to Newtonian mechanics, describe a perfect, closed ellipse with fixed perihelion (the point in the planet's orbit at which it comes closest to the Sun).

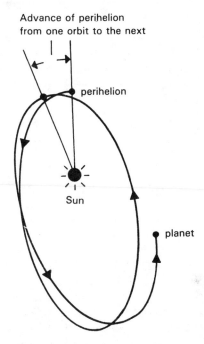

Figure 6. The advance of the perihelion of a planet orbiting the Sun. [Adapted from W. Rindler. *Essential Relativity*, p. 145.]

But the problem with Mercury's perihelion, indeed its entire orbit, is that it is *not* fixed. The gravitational forces exerted on Mercury by the other planets and by the asteroid belt in the solar system together provide a small additional influence which causes both the orbit to precess and the perihelion to 'advance' with the passage of time, completing one revolution over three million years (*see* Figure 6). But despite all such known gravitational influences on Mercury, there remained a completely unexplained – hence 'anomalous' – additional advance observed by astronomers in the perihelion to the tune of a mere

43 seconds of arc per century. Prior to Einstein's work, it was thought that this precession was caused by an undiscovered planet. But Einstein was able to explain the value exactly on the basis of the curvature of spacetime engendered by general relativity. Much more recently, the 'anomalies' in advance of the perihelia of certain other planets have been measured; within the observational uncertainties, their values also agree with those computed from general relativity.

The second result immediately computed by Einstein was the re-calculation of an effect he had anticipated before the completion of his general theory. This concerned the bending of the path of light by matter. One consequence of special relativity and the idea that the speed of light is the same for all observers, no matter what their speed, is the equivalence of energy and mass.[50] Thus there is a mass associated with the energy of a ray of light, and this should feel the gravitational pull of other matter; hence the path of the light ray will bend around a massive celestial body like a star.

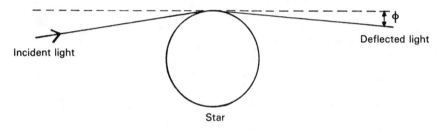

Figure 7. Incident light grazing the edge of a star and undergoing deflection through a total angle of 2ø. (The value of ø is 1″75 for the Sun and has been observed by comparison with known positions of stars during a total eclipse, and more recently by radio observations of quasars approximately in line with the Sun.)
[Adapted from W. Rindler. *Essential Relativity*, p. 147.]

Previously, Einstein had calculated the angle of deflection of distant starlight around the Sun on the basis of a kind of hybrid between special and general relativities, when spacetime was still assumed to be flat; repeating his calculations but now allowing for the curvature of spacetime, he found precisely twice his original result. Now the light must follow a geodesic in curved spacetime.

It was Arthur Eddington who helped test this second prediction of Einstein's theory. Eddington made his contribution in spite of the fact that Britain and Germany were at war when he first learnt of Einstein's work in Berlin through Willem de Sitter in the neutral Netherlands. A conscientious objector because of his Quaker beliefs, Eddington was granted an exemption from military service on the condition that he continue his scientific work, particularly the preparations to monitor a forthcoming solar eclipse in which the Moon was due to pass in front of the Sun. The eclipse of 1919 made it possible to observe the light from stars which passed close to the Sun, and to determine whether it was bent by the Sun's gravitational field. Normally, the glare of the Sun makes it impossible to observe that starlight. But, after returning from the Principe Island in the Gulf of Guinea where the eclipse could best be recorded – despite unexpected cloudy weather – Eddington told a Royal Astronomical Society dinner, in a parody of Omar Khayyám's *Rubáiyát*:

> Oh, leave the Wise our measures to collate.
> One thing at least is certain, light has weight;
> One thing is certain and the rest debate –
> Light rays, when near the Sun, do not go straight!

Later in his life, Eddington referred to this confirmation of general relativity – an observation that made Einstein an instant international celebrity – as his greatest moment.[51]

More recent confirmation of general relativity has come from studies of the binary *pulsar*, a pair of celestial objects believed to be tiny collapsed cores of old stars. Called a pulsar because it emits regular pulses of radio waves, the extremely rapid rotation of the pair round each other has to be modelled using relativistic as opposed to Newtonian mechanics. The advance of the perihelion is therefore much larger than for Mercury and the other planets. Ripples of spacetime curvature have also been predicted on the basis of Einstein's equations, and the existence of such gravitational radiation can be inferred from the orbital decay of the binary pulsar.[52] In addition, electromagnetic radiation emitted by distant *quasars* – the brightest objects in the universe – can sometimes be seen to undergo a gravitational lensing effect caused by the presence of intervening galaxies: the gravitational field of each galaxy acts as a peculiar sort of lens, producing multiple images of the original object in Earth-bound telescopes.

To summarise, general relativity entails, not by design but by necessity, a radical reconceptualisation of space and time, henceforth visualised mathematically as a single structure called spacetime. The geometry of this spacetime is determined by the distribution of matter; gravity itself no longer appears explicitly. That is one way of putting it, anyway. But for an altogether less dry précis, we turn to a send-up of Lewis Carroll's 'The Walrus and the Carpenter' written in 1924 by a

Professor W. Williams, a specialist in relativity theory. Entitled 'The Einstein and the Eddington', it contains the following verses:

'The time has come', said Eddington,
'To talk of many things;
Of cubes and clocks and meter-sticks,
And why a pendulum swings,
And how far space is out of plumb,
And whether time has wings.

'You hold that time is badly warped,
That even light is bent;
I think I get the idea there,
If this is what you meant;
The mail the postman brings today,
Tomorrow will be sent.'

'The shortest line', Einstein replied,
'Is not the one that's straight;
It curves around upon itself,
Much like a figure eight,
And if you go too rapidly,
You will arrive too late.

'But Easter Day is Christmas time,
And far away is near,
And two and two is more than four
And over there is here.'
'You may be right,' said Eddington,
'It seems a trifle queer.'[53]

Gravitational time dilation

How does time pass in curved spacetime? By 1911, Einstein had realised that clocks tick more slowly the stronger the gravitational field: the closer a clock is to a massive object, such as the Sun (or for greater effect a superdense object such as a black hole), the more slowly the clock ticks relative to a clock placed much further away. This conclusion is borne out by the full general theory of relativity and is known as gravitational time dilation; as distinct from the dilation effect we encountered in special relativity (p. 80).

This led to the third verification of Einstein's general theory. An atom can be viewed as a very simple sort of clock – it contains electrons which circulate round the nucleus with extremely precise frequencies, a phenomenon exploited in atomic clocks. This provided scientists with a marvellous opportunity for an experiment to show once and for all whether there is a 'universal time' that permeates the universe. However, there is no need to send atomic clocks into space at high speed or put them near the Sun's huge gravitational field to test relativity – they are already in place. According to Einstein's predictions, the electronic oscillations of atoms (more correctly ions, which are electrically charged atoms) in the Sun should take place at a slightly lower rate than on Earth. This slowing of the vibrations, as manifested in the radiation given off by the ions shifting to longer wavelengths, has indeed been verified. Although in the case of the Sun the effect is very small, light emitted from ions in White Dwarf stars (which are of similar mass but smaller size and hence much greater surface gravity) is received on Earth significantly reddened by this effect, dubbed the gravitational *redshift*.

Similar, albeit smaller, effects are even detectable from different geographical locations on Earth. For example, an atomic clock kept at the National Bureau of Standards in Boulder, Colorado, 1650 metres above sea level, gains about five microseconds (that is five millionths of a second) each year relative to an identical clock kept at the Royal Greenwich Observatory, which is only 25 metres above sea level.[54] This is because the nearer one is to the centre of the Earth, the stronger the gravitational field.

Experiments performed by Carroll Alley of the University of Maryland have directly shown this gravitational time dilation.[55] He used two groups of atomic clocks in a series of experiments which took place during the winter of 1975. In the course of an experiment, one group stayed on the ground while the other was taken on a flight round Chesapeake Bay at an altitude of about 9000 metres. Allowances were made for the special relativistic effects of motion mentioned earlier in connection with the twin paradox, and it was found that the time in the air advanced a few billionths of a second every hour.[56] A similar effect was noted when a hydrogen atomic clock was launched in June 1976 to an altitude of 9600 kilometres above sea level: now the clock ran fast to the tune of one billionth of a second every second, in agreement with the predictions of general relativity.

One can also imagine a gravitational analogue of the special relativistic twin 'paradox' mentioned earlier. If one of the twins whiles away his life on a very dense planet (such as a White Dwarf or neutron star), and the sibling remains comfortably wrapped up on the Earth, then with time passing the former ages more slowly than the latter. Note that again ageing is simply assumed to occur; the theory of relativity makes no distinction between the two possible directions of time, so that one may just as easily claim that the Earth-based twin gets younger more rapidly. As before, the phenomenon of ageing, associated with uni-directional, irreversible time, receives no explanation in Einstein's theories.

Cosmology and time

Our understanding of the nature of time has always been intimately linked with our knowledge of the structure of the universe. And according to the philosopher Karl Popper, the problem of *cosmology* is one 'in which all thinking men are interested'.[57] The large-scale structure of the universe – beyond the galactic level – is properly described in terms of Einstein's theory of general relativity. This is because it is only when we reach the truly vast distances which arise at this scale of size that the Newtonian picture of gravity breaks down. Although the average density of matter in the universe is extremely low[58] and thus so too is the average curvature of spacetime, the distances involved are so enormous that the small local curvature adds up to something highly significant. Insofar as they are correct, Einstein's equations of general relativity have permitted physicists for the first time to begin to probe consistently the behaviour of our world on the grandest of all scales, and to think in a dispassionate, scientific manner about the origin of the universe.

The ever-increasing sophistication of observational astronomy furnishes us with facts against which to test the validity of the cosmological models general relativity provides. However, we occupy an insignificant cosmological position. Our Sun is a typical star of middle age (some 4600 million years old) and of indifferent mass, located some 30,000 light years from the centre of our Milky Way galaxy, a vast disc-shaped, spiral swarm of matter 90 per cent of which is in the form of some 200,000 million stars (the rest is gaseous or dust).[59]

The solar system lies in a spur off one of the major arms of the spiral. Given our position and our relatively feeble means of probing the rest of the universe, the available astronomical data on the cosmos are severely restricted. This ignorance provides a rather free rein to cosmologists. Some scientists jest that there is speculation, pure speculation and cosmology. It is true that the universe presents us with the greatest of all scientific laboratories, yet it is a laboratory with a difference. The art of experimentation is to be able to control phenomena in a systematic way so as to be able to discern the regularities that would otherwise remain latent in nature. But the astronomer's laboratory is beyond our control; we must be prepared to rationalise whatever comes our way from the heavens.

The origin of time and the universe

Despite these difficulties, we already know with considerable confidence some important features of the universe. Among these, Edwin Hubble's discovery in 1929 that the universe is not static but is actually expanding with the passage of time really launched cosmology as a respectable scientific discipline. Whether the universe is destined either to go on expanding forever (an *open universe*) or the influence of the gravitational pull between all the cosmic matter is sufficient eventually to halt the expansion and bring about a contraction to a *Big Crunch* (a *closed universe*) still needs a definitive answer. Gravity implies a restless universe – even if all the matter in the cosmos were initially static, gravity would draw it inexorably together. The problem is that we are still uncertain about just how much matter there is out there in space, the key factor in deciding whether the universe will expand endlessly or collapse back on itself. At the time of writing, the observational data is not accurate enough to tell us what the fate of the universe will be.[60]

If the universe is expanding, does this mean that it had a beginning in time? One can estimate a minimum age for the universe on the basis of various pieces of observational data. For example, from the abundance of heavy elements produced in nucleosynthesis – the fusion of lighter nuclei into heavier ones – which are necessary for the existence of life as we know it, one can conclude that the universe is at least 10,000 million years old. By itself, this does not mean that there had to be a beginning to the universe.

However, even in the mid-1950s the measured value of the rate of expansion of the universe implied that the universe was younger than the Earth by 3.5 billion years.[61] This led to the rise in support during the 1950s and early 1960s among cosmologists for a so-called *steady-state* model of the universe.[62] Put forward by the cosmologists Hermann Bondi, Thomas Gold and Fred Hoyle, the model proposed a continual process of matter creation in such a way that, as the universe expands, it remains unchanged in its spatial and temporal properties.[63]

One immediate difficulty with this steady-state theory is that such matter creation is completely unexplained by general relativity. But the acid test – as ever – arose in confrontation with astronomical observation. In 1965 came the accidental discovery by Arno Penzias and Robert Wilson of an all-pervasive microwave radiation background, for which they were awarded the 1978 Nobel Prize for Physics. Having gone to great lengths to eliminate this extraneous microwave noise, including the eviction of a pair of pigeons roosting in the horn-like antenna used in their experiments, they were obliged to conclude from a detailed study of its properties that it was thermal electromagnetic radiation, whose origin lay beyond our galaxy, being received uniformly from all directions in the sky with an effective temperature of three degrees Kelvin above absolute zero.[64] It could only be interpreted as a relic from a much earlier stage in the evolution of the universe, when it was much hotter and denser than it is today – an echo of the birth of the universe.

The presence of this ubiquitous background radiation had been predicted by Ralph Alpher and Robert Herman in 1948 on the basis of a model developed by George Gamow, who had at one time been a student of the Russian physicist Alexander Friedmann. Gamow used Einstein's equations to study the behaviour of the universe at very early times. From this he concluded that the universe as it then was would have been very dense and exceedingly hot. It was the radiation from the fireball which constituted the early universe, and which had cooled during the subsequent expansion, that Penzias and Wilson had unwittingly discovered.

The substance of Alpher's doctoral thesis, which described how the 'elementary' particles present in the original cosmic stew evolved from hydrogen into helium by nuclear fusion of protons and neutrons, formed the classic 'Big Bang' paper that appeared on April Fool's Day 1948 in the American journal *Physical Review*. Its substance and date of publication were not the only notable points: Alpher's doctoral supervisor Gamow persuaded nuclear physicist Hans Bethe to add his name to the paper, so that its list of authors ran Alpher Bethe Gamow. Gamow wrote in his book *The Creation of the Universe*: 'It seemed unfair to the Greek alphabet to have the article signed by Alpher and Gamow only, and so the name of Dr Hans Bethe (in absentia) was inserted in preparing the manuscript for print. Dr Bethe, who received a copy of the manuscript, did not object . . . there was however a rumour that later, when the alpha beta gamma theory went temporarily on the rocks, Dr Bethe seriously considered changing his name to Zacharias.'[65]

Many years earlier, in his paper submitted in 1917 to the Berlin Academy of Sciences entitled 'Cosmological Considerations on the General Theory of Relativity', Einstein caught a glimpse of this cosmic story which showed a beginning, and perhaps an end, of time. In spite of the vision he showed in developing relativity, even Einstein could not accept its Genesis and Revelations: the *Big Bang* and the (possible) Big Crunch. He had at his disposal the means to predict Hubble's discovery of an expanding universe but was blinkered by the then prevailing notion of the universe as a static, time-independent entity. He therefore bent his theory to fit conventional wisdom by introducing a new constant of nature – the cosmological constant – rather in the manner that one would bludgeon a square peg into a round hole. The possibility of a dynamical cosmological model was probably the greatest prediction Einstein never made. Hawking calls it 'one of the great missed opportunities of theoretical physics'.[66] Later, Einstein eventually came to reject the need for this extra constant; Gamow wrote that Einstein felt it was the biggest blunder of his life.[67]

The full implications of relativity for the birth and death of the universe were conjured up by others, notably Friedmann, de Sitter and the Belgian cosmologist and priest Georges Lemaître. Friedmann himself had been the first person to accept general relativity as a self-contained theory and to work out some of its consequences for the universe. In fact, using general relativity, Friedmann had already been led to predict an expanding universe several years before Hubble's work. Friedmann's cosmological model implies that if at some time in the past the density of matter in the universe was infinite so too would have been the curvature of spacetime. The universe must have emerged from such an indescribably dense scenario in the form of some kind of Big Bang; prior to that there was literally nothing at all – no time, no space, no matter.[68]

Thus beyond the Big Bang the laws of physics lay silent and time itself stood still. In general relativity 'time is just a coordinate that labels events in the universe. It does not have any meaning outside the spacetime manifold,' in the view of Stephen Hawking. 'To ask what happened before the universe began is like asking for a point on the Earth at 91 degrees north latitude; it just is not defined. Instead of talking about the universe being created, and maybe coming to an end, one should say: "The universe is."'[69]

In many ways the Big Bang and the Big Crunch are repugnant. These are colourful terms for what are known as *singularities* by mathematicians – they are spacetime points which occupy zero 'volume' and have infinite mass. The whole mathematical edifice on which the theory of relativity is based becomes quite meaningless under such conditions of infinite matter density, showing that the theory is falling apart at the seams. It certainly suggests a major weakness in any claim that general relativity is the final theory of spacetime and gravitation – the limits of applicability of the theory are exceeded at the singularities. As Clifford Will of Washington University pointed out: 'The idea that there should exist singularities of spacetime, places where general relativity and all the other laws of physics should break down, is very disturbing.

Physicists are uncomfortable with the possibility of being unable to predict the future from some given initial data, but this was precisely what a singularity would do, because the physics that emerges from a singularity is not under any control.'[70]

There was and is still life in the theory of relativity. Seminal work by Stephen Hawking and Roger Penrose, between 1965 and 1970, demonstrated that if the behaviour of the universe is determined by the equations of general relativity, there must have been a Big Bang singularity of the kind described above, some time in the past.[71] Since the description of physics at these singularities is quite impossible – because the mathematics then falls apart – the indications are that we cannot really hope to use general relativity to deal with the origins of space and time. Nevertheless, according to Roger Penrose, the Oxford mathematician who was the first to undertake the study of singularities, this is far from meaning that we must discard relativity altogether. 'Some people say that singularities show you that general relativity is wrong. But it is the strength of relativity that it can tell you its own limitations.'[72] Roger Penrose, among others, would like to turn this vice into a virtue, as we shall see in Chapter Five.

Black holes, cosmic censorship and time warps

The pull of the gravitational attraction which all matter experiences can bring time to an end just as the Big Bang singularity is thought to have started it. For stars of large enough mass, the gravitational attraction can outweigh other forces tending to keep matter apart, and an inexorable collapse finally ensues.[73] The gravitational field may then become so enormous that even light itself cannot escape and time dilation could become so extreme that time might appear to stand still. The limits of such superdense objects, called *black holes*, are defined by an *event horizon*. This event horizon is not a physical surface but rather represents the point of no return for anything drawn towards it.

The American theoretical physicist John Wheeler coined the term 'black hole' to describe this one-way behaviour during a conference in New York in 1967.[74] But the concept of a black hole dates back to the eighteenth century, in the writings of a British amateur star-gazer, John Mitchell. He reasoned on the basis of the then favoured corpuscular theory that light would be attracted by gravity.[75] Black holes are now thought by many astronomers to be present at the centres of quasars and the hearts of other large galaxies. Stellar mass black holes are believed to have been detected in certain binary star systems which emit X-rays.

The latter provide the best contemporary observational evidence for black holes, although none has yet been proved to exist. After all, they cannot be directly observed, only indirectly detected through their gravitational pull on other objects. There is nothing remarkable about the event horizon of a black hole in terms of its appearance. A hapless space-traveller, along with anything else, might be drawn into a black hole without noticing that anything special had occurred; in particular, his own watch would continue to tick normally. But once inside the event horizon, nothing can ever leave again (if we ignore quantum mechanical effects). And the unstoppable effects of gravity would continue to operate, pulling our unsuspecting spaceman towards his own local Big Crunch, another nasty singularity of the Einstein equations. The differences in gravitational forces on his head and his feet would tear him apart.

Suppose our space-traveller, Tweedle-Dum, left Tweedle-Dee beyond the event horizon, which he entered at, say, 01.00 hours according to his own watch. As the fateful hour approached, Dum sent signals every second to Dee. The closer to the event horizon, the greater the interval would appear to Dee until it became infinite at the horizon. Then, in theory, Dee would have witnessed Dum dithering for ever at the horizon; as perceived by Dee, Dum's clock would never quite reach the hour of 01.00 because of extreme gravitational time dilation: time would appear to stand still. In fact, however, the intensity of the light from Dum that Dee saw would have become dimmer and dimmer and redder and redder as the light waves were stretched into longer wavelengths by the intense gravity.[76] So Dum just vanishes from sight into the black hole. Note that the singularity at the centre of the black hole is clothed from the view of outsiders by virtue of the event horizon, which prevents any light from escaping.

The singularity is where space and time come to an end. There are solutions of the equations of general relativity in which an astronaut could fall into a black hole, avoid hitting the singularity and go instead through a small passage, coming out of a *white hole*, the time reverse of a black hole. This feature arises because general relativity is a time-symmetric theory: 'The white hole might be in another universe or another part of our universe. In the latter case one could use black holes to travel to distant galaxies. Indeed, one would need something like that if intergalactic travel is to be a practical possibility,' commented Stephen Hawking.[77] Unfortunately for space-travellers, 'the least disturbance, such as a space ship going through, will cause the passage between a black hole and a white hole to pinch off.'[78] Such a white hole would describe exactly the opposite behaviour in time in which infinitely dense matter within a singularity would explode into life along with a blinding release of light – like the Big Bang itself on a local scale. And then the singularity would be naked, exposed to full view.[79] Physicists generally feel that white holes are unrealistic and would lead to untenable physical consequences of the kind described by Hawking. To deal with them, Roger Penrose introduced the 'Cosmic Censorship' hypothesis, an edict without theoretical justification stating that naked singularities are forbidden in our universe. According to this censorship, all singularities should be shrouded by event horizons. It is another illustration of an *ad hoc* postulate invoked to rule out unpleasant features of a time-symmetric reversible theory.

The anthropic principle

An intriguing crutch of support for our cosmological gropings-in-the-dark is provided by the *anthropic principle*. This draws attention to the fact that our very existence determines to a considerable extent the properties of the universe we observe out there. Why is the universe so big and life apparently so rare? The visible universe is truly of an enormous size, some 13,000 million light years across. Because of its expansion, this means that it must be about 13,000 million years old. On the other hand, life as we know it depends on the presence alongside hydrogen of certain elements, most notably carbon, oxygen, nitrogen and phosphorus. These could not have been made in the original Big Bang during which only hydrogen and helium nuclei were formed in significant quantities. The heavier elements had to wait for the forma-tion of galaxies and stars whose interiors, resembling vast furnaces, could achieve nucleosynthesis by fusing together the lighter elements. And after that, up to a few thousand million years of heating is required to generate such elements.[80] Therefore, in order for us to exist, at least this much time must have passed since the universe began. As John Barrow puts it: 'We should not be surprised to discover that the Universe is so vast in scale, because we could not live in one that was significantly smaller . . . It is a sobering thought that the global and possibly infinite structure of the Universe is so linked to the conditions necessary for the evolution of life on a planet like Earth.'[81]

The weak anthropic principle, whose name was first put forward by Brandon Carter in 1973, is nothing more than a statement that the existence of life (ourselves) may determine some of the properties of the universe we observe.[82] It followed the pioneering work of Gerald

Whitrow, who in 1955 used the special features of much of three-dimensional mathematical physics to argue that the fact that we live in a three-dimensional space is related to the nature of ourselves as rational information-processing observers. Whitrow then went on to relate the need for a very large universe with the conditions necessary for life.[83] Controversy surrounds the principle owing to the fact that other, more speculative, versions have been proposed in recent years. The strong anthropic principle contends that the universe must be such as to allow the existence of life, while the final anthropic principle maintains in addition that once life has come into existence in the universe, it will never die out. These two statements seem at first sight to be more metaphysical than scientific, having more in common with the teleological mode of explanation (namely that the universe has a purpose) favoured by theologians than that used by scientists.[84]

Time travel

There is another fascinating consequence of general relativity: genuine time travel. As we have seen with the help of the twin paradoxes, in a very limited sense both special and general relativistic time dilation permit 'time travel' of one observer relative to another. Nevertheless, the temporal ordering of events is the same for all observers, even if they cannot agree on a universal 'now': no observer, in any state of motion, will see light arrive on the Earth before it shines out of stars.

However, the famous logician Kurt Gödel showed in 1949 that the tricky feat of travel into the past is possible according to Einstein's equations of general relativity.[85] He found a cosmological model of a rotating universe satisfying the Einstein equations in which journeys into history are permissible. There are worrying implications here. As Gödel put it: 'This state of affairs seems to imply an absurdity, for it enables one, e.g., to travel back into the near past of those places where he himself has lived. There he would find a person who would be himself at some earlier period of his life. Now he could do something to this person which, by his memory, he knows has not happened to him.'[86] Time travel, were it possible, would raise bizarre possibilities indeed. For consider the most lurid implication of Gödel's idea – auto-infanticide. If such a deed were enacted, then a time-traveller would not have lived in order to perform the deed. This furnishes a beautiful example of a form of logical argument known as *reductio ad absurdum*.

The model discovered by Gödel bears no resemblance to the universe we inhabit, so one can try to dismiss it as a physically meaningless

103

solution of the Einstein equations.[87] It would seem, on the grounds of *reductio ad absurdum*, that time travel must be ruled out in our universe. As the Oxford astronomer Cedric Lacey puts it: 'Time travel is not permitted in any reasonable cosmological model (but I guess that's the definition of reasonable!)'

On the other hand, cosmologists have unearthed a novel feature of the landscape of spacetime known as a 'wormhole'. Wormholes, first mooted by John Wheeler, are solutions of Einstein's equations which connect otherwise distant parts of a single universe, or even what would otherwise be separate universes (*see* Figure 8).

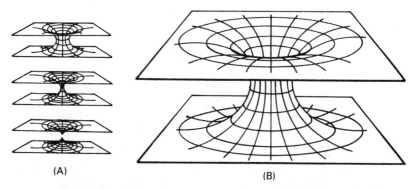

(A) (B)

Figure 8. Two possible kinds of wormhole in spacetime. In the classical (Newtonian or Einsteinian) case (A) the wormhole pinches off before a traveller can pass through. In the quantum case (B) it may be possible to avert the collapse, thereby permitting auto-infanticide and much more besides.
[Adapted from R. L. Forward. *J. Brit. Interplan-Soc.* <u>42</u>, 533 (1989).]

A traveller who fell through a suitable wormhole could end up at the same place in his 'past'. However, matter as described in Newtonian or Einsteinian mechanics cannot support the existence of wormholes for sufficiently long to enable the intrepid traveller to go back into the past – he would be crushed beyond all recognition in the attempt to reach it. Yet, remarkably enough, there is a recent speculative claim by three cosmologists, Michael Morris, Kip Thorne and Ulvi Yurtsever that if allowance is made for the quantum properties of matter, the collapse of these wormholes might be averted and auto-infanticide, along with much else, could after all be a possibility.[88] Their work was inspired by a science-fiction novel, *Contact*, in which the author, Carl Sagan, describes a wormhole constructed by an ancient civilisation to permit ultra-fast space travel.[89] Nevertheless, one must ask whether some-where along the way we have not 'missed a fundamental constraint that maintains the consistency of physics and prevents us from journeying back in time'.[90]

The problem with relativity

We are back to our main problem: Einstein's failure to explain time's arrow. The reason for this is closely linked to the concept of causality, the notion that effects never precede their causes. Consider any process whereby in a given frame of reference, some event – such as a batsman hitting a ball – at a given location in space, X, and at a given time causes an event elsewhere at spatial coordinates Y and a *later* time – say the ball falling into the hands of a fielder. In order for the event at Y to happen, information must pass from X to Y at some speed, in this case the speed of the cricket ball. As we saw in the last chapter, this is of fundamental significance in Newtonian mechanics. It was fully accepted by Einstein as an unassailable guiding principle when he constructed his theory of relativity.

Einstein insisted on the invariance of causality – X must precede Y for all observers in inertial frames of reference. This seems to be reasonable on the basis of the fact that we know batsmen are not caught out before they have struck a ball. However, an analysis of the Lorentz transformations, together with insistence on causality, shows that no influences can travel faster than the speed of light. Light therefore becomes a kind of universal umbilical cord connecting events. Otherwise there would exist frames of reference, be they cricket pitches, china shops or stars, in which observers could see Y preceding X in time if signals from Y travel faster than those from X. Put another way, Einstein's second postulate of special relativity, that there *is* a limit fixed by the speed of light, guarantees that causality is maintained whatever the circumstances of an observer.

It is difficult to overplay Einstein's commitment to the concept of causality as the bedrock of physics. The reason is readily understandable if we pause for a moment to consider a world in which causality is violated. It might mean that a pebble could levitate off the ground so that you could grasp it; worse than that, you might be struck down by a stone before it fell or kill your own grandmother before you were born. Causality is closely related to determinism; it will be recalled that Newtonian mechanics is deterministic – specification of the dynamical conditions at one instant of time is enough to predict all subsequent behaviour. Einstein's modification of dynamics left this deterministic structure intact. However, considerations of causality play a more dominant role in relativity because for the first time the finite speed of signals is taken into account. In Newtonian physics, it had been tacitly assumed that effects such as the gravitational attraction between massive bodies – action-at-a-distance – propagated instantaneously.

But the odd thing about Einstein's insistence on the common-sense notion of causality is that it self-destructs. We saw in Chapter Two that Newton's description of the world is a rigorously deterministic one, where the future and the past can be constructed if enough information is provided. We also saw that his description is reversible – there is no physical distinction between time running forwards or backwards. Precise specification of a system at any instant in time at once provides knowledge of its entire past and future behaviour. This is due to the causal structure of the theory: an event at some point in time leads unequivocally to an effect at a later instant; moreover, because of the time-symmetry of the equations of motion, the later event can also be said to *cause* the earlier one, since both processes are mathematical solutions of Newton's equations and are thus physically possible.

Exactly the same problem afflicted Einstein. All his equations of relativity (whether in the special or general theories) possess in broad terms the same deterministic structure and time-reversible features as Newton's mechanics. There is no distinction between time running forwards and backwards. This is why one has to introduce in an *ad hoc* way assertions such as the Cosmic Censorship Hypothesis and the banning of time travel. Thus, for example, the metric structure (which, it will be recalled, determines the geometrical properties of spacetime) as well as the equations of motion which it spawns are unchanged if one replaces a positive value for time in the equations by a negative one. Once again, given the conditions at any single spacetime coordinate, the entire history and future of the universe can be computed.[91] One need go no further to conclude that just as Newton's physics inherently lacks the ability to explain many kinds of irreversible processes with which we are familiar (melting snowmen, cooling cups of coffee, crumbling statues and the ageing process) the same can be said of Einstein's relativity.

In the other remarkably successful twentieth-century discovery of theoretical physics, quantum theory, the whole concept of causality has undergone a radical reassessment. According to the conventional interpretation of quantum theory, naïve acceptance of the principle of causality is no longer possible. Einstein never accepted the quantum mechanical renunciation of causality as final, but it could offer a way out of these difficulties.

Retracing the evolution of the universe back in time from the present moment, we find that we approach the Big Bang, in which all matter in the universe is fantastically compressed together and spacetime warped beyond all belief. Close to the moment of the Big Bang singularity we are dealing with a scenario in which Einstein's theory must start to fail. For there *are* certain built-in suppositions, even to Einstein's powerful theory, which are known to be incorrect over very short distances. As we have stressed before, general relativity is valid on the very large, cosmological scale where it replaces Newtonian physics; but in the realm of elementary particles, atoms, and molecules, quantum theory replaces classical physics. The singularity is sufficiently troublesome for scientists to seek a synthesis of quantum and relativity theories as a way around it.

Could such a synthesis hold the key to the origin of the arrow of time? Roger Penrose thinks so, for as we shall see in Chapter Five he believes that 'quantum gravity' must be an intrinsically time-asymmetric theory. Nevertheless, 'It is one of those things that worry me. The answer seems to be almost obvious, but when I say these things no one seems to agree.'[92] Perhaps a more satisfactory concept of time may arise by merging relativity with the quantum mechanics that governs the enchanted microscopic world to which we now turn.

A quantum leap for time

If [atomic research] cannot be fitted into space and time, then it fails in its
whole aim and one does not know what purpose it really serves.

Erwin Schrödinger
Letter to Willy Wein, 25 August 1926

QUANTUM mechanics gives us tantalising hints of the presence of
time's arrow. As we shall see, it suggests that the passage of time is
guaranteed by something very simple: our own observation of change.
And by opening up the secret world of the *atom*, it reveals one tiny
particle whose existence indicates that time is irreversible. Physicists
dispute whether these are fundamental insights or simply red herrings.
One thing is certain – the quantum world is full of problems and
paradoxes. For example, in many respects this new theory follows its
predecessors by appearing to imply that time can go backwards as well
as forwards. But in addition to making the claim that events are doomed
to endless repetition, it also seems to support the notion that a watched
pot never boils. It suggests that a cat can be alive and dead at the same
time, and that something can be everywhere and nowhere simultaneously. It is such a strange theory that many of the scientists who helped
to invent it – including Einstein – later tried to disown it. And decades
after their original work, there is still great dissent about what quantum
theory actually means.

Quantum theory deals with the properties of matter on the tiniest of
scales, including atoms, the smallest units of a chemical element. In
attempting to describe the world at this microscopic level, we find that
Newton's mechanics breaks down. Newton's failure here is even more
spectacular than his failure to cope with the fast and highly massive
objects dealt with by relativity. Quantum theory, by contrast, is
extraordinarily successful at the atomic level. We depend on it for our
detailed understanding of chemical reactions, lasers and the transistors
and diodes that underpin modern computer technology. Today, the
existence of atoms appears indisputable – images of atoms and
molecules can even be seen with the aid of field-ion, electron or scanning
tunnelling microscopes (*see* colour plates). But it is easy to forget how
recently it was that the very existence of atoms was a controversial issue.

Although the idea originated in antiquity, it was suppressed for centuries.

The story of the atom probably started at Abdera, a seaport in the Aegean, around 500 BC. The two pioneers of *atomism* were the philosopher Leucippus and his pupil Democritus of Abdera, whose views were not too distinct from that of scientists today. They thought that the world was made up of minute, invisible, irreducible bodies – differing only in shape and size – in a state of perpetual motion through a limitless void. They called these entities *atomos*, meaning indivisible, and suggested that everything from tables to turtles was formed by their random collisions. The atomists also used atoms to explain such sensory phenomena as taste and smell. Sadly, their ideas were cast into oblivion by the baleful influence of Plato and Aristotle.[1] These fathers of Western philosophy argued that matter could be infinitely subdivided, there being no smallest unit which could not be broken down further. Atomism had been defeated and remained in the shadows for two and a half thousand years.

Why did it suddenly reappear? Much of the credit belongs to a Quaker teacher called John Dalton who was born in 1766 in Eaglesfield, Cumberland. His treatise entitled *New System of Chemical Philosophy*, written in two volumes between 1808 and 1827, marked the rebirth of atomism and laid the foundation for modern chemistry. Dalton recognised that atoms could help to explain a growing wealth of scientific knowledge, including the behaviour of gases and the chemical transformation of one substance into another. To Dalton, an atom was the smallest indivisible unit of a substance that still possessed its chemical properties. He maintained that chemical reactions did no more than separate and unite these fundamental building blocks of matter. Today we would usually describe these building blocks as *molecules* – the smallest combination of atoms that can take part in a chemical reaction. For example, a water molecule consists of two hydrogen atoms and one oxygen atom.

Initially, other chemists were wary of Dalton's ideas. They realised that he was on to something but held back from his conviction that atoms really existed, preferring to treat them as handy devices to interpret and rationalise their data. The French chemist Jean Baptiste Dumas even said: 'If I were master of the situation, I would efface the word atom from science.'[2] It was some time before chemists and physicists began to recognise that they had accumulated independent evidence pointing unequivocally towards atomism. Much of the intellectual debate focused on gases and the so-called kinetic theory which sought to explain their properties in terms of atoms and molecules.

Physicists such as James Clerk Maxwell and Ludwig Boltzmann put forward simple models to explain the pressure exerted by a gas on a container, portraying the gas as a collection of hard spheres like billiard balls which rattled against the container's sides in accordance with Newtonian mechanics. The properties of gases were interpreted in terms of the movements of their component atoms and molecules. Pressure could be easily calculated from the rate at which these spheres struck the walls of a container. Heat was the result of the rapid and random motions of molecules: the hotter the gas, the faster the molecules would move.

But this was still not enough for the die-hard opponents of atomism like Ernst Mach or the German physical chemist Wilhelm Ostwald. As positivists, they contended that statements about a world which could not be directly observed were meaningless. What the atomists needed was an example where molecular effects happened right in front of the sceptics' eyes. In 1905 they realised that one had been available from the days of Dalton himself. It was called *Brownian motion* – the random dance of small particles of pollen (or, indeed, dust or soot) suspended in water. The effect had been watched under a microscope by the Scottish botanist Robert Brown as early as 1827 but no convincing explanation was offered until Einstein provided a characteristically brilliant break-through. He explained that these Brownian motions were due to random collisions of the suspended particles with the invisible water molecules surrounding them.[3]

It was a powerful vindication of the atomic theory of matter. But by then the classical conception of the atom, as envisaged by Leucippus and Democritus, was already out of date. It had been superseded at the very end of the nineteenth century by the discovery of radioactivity. In 1895, the German physicist Wilhelm Röntgen had stumbled upon a myster-ious form of radiation which he called X-rays. The following year, while investigating X-rays, the Frenchman Henri Becquerel detected strong radiation emanating from uranium compounds. Subsequent work by the Polish chemist Marie Curie and others helped to piece together what was happening. They found that atoms of certain elements could decay into other elements with quite different chemical properties. This transmutation of the radioactive elements – almost like something out of the dreams of the medieval alchemists – was made explicit in a law enunciated by Ernest Rutherford and Frederick Soddy in 1902. At this point, the modern atomic theory of matter parted company with that handed down by antiquity, for it implied that the atom itself had structure and could be further subdivided.

Atomic structure and the fall of classical physics

In the first decade of the twentieth century, the race to discover the structure of the atom was on. Undoubtedly the most important experiment was performed by Hans Geiger and Ernest Marsden under the direction of Ernest Rutherford in 1909 at the University of Manchester in England. The result was extraordinary, providing the first glimpse of what an atom looks like and opening the way to a new view of matter. The two men used radioactive material to bombard gold foil with beams of alpha-particles (positively charged particles emitted during radioactive decay). Most of the particles passed straight through the thin sheets of foil, suffering only small deflections, but very rarely one bounced right back in the direction of the beam. Rutherford said that it was the most incredible thing that ever happened to him: 'It was almost as if you fired a fifteen-inch naval shell at a piece of tissue paper and it came right back and hit you.'[4] He made sense of these scattering results by postulating that the alpha-particles were deflected by a massive but very small positively charged nucleus located at the centre of each gold atom in the foil.

In Rutherford's famous model of the atom, which made its first appearance at Christmas 1910,[5] the nucleus was orbited by negatively charged electrons just like planets in a miniature solar system. Atoms were mostly empty space and their size was defined by the orbit of the outermost electrons. The radius of the atomic nucleus (equivalent to that of the Sun) was about one-thousandth of a millionth of a millionth of a metre and the electrons roamed in a region of radius 100 millionths of a millionth of a metre. It was soon recognised that radioactivity was caused by the emission of particles from inside the nucleus, which in turn implied that it too had an internal structure.

With all the successes of Newtonian physics (discussed in Chapter Two) it seemed natural to make the leap of faith that the motions of electrons obeyed Newton's deterministic equations. If Newton's mechanics worked for cricket balls, why not for atoms and the electrons within them? There were plenty of predictions that could be made in this way, many of which could be tested by experiment.

However, the classical approach broke down when applied to the way atoms absorbed and emitted light. According to Maxwell's theory of electromagnetism and Newton's mechanics, the negatively charged electrons orbiting the nucleus should emit a continuous range of light frequencies, shading into each other like the colours of a rainbow. In fact, scientists saw a series of discrete, utterly distinct lines, very similar to a supermarket 'bar code', only in Technicolor. Worse still, if the electrons emitted light in the manner predicted by classical electrodynamics, they would soon lose their energy and spiral into the nucleus – like water flowing down a plug hole. The inevitable implication of classical theory was that the atom – the very building block of matter – was unstable.

The birth of the quantum

Quantum mechanics came to the rescue of these self-destructive classical atoms by a roundabout route, starting with a rather reluctant revolutionary, the German physicist Max Planck. His work actually pre-dated Rutherford's atom by ten years. It centred on the inability of existing theory to explain the relationship between the temperature of a body and the amount of electromagnetic radiation it emits: for instance, why a red-hot poker becomes white when heated. The theorists of the time based their predictions on an idealised example, a perfect emitter and absorber of radiation called a 'black body'. For low frequencies at the red end of the spectrum, they achieved reasonable agreement with experiments. For high frequencies, they predicted that the body would send out an infinite amount of energy. This absurd result was dubbed the 'ultraviolet catastrophe'. In fact, observations showed that the density of radiation was small at both high and low frequencies, with a peak somewhere in the middle which depended on the temperature of the radiating body.

By the late 1890s some approximate laws were cobbled together to fit experimental measurements on black body radiation. But a completely satisfactory explanation for the way in which the radiation density varied with frequency had to wait until 19 October 1900 and an announcement by Planck at a meeting of the German Physical Society. The roots of his historic announcement reach back to an argument he had had with Boltzmann in 1897, when the latter suggested that he use a statistical approach for the resolution of his difficulties.[6] Planck was one of the foremost thermodynamicists of his day and he naturally hoped that thermodynamics would come to the rescue of the black body problem. Boltzmann had pioneered a statistical mechanical framework to underpin the subject of thermodynamics, based on the assumed existence of atoms and molecules. As an opponent of atomism, Planck was obliged to reject it. During one encounter in 1891 Boltzmann remarked to Planck and Ostwald that he saw 'no reason why energy shouldn't also be regarded as divided atomically'.[7] Planck, a physicist of the old school, eventually relented and took up Boltzmann's suggestion: his resulting law gave a beautiful description of black body radiation.

To derive his law, Planck had indeed to assume that electromagnetic radiation comes in packets of energy, called *quanta*. He found that, like matter, it can only be divided into a finite number of pieces rather than an infinite number. The mathematical relationship lying at the heart of this work showed that the energy of the quantum could be calculated by multiplying the frequency of the radiation with a new fundamental constant of nature, now known as Planck's constant. This simple 'Planck relationship' between energy and light frequency in effect said that energy and frequency are the same thing measured in different units.

Enter Einstein

Planck thought his explanation of black body radiation was grotesque, for it contradicted the tenets of classical electrodynamics. Its far-reaching implications were lost on him, a conservative scientist who treated his theory more as a convenient assumption than as a profound truth. Yet it worked wonderfully well and, being a pragmatic man, he accepted it in that spirit. That did not prevent him from being shocked when the full implications of his ideas were later spelled out by others. It has been said that after he let the spirit of the quantum out of the bottle, Planck was scared to death by it.[8]

Once again it was Einstein who took the big step forward. His remarkable contribution to quantum theory came in 1905, the same year that his papers on relativity and Brownian motion appeared in *Annalen der Physik*. Indeed, it was for this breakthrough – not relativity – that he was awarded the 1921 Nobel Prize for Physics (announced in 1922). His achievement centred on something called the photoelectric effect. Experiments had shown that light cast upon a solid metal surface caused the ejection of electrons from the metal. The energy of these electrons varied, not with light intensity, but with its colour. Such behaviour is inexplicable by classical electrodynamic theory, which predicted that the more intense the light, the greater the speed with which electrons would be swept out of the metal. Instead, it was observed that increasing the intensity of light of a given colour simply swept out more electrons of the same energy. To explain this behaviour Einstein suggested that the energy associated with the light beam came in microscopic packets which he called 'light quanta'. Brighter light means more quanta – hence more electrons swept out of the metal.

Higher-frequency light means bigger quanta and a greater speed for the escaping electrons. Below a certain size of quantum, the electrons do not have enough energy to leave the metal surface at all.

This *ad hoc* explanation (based purely on deduction) was greeted with scepticism. For despite Planck's earlier work, it was still generally believed that the energy of electromagnetic radiation came in continuous quantities. Einstein was proposing that in a sense light consisted of corpuscles, a notion favoured by Newton which had long ago been overtaken by the wave theory of light proposed by the Dutchman Christiaan Huygens in 1678. The wave theory seemed too good to give up, as it offered a very clear explanation of optical phenomena such as refraction, reflection and interference. (Interference can occur when light from two sources combines to produce alternating light and dark bands – an *interference pattern*).

Einstein's theory of the photoelectric effect was subjected to detailed scrutiny by experimentalists over several years; by 1916 it had been completely confirmed. The remarkable success of this theory eventually forced scientists in the 1920s to reconsider the nature of light. Yet, ironically, this conflict with the wave theory caused Einstein anxiety throughout the rest of his life. He always stressed the provisional nature of the light-quantum postulate. On 12 December 1951, near the end of his life, Einstein wrote to his friend Besso: 'All these fifty years of pondering have not brought me any closer to answering the question, What are light quanta? Nowadays every Tom, Dick and Harry thinks he knows it, but he is mistaken.'[9] Today, despite Einstein's own reservations, the particulate aspect of light has been unambiguously demonstrated. The light-quantum goes under the name of the *photon*, first suggested by the physical chemist Gilbert Lewis in 1926. This does not, however, mean that light is not also a wave motion. For the photon has the properties of a wave as well as those of a particle. Sometimes it appears in one guise, sometimes in another. This is our first encounter with one of the strangest features of the quantum world: the duality of particles and waves.

Waves and particles

Wave–particle duality gives a bizarre twist to a classic experiment demonstrating the wave nature of light which was performed and analysed by Thomas Young during the early years of the nineteenth century. He shone light from a single source upon an opaque barrier in which there were two slits. Beyond this obstruction the light spread out from the slits – which behave like secondary light sources – into alternating light and dark bands visible on a screen, an effect typical of an interference pattern. It is a good demonstration of the wave nature of light: a similar pattern can be produced by passing parallel waves in water through two slits in a barrier. Then the pattern of alternating constructive and destructive interference arises as water waves from the adjacent slits add or cancel respectively.

Now imagine the same experiment that Young carried out but using only a single particle of light. A particle has its mass concentrated at a single point, while a wave is an entity without mass that is smeared over a region of finite extent. Surely the particle goes through either one slit or the other? It turns out, however, that if individual photons are fired at the two slits one at a time and their point of arrival at the screen is recorded, an interference pattern of photon intensity identical to the original one builds up. Thus it seems that an individual particle senses both slits owing to its wave character.

This duality does not just apply to light – it has much wider significance. This became clear following a flash of insight by a young Frenchman named Louis de Broglie. De Broglie was an aristocrat whose wartime involvement in radio work directed his intellectual interests away from medieval church history and into physics. He came to dominate the subject in France for over a generation (some have said deleteriously). His contribution to quantum theory was so radical that others even mocked his ideas as 'la Comédie Française'.[10] De Broglie recognised, through the encounter with the light quantum, that just as light waves can behave like particles, so too can particles behave like waves. He put this apparently straightforward idea on a firm foundation by proposing a simple mathematical relationship between the momentum of a moving body and its associated wave: the wavelength of the 'particle' is inversely proportional to its momentum (the product of the particle's mass and velocity), the constant of proportionality again being Planck's constant. Thus the greater the speed or mass of the particle, the shorter its de Broglie wavelength.

De Broglie first broached these ideas in three short publications in the *Comptes Rendus de L'Académie des Sciences de Paris* in 1923. Then he set about writing his doctoral thesis, which was completed in 1924. One of his examiners was Langevin, who was apparently astounded by his ideas; he sent a copy of the thesis to Einstein, who, immediately recognising its significance, wrote back that de Broglie 'has lifted a corner of the great veil'.[11] Undoubtedly, Einstein's opinion influenced the positive outcome of the *viva voce* examination. But de Broglie did not wish to rely on Einstein's authority – he suggested himself how his ideas might be tested. In 1927 their validity was unambiguously demonstrated by Clinton Davisson and Lester Germer at the US Bell Telephone Laboratories and by Alexander Reid, under the direction of George Thomson, at the University of Aberdeen, Scotland. It is amusing to note that Joseph J. Thomson had earlier received the Nobel Prize for proving that the electron was a particle, while his son received the same award for demonstrating that it was a wave.

The schizophrenic particle-and-wave behaviour of matter has strange consequences. Imagine a game of 'quantum snooker'. The wave properties of the snooker balls would lead to a number of astounding effects. No matter how well a quantum player aimed the cue ball – for instance, at a red ball teetering on the edge of a pocket – there would always be a chance that the red would end up in a pocket at the other end of the green baize (as a wave, the red ball would extend across the table). It would also be possible to overcome a 'snooker' by a quantum phenomenon called 'tunnelling', in which the cue ball would pass right through an intervening ball. Again, the wave-like properties of the cue ball would allow it to 'spread' through the obstacle.

If de Broglie's idea is right, why don't we see such wave-like effects and other quantum phenomena in everyday snooker? The reason is revealed by de Broglie's relationship. It shows that the wave properties of particles depend on their mass – the larger the mass the smaller the associated wavelength. For atoms, this wavelength is large relative to their 'size' in the classical sense, whereas for ordinary (macroscopic) snooker balls it is vanishingly small. Only by shrinking the game of snooker to the microscopic level could we observe these curious quantum effects.

Bohr's atom and quantum theory

How did all this relate to the atom? To understand, we must first go back to 1913 and the work of the Danish theoretical physicist, Niels Bohr. Opinions on Bohr's intellectual abilities varied enormously: Rutherford described him as 'the most intelligent chap I have ever met', while George Gamow maintained that Bohr's most characteristic quality was 'the slowness of his thinking and comprehension'.[12] This observation was repeated by Winston Churchill who thought that Bohr was an immense bore (*sic*). Yet he was to inspire and dominate the new way of thinking about the world engendered by quantum theory.[13]

Niels Bohr proposed a radical though inelegant solution to the failure of classical mechanics to cope with Rutherford's self-destructive atoms, a halfway house between classical mechanics and modern quantum mechanics. He simply asserted as an *ad hoc* postulate – solely for electrons revolving around nuclei – that Newton's mechanics be replaced by new quantum rules. These rules could not really be called laws of motion since they lacked any theoretical basis. They just stated baldly that electrons occupy fixed orbits in which they emit no radiation.[14] Bohr called these orbits 'stationary quantum states'. As an idea almost plucked from nowhere, it had a lot in common with Einstein's response to the photoelectric effect.

As before, the electrons moved in different orbits, rather like the planets round the Sun. But when the atom received electromagnetic energy, for example by absorbing a photon of light, one of the electrons jumped instantaneously into another orbit further from the nucleus. This explained the characteristic discrete spectrum, because light is only emitted when an electron makes a fixed jump back to its original orbit, or 'stationary quantum state'. Here we can see one of the principal features of quantum systems: rather than being absorbed or emitted in a continuous way, energy can only vary at the atomic level in abrupt quantum jumps.

Bohr applied his model to the simplest of all atoms, hydrogen. This consists of a single electron orbiting about an equally yet oppositely charged nucleus of far greater mass, which was called a proton when it was discovered by Rutherford in 1919. Using his quantum rules, Bohr was able to explain for the first time the electromagnetic spectrum emitted by hydrogen atoms. Not only could physicists now understand why the hydrogen spectrum had a 'bar code' structure consisting of discrete frequencies (corresponding to quantum jumps by the electron), but the exact frequencies at which the lines of light appeared could be predicted (*see* Figure 9).

Figure 9. Examples of atomic spectra. The absorption/emission lines occur at discrete values of light frequency, providing powerful evidence for the quantum nature of atoms.
[Adapted from W. J. Moore. *Physical Chemistry*, p. 587.]

Although the new theory was very exciting, it did not account for the more complicated atomic spectrum of helium, the next simplest atom, which merely possessed two electrons revolving round a doubly positively charged nucleus. More complex atoms were even further beyond its scope. Moreover, the Bohr model was fundamentally flawed insofar as no explanation was offered for the *ad hoc* quantum postulates. Thus its status remained that of a partial or stop-gap theory of atomic structure.

The new mechanics

The next breakthrough took a while in coming. When it eventually arrived it did so nearly simultaneously via two separate intellectual pathways, one associated with the young German Werner Heisenberg, the other due to the Austrian Erwin Schrödinger.

In 1925, Heisenberg was the first to reach a consistent quantum formulation of the microscopic world, which he called 'matrix mechanics'. He invented the world's first fully fledged quantum theory while recovering from a hay fever attack on the North Sea island of Helgoland. His theory gets its name because it expressed the microscopic world in terms of mathematical objects called matrices. The algebra of matrices is similar to that of ordinary numbers, but with one very important exception. In normal multiplication, two times three is the same as three times two. Matrix A multiplied by matrix B does not equal matrix B multiplied by matrix A. This lopsided mathematical feature was later recognised to be related to the fact that the microscopic world could yield different results depending merely on the order in which measurements are carried out. This is one of many curious features displayed by quantum theory.

Matrix mechanics is rather abstract and it is easier to concentrate on the version developed by Schrödinger. He had been deeply influenced by Boltzmann, whose 'range of ideas played the role of a scientific young love, and no other has ever again held me so spellbound'. [15] Schrödinger's solution to the problem, praised by Arnold Sommerfeld as 'the most astonishing among all the astonishing discoveries of the twentieth century', appeared in 1926. [16]

Schrödinger was impressed by de Broglie's work and committed himself totally to the new wave-particle ideas. [17] He developed wave mechanics in Arosa, an Alpine *Kurort* near the ski resort of Davos, in the company of a mistress, when his life-long marriage was at an all-time low. Walter Moore wrote in his biography of Schrödinger: 'Like the dark lady who inspired Shakespeare's sonnets, the lady of Arosa may remain forever mysterious . . . Whoever may have been his inspiration, the increase in Erwin's powers was dramatic, and he began a twelve-month period of sustained creative activity that is without a parallel in the history of science.' In December 1925 Schrödinger described in a letter how he was struggling with a new atomic theory, one that when solved would be 'very beautiful'.[18]

In 1926 Schrödinger's creativity peaked. To ensure that he was not distracted by noise, he often resorted to using pearls as earplugs.[19] That year, he produced six major publications on his wave mechanics for the German journal *Annalen der Physik*.

Schrödinger's theory, being written in terms of differential equations, is amenable to a more directly pictorial description than Heisenberg's. Mathematically, the formulations of Schrödinger and Heisenberg looked totally different but, as Schrödinger himself soon showed, led to exactly the same results. 'It was as if America was discovered by Columbus, sailing westwards across the Atlantic ocean, and by some equally daring Japanese, sailing eastward across the Pacific ocean.'[20] Owing to its more intuitive and flexible basis, Schrödinger's theory has become the favoured way of thinking about the subject of quantum mechanics.

De Broglie's work indicated that Bohr's quantum rules for the hydrogen atom, which dictated a set of fixed electron orbits, could be understood in terms of electron waves around the nucleus. These are called *standing waves* and are similar to those set up in an organ pipe when it resonates, or by a violin string when stroked with a bow. The resonance occurs because an exact number of waves can be fitted in the pipe or along the string. Similarly, only certain values of the de Broglie wavelength could produce standing waves around the atom. Each value of the wavelength corresponds to an 'orbit' of the electrons (*see* Figure 10).

By this line of reasoning, Schrödinger was led to write down an equation describing the way matter-waves evolve in time at the microscopic level.[21] In essence, it was the same task as had been tackled by Newton and Einstein before him. For them, the key to dynamics was to describe directly how the position of a body like a snooker ball changes in time. Schrödinger, however, was working at the atomic level, where a particle is at one and the same time a wave. This made the analysis much more subtle. Accordingly, his equation includes a completely new mathematical quantity called the *wavefunction* which takes into account the Jekyll and Hyde character of tiny particles and maps out all the possibilities of their behaviour. Schrödinger connected his wave

119

mechanics to the everyday world by constructing it in such a way that for macroscopic objects like snooker balls it reduces precisely to equations familiar from Newtonian mechanics.[22] *Schrödinger's equation*, as it is called, has since become one of the most fundamental of all the equations used in theoretical physics and chemistry. The paper that unveiled it was later described by one of the leading exponents of quantum theory as containing the whole of chemistry and a large part of physics.[23]

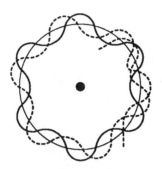

Figure 10. Standing electron wave (full line) in the hydrogen atom as envisaged by de Broglie. The dashed line shows how non-standing waves would be destroyed by interference.
[Adapted from W. J. Moore. *Physical Chemistry*, p. 595.]

Schrödinger tested the validity of his equation by applying it to the ripple of electron waves around the hydrogen atom. As a piece of applied mathematics, the equation was not new. Schrödinger solved it with standard techniques and found, as had Bohr, exact agreement with the energies of the radiation emitted by hydrogen atoms in spectroscopic experiments. But now the quantum rules dictating that electrons only occupy fixed orbits no longer appeared out of thin air. And unlike Bohr, Schrödinger was able to go on to analyse and successfully explain the two-electron helium atom.

Despite this success, Schrödinger was uncertain about what the wavefunction meant in physical terms. Evidently, he hoped at this early stage to lead a return to the crystal-clear perspectives of the old classical concepts, which were rooted in everyday images. But the wavefunction was to spoil this dream. About six months later, while applying Schrödinger's equation to the study of collisions between microscopic particles such as atoms or electrons, the German physicist Max Born made a far-reaching conjecture. He proposed to interpret the wavefunction as a 'probability amplitude', a quantity that can be used to calculate the probability of finding a particle in a region of space. He maintained that the square of the wavefunction gives the probability of finding the particle in a specified place at an instant in time.

Acceptance of the Born interpretation was by no means immediate, and disputes over it continue to this day. They are central to the meaning of quantum mechanics, for either one of two consequences has to follow. On the one hand, Schrödinger's equation is accepted as the basis for a new, fundamental mechanics, in which case Born's interpretation means that instead of exactly predicting phenomena at the atomic or sub-atomic level we can only talk about the probabilities of their occurrence. This has profound consequences for determinism and causality. The only alternative is to deny that quantum mechanics is truly fundamental; instead it is simply a method of making statistical statements about events of which we have an incomplete understanding – but which at a deeper yet unknown level remain strictly deterministic and causal, just like Newton's and Einstein's physics.

It was at this point that Einstein parted company with the new generation of quantum physicists. The latter preferred the first choice while Einstein doggedly adhered to the second: 'God does not play dice with the world,' was his famous riposte.[24] Einstein realised that the commonly favoured interpretation of the wavefunction undermined causality, the centuries-old notion linking cause with effect. He wrote in a message for the bicentenary of Newton's death: 'May the spirit of Newton's method give us the power to restore unison between physical reality and the profoundest characteristic of Newton's teaching – strict causality.'[25] It is a remarkable fact that both de Broglie and Schrödinger also held profound reservations about the direction the newly born physics was taking. In a famous exchange with Niels Bohr, Schrödinger remarked: 'If we are still going to have to put up with these damn quantum jumps, I am sorry that I ever had anything to do with quantum theory.'[26] But Bohr, Born, Heisenberg and their disciples gave short shrift to such objections. In so doing they laid the basis for the 'orthodox' version of the subject which is widely accepted today, although in truth it poses as many questions as it answers.

Everywhere and nowhere at the same time

The shocking picture of the world described by quantum theory can be illustrated by the very simple experiment shown in Figure 11(b), an electronic version of Young's slit experiment. The source is designed to fire one electron at a time at the double slit in the barrier, beyond which

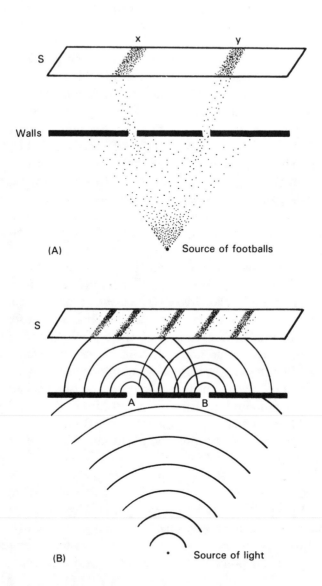

Figure 11. (A) The double slit experiment with classical particles – footballs – kicked at a screen S from behind a wall with two slits. Balls can only arrive at the positions X and Y. (B) The intensity pattern built up on the screen S from a wave source sending waves through a barrier with two slits. Dark regions on the screen correspond to high wave intensity, light regions to low intensity. Slits A and B act as secondary wave sources.

[Adapted from J. D. Barrow. *The World Within the World*, p. 134.]

122

there is a fluorescent screen S which announces the electron's arrival by a flash of light. In the experiment, an interference pattern builds up on the screen, wherein the number of electrons recorded waxes and wanes depending on the position along the screen.

We recall that the electron is neither a particle nor a wave, but has attributes of both. If the electron were a particle, it would either hit the barrier and be deflected or pass through one or other of the slits to make a flash on the screen at either the point X or the point Y (see Figure 11(a)). Repetition of the experiment would lead to as many flashes at X as at Y. If, on the other hand, the electron were a wave, spread throughout space, the wavelike disturbance would pass through both slits simultaneously (see Figure 11(b)). Just as overlapping ripples of water in a pond interfere with one another, so these electron waves would also produce a characteristic interference pattern on the screen, consisting of peaks and troughs of intensity as shown.

What does quantum mechanics predict? Like a bookie, it only gives odds, even when a single electron is involved. Essentially, it provides the odds for the electron to arrive at each point on the screen. It can only tell us with absolute certainty where the electron will *not* be found. In this case, it predicts that the electron, if conceived of as a point particle, can arrive anywhere on the screen except where there are zeros in the intensities in the wave description, when the probability of arrival is zero. The probability of spotting the electron varies with its position on the screen in exactly the same way as the wave intensity predicted from wave interference. However, the electron finally arrives at a fixed position – it is not spread out over the whole screen. Only with the passage of enough electrons through the apparatus does the interference pattern build up.

An elegant formulation of quantum mechanics which enables one to envisage this was put forward by the American theoretical physicist Richard Feynman in the 1940s while he was a graduate student at Princeton University. Popularly known as the 'sum over histories technique', it shows how the quantum description of the world turns smoothly into Newton's as one moves from the microscopic regime into the macroscopic.[27] In Feynman's description, a quantum particle, like an electron or a photon, is regarded as trying out every possible classical path or trajectory between the source and a given position of arrival on the screen shown in Figure 11. Because the particles have an associated de Broglie wavelength, each of these classical 'histories' interferes with all others, just as we have seen that water waves interfere with one another. This leads to the probability distribution characteristic of the interference pattern shown in Figure 11 for tiny particles. But if one boosts the mass of the particle, to that of a football for example, the mathematics shows that destructive interference then occurs along almost all paths or histories of the particle, except precisely along the paths (trajectories) that would be predicted by Newtonian mechanics.

As we have said, quantum theory cannot predict where the electron will arrive: it only delivers a probability for the event to occur at a given point. Thus all quantum mechanics can say, if a single electron is fired at the slits, is that the electron will cause a flash somewhere on the screen, save for the taboo regions where there is zero wave intensity. How is this possibility turned into a reality? The flash of light could appear in any number of locations on the screen. But probability is converted into certainty by a measurement: once the electron has been found at a particular point, the probability of it being found anywhere else quite properly drops to zero. Only if the experiment is repeated a great many times will the probability distribution become meaningful and the interference pattern build up.

To the extent that quantum mechanics provides the ultimate description, no more can be asked about the whereabouts of the electron until it impinges on the screen. We are therefore forced to conclude that the electron is somehow spread or delocalised throughout space and time, passing through *both* slits and interfering with itself before miraculously collapsing instantaneously to some spot on the screen in an entirely random fashion. It is everywhere and nowhere at the same time.

The Copenhagen interpretation

Niels Bohr confronted the dilemmas posed by measurement head-on. His approach, which is adhered to by the majority of working physicists, has come to be known as the *Copenhagen interpretation* of quantum mechanics, after the city where he lived and worked. Its basic premise is that our description of the microscopic world is cramped by the paucity of our language, which is based on the information delivered by our senses. The world has a classical part, consisting of the act of measurement, and a quantum part, the thing we are measuring. In other words, the world we observe seems to be independently real but is nevertheless suspended on an 'unreal', microscopic world. This limitation is inescapable; accordingly, we cannot hope to give a description of quantum processes as they really are. Bohr's close collaborator Werner Heisenberg maintained: 'The hope that new experiments will lead us back to objective events in time and space is about as well founded as the hope of discovering the end of the world in the unexplored regions of the Antarctic.'[28] Instead, our discussions can only be expressed in terms of the necessarily macroscopic observations we make of the world. Pushed to the limit advocated by Bohr himself, this interpretation of quantum mechanics denies the objective reality of phenomena which are not subjected to measurement: 'There is no quantum world. There is only an abstract quantum description.'[29]

The mere suggestion that 'elementary particles', atoms and molecules have no independent existence is anathema to most physicists, chemists and molecular biologists. Among the chemists and molecular biologists, who have convinced themselves of atomic reality, it would be regarded as pure heresy. Yet the quantum theory, on which all their models are ultimately based, provides little support for such notions. It is the modern version of Bishop Berkeley's famous denial of the existence of the tree in the quadrangle at New College, Oxford when he turned his back on it.[30]

According to the Copenhagen interpretation, every observation must be referred back to its macroscopic context. The very process of measurement evidently assumes considerable importance in quantum theory, whereas it had been completely ignored in classical physics.[31]

Heisenberg's uncertainty principle

In Heisenberg's development of quantum mechanics, another extraordinary feature arose. This was the famous *uncertainty principle* of 1927 which asserts that there is a limitation on the accuracy of measurements built into nature. Imagine an object like an electron flying through space. According to classical physics, it has a position and a momentum, both of which can be measured simultaneously. Simply put, Heisenberg's principle says that in the sub-atomic realm it is impossible to know simultaneously the position and the momentum of that electron with any precision. If one wishes to measure the exact position at some instant, then its momentum (or equivalently its velocity) cannot be known with any certainty whatsoever and *vice versa*.

The principle reflects the paradox of wave–particle duality: position is very much a particle property yet waves have no precise location. The more that is known about wave character, the less that can be said about particulate attributes. One way to envisage this is to imagine what takes place when a measurement is made of the position of an electron. We could, for instance, do this by bouncing a photon off it. But although we could deduce the position of the electron with some certainty from the resulting trajectory of the photon, we have transferred an unknown amount of momentum from the photon to the electron in the process. Somewhat analogous effects are familiar in the everyday world: the very act of measuring the pressure of a tyre allows some air to escape and changes that pressure.

In general terms, the uncertainty principle means that the more precisely we know the measured value of one quantity, the greater the uncertainty in another 'conjugate' quantity. The proportionality constant connecting the two uncertainties is our old friend Planck's constant, which is so small that for macroscopic objects, such as snooker balls or Newton's apple, it is effectively zero. Hence for the objects described by Newtonian (and relativistic) mechanics, there is no limitation on the simultaneous measurement of position and velocity.[32]

Heisenberg's principle has consequences for the measurement of time. Just as we cannot know simultaneously the position and momentum of a sub-atomic particle with any precision, so there is a limitation to the accuracy with which we can measure energy within a given interval of time. The principle connects these two quantities in the same way as it links the mutual uncertainties in position and momentum. A precise measurement of an atom's energy in a particular quantum state can only be performed at the expense of uncertainty about the time it spends in that state, in other words its lifetime. But if its lifetime is known very precisely then its energy cannot be known with any certainty. The energy–time uncertainty principle has important repercussions in cosmology, as we shall see later; some believe that it may hold the key to how time started ticking in the first place.

The Bohr–Einstein debate

Einstein could not accept the uncertainty principle as a fundamental fact of nature. As we have already remarked, he preferred to believe that quantum mechanics was really a mathematical recipe for predicting in a statistical sense the outcome of a large number of experiments, rather than the best account possible of a single one. This led to a famous debate between Einstein and Bohr about the foundations of quantum mechanics, which was to mark each of them for the rest of their lives. On one occasion, at the Sixth Solvay Conference in Brussels in 1930, Einstein announced a *Gedankenexperiment*, a 'thought experiment' rather than one conducted in the laboratory. He fiendishly designed it to contradict the uncertainty relationship linking energy and time; but after a sleepless night Bohr managed to defeat it with the help of one of

Einstein's most important discoveries – relativity. That was far from the end of the story. On the morning after Bohr died in 1962, a drawing of Einstein's 1930 *Gedankenexperiment* was found on his blackboard. It seems that Bohr struggled with Einstein's ideas to the end.[33]

At the outset, Einstein thought that the new quantum theory was simply incorrect (by being inconsistent). But his repeatedly unsuccessful skirmishes with Bohr eventually led him to change his critical line of attack: he came to regard quantum theory not as incorrect but rather as incomplete. His objections became more firmly based on the apparent lack of causality in the theory and its concomitant incompatibility with the principles of relativity. But although their mutual respect was unfaltering, Einstein could never persuade Bohr of the validity of his arguments. It was a battle that tormented both men. A colleague, Abraham Pais, described Bohr's anguish on one occasion at the Institute for Advanced Study in Princeton: 'He [Bohr] was in a state of angry despair and kept saying "I am sick of myself." . . . As always, they had become involved in an argument about the meaning of quantum mechanics, and, as remained true to the end, Bohr had been unable to convince Einstein of his views.'[34]

This great debate – with Einstein in the role of old reactionary – is full of ironies. We have already noted how Einstein was isolated at the turn of the twentieth century by his total confidence in his own approach to physics, even when it was completely at odds with established Newtonian thinking. With his explanation of the photoelectric effect he led the way in quantum theory. But when the new mechanics burst onto the scene in the mid-1920s, Einstein was no longer in the vanguard. The whole structure of the theory was completely at odds with his thinking. True to his own convictions until the end, the man who had done so much to usher the theory into existence remained aloof from the great strides in modern physics which flowed from its discovery. As Pais points out, his view on quantum theory 'changed Einstein's apartness from that of a figure far ahead of his time to that of a figure on the sidelines'.[35]

Today, perhaps, Einstein is having his revenge, for an increasing number of physicists are now prepared to question the foundations of quantum theory. It may be that removed by more than a generation from the founding fathers and propagandists of quantum mechanics, many first-rate minds are no longer prepared simply to accept the remarkable traditional dogma which has been handed down to them.[36] And some of their qualms revolve around the search for an adequate understanding of time's role in physical processes.

Time: lost and found

At first sight, quantum mechanics does as much to undermine the arrow of time as the theories that preceded it. Compared with quantities like momentum and energy, time has a second-class status in quantum mechanics. It is possible to fish out observable information from the wavefunction by using special mathematical objects, called 'operators'. For instance, from a wavefunction describing an electron, such operators can reel in values for position or momentum. But there is no operator for time because time is not regarded as an 'observable' (in other words, a measurable quantity) in quantum mechanics. For this reason, the status of the energy–time uncertainty relationship is somewhat unclear. However, as we shall see in Chapter Eight, if an arrow of time is introduced into a generalised version of quantum theory, it becomes possible to define such an operator for time.

Schrödinger's wave equation is deterministic, just like the equations of motion of Newton and Einstein. Given a value for the wavefunction at one moment in time, its value can be strictly inferred at any earlier or later instant. The equation describes a behaviour that is totally reversible in time. Imagine one particular wavefunction, representing mathematically the behaviour of an unobserved electron. The function stores all the fates that can befall the electron at a moment in time if we observe it with a measuring device, for example a phosphorescent screen. Equally, the equation allows us to predict all the possible fates that could befall the electron if we observed it at some time in the future. And most importantly, it allows us to retrodict all the possible fates that could have befallen the electron had we observed it at some time in the past. So long as we are only talking about probabilities and possibilities – the theory's stock-in-trade – quantum mechanics is perfectly time-symmetric.

There is also a quantum mechanical version of the Stoics' eternal return. Quantum theory, applied to an isolated system, suffers from a strong form of Poincaré's return (described in Chapter Two) which would appear to support the notion of cyclic time. Given enough time, a wavefunction of an isolated system, for example that of the universe, could return to its initial state. Thus it does not seem to offer a satisfactory foundation for an objective flow of time or even for a meaningful division of time into past, present and future.

One naturally needs to go beyond the wavefunction, which contains all the potential behaviour of a system, to find out what actually happens in any experiment: one has to make measurements. And it is at this point that quantum mechanics needs the arrow of time. Let us go back

128

to that apparently time-symmetric wavefunction. When a particular measurement is made, the electron will be recorded as having arrived at one, and only one, position. Thus, the wavefunction – and also the system itself – must undergo an instantaneous transformation when the measurement is made.[37] It is a discontinuous contraction from a form reflecting all possible outcomes to one which corresponds to the single value recorded in the experiment.

This transformation, from a plethora of potential outcomes to the one observed, is known as the 'reduction' or the 'collapse' of the wavefunction.[38] If we adopt the Copenhagen interpretation of quantum mechanics, there are an infinite number of possible outcomes of which just one becomes actuality when we 'pop' the wavefunction.[39] Imagine you are sitting in a quantum theatre. There is an endless profusion of possible plays which might take place, from Shakespeare to Coward to Ibsen to Wilde. But as soon as the curtains swing apart the theatre wavefunction collapses to reveal Agatha Christie's *The Mousetrap*.[40]

It thus seems that if we turn our backs and do not peek, a wavefunction evolves in a reversible and deterministic manner. Yet one measurement of the location of the electron on the screen alters the behaviour irreversibly. When the wavefunction collapse occurs, all the many possibilities reduce to a single real event. This removes the symmetry between the past state of the system (potentiality) and the present state (actuality). Indeed, if one tries to use the method to retrodict the past from a given measurement result, one gets incorrect results.[41] Thus the very act of measurement introduces an arrow of time into the phenomena described by quantum mechanics.

However, neither the collapsed wavefunction nor its precursor says anything about the direction of time. The majority of working physicists simply accept the additional postulate proposed by the Hungarian-born mathematician John von Neumann, that the wavefunction collapses on observation.[42] No mechanism for the collapse is proposed. Indeed, it is clear that the collapse cannot be described by the Schrödinger equation itself, for that equation is reversible and deterministic, while the collapse is irreversible and random. This is the nub of the measurement problem, which has great significance for the arrow of time, and has spawned a number of paradoxes.

The quantum cat paradox

The most famous of these is Schrödinger's 'cat paradox'. Many physicists view the paradox with irritation because they believe that it does not have any 'real' consequences. Stephen Hawking, for instance, once remarked: 'When I hear of Schrödinger's cat, I reach for my gun.'[43] Such a dismissive attitude is often prevalent when problems of a profound nature are considered: a thin layer of ice suffices for many to skate happily over, oblivious to the perils which lurk below. Thus the cat paradox has been regarded as a scrap to fall off the quantum table for the philosophers to fight over, as indeed they have. Yet it may be possible to design experiments to test the paradox.[44]

Figure 12. Schrödinger's cat and the diabolical apparatus containing her. The walls are supposed to be opaque. If they are transparent, the radioactive material can never decay and the cat will never die, according to the quantum Zeno effect.
[Adapted from B. S. deWitt and N. Graham (eds). *The Many Worlds Interpretation of Quantum Mechanics*, p. 156.]

Schrödinger envisaged a *Gedankenexperiment* consisting of the apparatus shown in Figure 12. A cat is curled up inside a box containing a sample of some radioactive material and a vial containing deadly hydrogen cyanide. The process of radioactive decay is itself quantum mechanical and accordingly can only be predicted to occur in a probabilistic sense. By a suitable hook-up of gadgetry, it is arranged that when an atom within the radioactive sample decays, a signal causes a judiciously placed hammer to drop on the vial, releasing the toxic gas and killing the cat.[45] According to common sense, the cat is either dead or alive, but according to the tenets of quantum theory, the system comprising the box and its contents are described by a wavefunction. Making the simplification that the cat can only exist in two quantum mechanical states – alive or dead – the wavefunction for the system involves a combination of these two possible and mutually exclusive observational outcomes. The cat is both alive and dead at the same time. If the good Professor Schrödinger does not open the lid of the box to gaze on the cat, his own equation says that the time evolution of the cat is described mathematically by a physically (and physiologically) indescribable combination of these two states. Just as the electron is neither a wave nor a particle until a measurement is made on it, so the unfortunate cat is neither alive nor dead until finally the professor chooses to peek inside.

It was through the invention of this 'hellish contraption' that Schrödinger attacked the indeterminacy of quantum mechanics by transferring it from the microscopic level of radioactive decay to the macroscopic level of a dead cat. The act of observation not only apparently injects a subjective element into the phenomenon – someone has to open the box and look at the beast – but it forces the cat irreversibly to adopt one or other of the two possibilities: either the vial is intact and the cat alive and well, or the vial is broken and the cat lies dead.

Schrödinger's cat brings the measurement problem home in a graphic way. We are apparently expected to believe that the state of the system is modified by the very act of observation, a notion that seems too crackpot to be true. As Einstein remarked: 'I cannot imagine that a mouse could drastically change the universe by merely looking at it.'[46] Yet even the pioneers of quantum mechanics did not regard this as necessarily a bad thing, as is illustrated by an exchange between the Swiss theoretical physicist Wolfgang Pauli and Niels Bohr in which the latter told the former that one of his ideas 'was crazy, but not crazy enough'.[47]

Wigner's friend

There are two popular ways of attempting to meet Einstein's objection. One is to say that measurements on quantum systems are not made by cats or mice, but by human beings endowed with the attribute of consciousness. Proponents of this line of thought, including most notably John von Neumann and Eugene Wigner, have maintained that the intervention of a conscious observer is required 'to do the looking' and thereby effect wavefunction collapse; they argue that a cat would not qualify as an observer capable of collapsing the wavefunction to the actuality of life or death, because it is not clever enough to distinguish between these states. According to this point of view, mind and matter are totally distinct notions, and quantum theory applies only to the latter. Only conscious minds pop wavefunctions.

A consequence is that Schrödinger's poor cat does not know whether it is alive or dead (even though most cat lovers claim human-like intelligence for their feline companions). However, whether the beasts are conscious or not is a matter of much dispute, not least because the word itself is difficult to define scientifically. It is usually taken to mean the possession of human intelligence, although such a black-and-white notion is probably unhelpful.

So we come to 'Wigner's friend', named after the Nobel laureate's imaginary acquaintance who, equipped with a gas mask, has the thankless task of crouching with the cat in the box. Once he opens his eyes the cat's wavefunction collapses. He can be asked what it felt like to be in the box before Schrödinger peeked inside. The friend will be able to report what happened in conventional language – there is no question of his having been scrambled in a quantum superposition of all possible experimental outcomes until the box was opened. Thus, according to Wigner, when a human mind is involved within the system under study, the normal kind of quantum description cannot apply. Proponents of this consciousness-based interpretation of quantum theory have also maintained that the passage of time is merely a mental impression corresponding to the repeated popping of the wavefunction.[48]

One could ask how many pints of beer have to be consumed by Wigner's friend before he loses consciousness and the concomitant ability to pop wavefunctions. We are in danger of returning to a very subjective picture of the world. As Alastair Rae put it in his book, *Quantum Physics: Illusion or Reality?*: 'Ever since the beginning of modern science four or five hundred years ago, scientific thought seems to have moved man and consciousness further from the centre of things. More and more of the universe has become explicable in mechanical, objective terms, and even human beings are becoming understood scientifically by biologists and behavioural scientists. Now we find that physics, previously considered the most objective of all sciences, is reinventing the need for the human soul and putting it right at the centre of our understanding of the universe!'[49] With all the evidence amassed by science in favour of an independent reality 'out there', it is hard to take the consciousness-based approach to quantum theory very seriously.[50] Indeed, it may one day even be possible to understand consciousness in physical terms.

Parallel universes

The most daring attempt to 'solve' the problem of measurement in quantum mechanics was proposed by Hugh Everett III in 1957.[51] His idea, which has often been exploited by science-fiction writers,[52] arose in the course of his doctoral work at Princeton University, under the enthusiastic supervision of Professor John Wheeler. Let us return to Young's slit experiment and consider the dilemma posed by asking which of the two slits the photon passes through. The wavefunction contains all the possible outcomes of the passage of the photon through both of the slits, yet at the most basic level there is a choice of slit for the photon to traverse. The Copenhagen interpretation says that the choice is made on the irreversible collapse of the wavefunction, according to a probabilistic rule.

Everett's interpretation is that the electron does not make a choice of slits but of universes. In choosing one slit in preference to the other, the universe splits into two. The slit used depends on which universe we occupy. After that point the two universes are completely separate and continue to branch into further parallel universes with each subsequent measurement. The crux of Everett's idea is that the universe itself is described by a wavefunction which contains the ingredients of any outcome. His interpretation carries with it a bizarre implication – that innumerable 'parallel' universes, each as real as our own, all exist independently. Your wildest dreams may be fulfilled within these other worlds. With every measurement made by an observer, who is by definition within a universe, the entire universe buds off an uncountable multitude of new universes (the 'many worlds'), each of which represents a different possible outcome of the observation (for example a living or a dead cat). No collapse of the wavefunction takes place, only the infinite propagation and budding of new branching universes: there is no need for an observer external to the universe. Everett's thesis bears a certain similarity to the ninth-century teachings of the *Kalām* School of Islamic scholastic theology, according to which with every event the world is born anew.[53]

The cosmologist Bryce DeWitt described the head-reeling effect induced on first coming across Everett's proposal: 'I still recall vividly the shock I experienced on first encountering this multiworld concept. The idea of 10^{100+} slightly imperfect copies of oneself all constantly splitting into further copies, which ultimately become unrecognisable, is not easy to reconcile with common sense. Here is schizophrenia with a vengeance . . .'[54]

Everett's many-worlds interpretation of quantum mechanics has found favour with many cosmologists because it removes the apparent necessity for an external observer.[55] This is thought to be important, for if the consciousness-based interpretation *à la* Wigner is adopted, the only observer who could collapse a conventional wavefunction of the universe must be God. Compared with the latter approach, Everett's seems to be 'cheap on assumptions but expensive on universes'.[56] It is intriguing to note that David Deutsch of Oxford University believes that his proposed model for a quantum computer, if and when eventually constructed, would provide an experimental test of the validity of Everett's conjecture.[57] Such a computer would possess many novel properties, including a 'quantum parallelism' by means of which many calculations could be performed simultaneously far more swiftly than is possible using a conventional computer. Deutsch claims that such miraculously fast computation would require that different parts of the calculation be done in parallel universes, although his argument is open to debate. For there is no scientific proof favouring Everett's conjecture, and even if such a computer were built it is unlikely to favour many-worlds over other interpretations. Moreover, the many-worlds interpretation suffers from many technical difficulties; in particular, it does nothing to explain what is so special about measurement processes that they should lead to splitting universes. One is left with the sensation of having chartered an aircraft to cross the road.

Caveat lector

The far-fetched nature of both the consciousness-based and many-worlds interpretations of quantum mechanics is typical of the difficulties posed by the measurement problem within the existing framework of quantum theory. The fundamental problem is obscured owing to the widespread yet fallacious belief that the conventional Copenhagen interpretation requires the existence of 'observers' — assumed to be human beings — thus opening the door for subjective elements to creep in. But there is no need for 'conscious' beings to be present: it is enough for there to be a measuring apparatus in place which records the outcome of an experiment. For example, it might consist of a computer, a phosphorescent screen, a photographic emulsion or a bubble chamber. Human minds are irrelevant. Wavefunction collapse is irreversible and fully objective.

The EPR paradox

If we reject the consciousness-based and parallel universe interpretations, we are still left facing a major difficulty in accounting for the way in which wavefunctions collapse. This collapse occurs in apparent defiance of causality, the common-sense idea that effects follow from their causes and must be near enough in space to be causally connected to them.

Take a system ruled by quantum lore, for example an electron flying towards a phosphorescent screen. Its evolution in time is described by a quantum mechanical wavefunction according to the Schrödinger equation. When a measurement is performed at the screen to find out where the electron is, the wavefunction collapses. But although a measurement is made locally, the collapse of the wavefunction alters the physical state of the electron *everywhere* at the same instant – remember that the wavefunction contains every possibility for that electron in space and time, not just the one that emerges when the wavefunction collapses.

This non-local collapse implies a violation of causality. There will be regions of spacetime which cannot be connected with the point on the screen where the electron is observed – even via speed-of-light signals – yet which nevertheless correlate instantaneously with the measurements made there. The mere observation of that electron sends out instantaneous repercussions to the farthest reaches of the universe, with no more explanation than of how a voodooist may hurt someone located miles away by sticking pins into a wax effigy. The reason for this peculiar property is that the wavefunction is actually a kind of abstract field which pervades the whole of space.[58] An observation at any given point causes the collapse of a state of potentiality (all possible outcomes) to a state of actuality (the outcome observed), which is then fixed everywhere in the universe at one and the same instant.

These extraordinary non-causal correlations at a distance were first highlighted by Einstein in a collaborative paper written in 1935 with the Russian Boris Podolsky and Brooklyn-born Nathan Rosen, then two members of the Institute for Advanced Study at Princeton.[59] Einstein had been appointed to a permanent position at the Institute in 1932, a post which had enabled him to escape from the rising tide of anti-Semitism in pre-war Germany. By the time he was settling down at Princeton, Einstein's special theory of relativity was almost thirty years old. But he was still tormented by quantum theory, being irritated by its intrinsic randomness and the fact that reality is apparently 'created' by an observer. To Einstein, this statement seemed to put man back on the pedestal from which Copernicus had knocked him five hundred years previously. Einstein was as eager as ever to put a spanner in the works of quantum mechanics.

The *Gedankenexperiment* conceived by Einstein and his colleagues – named the EPR paradox after their initials – may be illustrated by two spinning particles that interact and then separate to a vast distance. We know from the manner in which they interact that each one will have an equal yet opposite value of the property called 'spin': if the spin of one is 'up' then that of the other is 'down'. Thus if we observe the spin of particle A, we can deduce the spin of particle B. But in quantum theory, both particles are spinning in a limbo state – 'up and down' – until the moment a measurement is made. The measurement of the spin value of particle A obliges it to become either 'up' or 'down' at that moment, but because of the quantum correlations particle B must *immediately* also be forced to adopt a definite spin, the opposite of A's. This result ensues, no matter how many light years away from A particle B has become. There is no need for us to make a measurement of the spin of B – it is known as soon as is that of A. This ghostly action at a distance seems to imply that the two particles communicate by a physical effect which propagates faster than light.

The contention of the EPR paradox was to show from this analysis that quantum theory cannot support an objective description of reality, and that it is incomplete since it involves such unphysical 'non-local' correlations. Much work has been done since then in the search for some underlying substratum to the quantum description – a 'hidden variable theory' – which might account for these correlations in a deterministic manner and, moreover, which could be subjected to the final arbiter, experimental test.[60] Such a test has been developed along the lines of an actual EPR experiment, whose outcome would be different depending on whether hidden variable or quantum theory held sway.

Unfortunately for Einstein and his followers, quantum mechanics won the day. That there do indeed seem to be faster-than-light connections between distant regions of spacetime was confirmed in 1982 by Alain Aspect and his colleagues at the Institut d'Optique Théorique et Appliquée in Paris.[61] Two quantum particles, in widely separated regions of the universe, somehow constitute a single physical entity.[62] It would therefore appear that the potty and peculiar universe suggested by quantum uncertainty is upheld: God does play dice with the universe. And we must conclude that Einstein's vision of a deterministic reality fully described by science is an elusive chimera induced by our 'common-sense' view of the world.

When time stands still

The irreversibility of wavefunction collapse provides firm evidence for the objective existence of the arrow of time. Nevertheless, objective wavefunction collapse in its present *ad hoc* form is unable to provide the full explanation of transience that we are looking for. This is clear from a remarkable consequence of the theory which implies that time effectively 'stands still' when a quantum system is continuously observed. Consider the case of an unstable atomic nucleus which can undergo radioactive decay, as happens in the cat paradox. Often in quantum theory measurements are idealised as occurring in a discontinuous fashion – by glancing at something – at a fixed single instant in time. Suppose instead that the nucleus is subjected to continuous measurement in order to determine at what moment in time it decays.

Baidyanath Misra and George Sudarshan working at the University of Texas at Austin in the mid-1970s showed that, under such conditions, the nucleus will never decay.[63] It is the quantum-mechanical analogue of the proverbial adage that 'a watched kettle never boils'. Continuous measurement forces the atom to remain in the undecayed state from which it cannot undergo a transition to the decay products. In order to draw attention to their work, Misra and Sudarshan called the phenomenon the 'quantum Zeno paradox', because it bears some resemblance to Zeno's paradox involving an arrow in flight, his analysis of which implied that the arrow cannot be moving. However, the quantum version is not really a paradox at all, being simply a consequence of orthodox quantum theory.[64]

Experimental support for the quantum Zeno effect came in an elegant experiment carried out by a team at the National Institute of Standards and Technology in Boulder, Colorado in 1989.[65] The researchers observed the behaviour of 5000 charged beryllium atoms trapped in a magnetic field. The atoms all started out in one energy level, although they could be 'boiled' by exposure to a radio-frequency field for 256 thousandths of a second. After exposure to this field, the atoms all occupied a higher energy state – so long as no measurements were made in the interim. By probing with a laser at intermediate moments, the team found that the more often they recorded the state of the atoms, the fewer reached the higher energy level: with observations made every four-thousandths of a second, no atoms boiled at all. It seems that a watched quantum kettle never does boil.[66]

The notion of a continuous measurement is as much an idealisation as is a discontinuous one. Any real measurement can only extend over a period of finite duration and so, eventually, quantum transitions will take place. Yet these two extreme caricatures of measurement are equally disconcerting: apparently, either something is induced to take place or nothing happens at all. Consider the diabolical apparatus of the cat paradox with one difference – the walls of the box are made transparent instead of opaque. Now we can see what is going on inside the box at all instants, and the quantum Zeno effect then implies that the cat must remain alive for as long as we keep looking (although recall that 'we' can be replaced by inanimate measuring apparatus). Something is surely amiss with our so-called 'greatest theory'.

Schrödinger's cat, Wigner's friend and the other paradoxes underline how urgently wavefunction collapse demands an explanation. In our view, all the difficulties arise from the fact that the existing quantum formalism is based on time-reversible foundations, while measurement processes are intrinsically irreversible. The consciousness-based and many-worlds interpretations are but manifestations of the time-honoured technique of attempting to drive a square peg into a round hole. For quantum theory as it is known today is incomplete. The problems posed by measurements can only be properly resolved with the help of a framework incorporating the arrow of time. In Chapter Eight, we will show how this can be done.

Matter and antimatter

Until then, our quest for time's arrow must continue. One fascinating place to look for it is in the strange world of sub-atomic particles. The foundations for this search came with the work of the British mathematical physicist, Paul Adrien Maurice Dirac, who played a key role in combining quantum theory with special relativity in the 1930s. He was originally trained as an electrical engineer but soon found that his inclinations drew him towards physics. He had an awesome reputation for mental gymnastics, and a shy, solitary nature. After one lecture he gave at the University of Toronto, a Canadian professor in the audience asked: 'Dr Dirac, I do not understand how you derived this formula at the upper left corner of the blackboard.' After a long silence, and a prompt from the chairman, came Dirac's reply: 'It was not a question; it was a statement.'[67]

Schrödinger himself had realised from the start the limitations of his own equation – it did not satisfy the requirements of Einstein's special relativity (specifically, it was not Lorentz invariant[68]) and as a result could not account for the finer details of atomic spectra. Dirac attacked this shortcoming and in 1928 had written down a relativistic equation which implied that the electron must be spinning on its own axis. In addition, the mathematical properties of Dirac's equation led him to propose the existence of the *positron*, a unit of 'antimatter'. The positron has the same mass as an electron but is of opposite electric charge: a collision between the two results in their mutual annihilation, together with the production of a burst of radiation. An experiment by Carl Anderson in 1932 confirmed the existence of the positron and therefore the reality of antimatter, a discovery that radically altered the conceptual foundations of elementary particle physics. Prior to that time, physicists had firmly believed in the Greek notion of the immutability of matter. From then on, it was recognised that matter could be created and destroyed at will.

On the basis of the mathematical description provided by Dirac, the positron has been interpreted as its antiparticle – an electron – moving backwards in time.[69] Feynman even went so far as to propose that there is only one electron in the entire universe, moving forwards and backwards in time on an elaborate trajectory so that at any instant we merely think that we see a vast number of separate electrons. Like the Gödelian notion of time travel which we met in Chapter Three, Feynman's 'one-electron theory', requiring the co-existence of different time directions, leads into direct conflict with the notion of causality and inevitably raises many more problems than it solves.

An intricate relationship exists between matter, antimatter, spatial symmetry and the two directions of time. It appears in the remarkable *CPT theorem*, which is a consequence of the mathematical form of the microscopic laws of physics. The roots of the CPT theorem lie in the symmetry of these laws, which remain unchanged (invariant) if in any process particles are swopped with antiparticles, the process being exchanged with its mirror image, and the direction of time being reversed. The theorem was developed by G. Lüders (1954) and Wolfgang Pauli (1955). Its name arises because symmetry is tested by the combined sequence of three abstract operations:

- **C** Charge conjugation, by means of which matter is converted into antimatter;
- **P** Parity inversion, which converts spatial coordinates into their mirror images;
- **T** Time reversal, which reverses the direction of time.

The effect of the threefold CPT operation on a process produces another equally permissible physical process which is also described by the same theoretical framework. Loosely speaking, the CPT theorem contends that the laws of physics predict equal but opposite events in a kind of 'generalised mirror image' world. They can also tell us how time symmetry can be broken to give an arrow of time.

Symmetry – in the guise of the CPT theorem – can be used to make deductions about physics. Thus we find that the quantum laws would say the same things for a cricket ball sent flying over a stadium as for its antimatter mirror image whizzing back in time to the batsman before bouncing into the bowler's hand. The most important facet of the CPT theorem is the statement that if the CP part is symmetrical then the process under discussion must also possess time-reversal symmetry (T). Here CPT is essentially being employed in the same way as we use considerations of symmetry to deduce the numbers of knives in a set of cutlery from the number of forks. CP symmetry also implies that if an antimatter left hand holding an antimatter stopwatch exists, then so too does an ordinary matter right hand clasping an ordinary matter watch; the T symmetry then means that each watch can go backwards and forwards in time. Conversely, if CP is violated – the antimatter left hand does not have an associated ordinary matter right hand – then T is also violated: the watch in the left hand can only go forwards, not backwards, in time. This appears to offer a theoretical foundation for time's arrow in the reductionist spirit; it is apparently quite independent from the measurement problem encountered above. However, almost all known 'elementary-particle' interactions satisfy the restricted CP symmetry, and so are oblivious to the march of time because they are symmetric under time reversal.

The curious case of the long-lived kaon

There is one, and only one, extraordinary phenomenon within the world of microscopic 'elementary' particle physics which is believed to break the symmetry between the two directions of time. It arises in certain decays of an unstable particle known as the long-lived *kaon* (K^0), discovered at Brookhaven in the United States by J. H. Christenson, J. W. Cronin, V. L. Fitch and R. Turlay.[70] Their work led to the award of the Nobel Prize for Physics in 1980 to Cronin and Fitch. In most decays, kaons form a negative *pion*, a positron and a *neutrino*, a process in which the CP symmetry can be shown to be preserved. But they can also decay (about once in 1000 million decays) to form a positive pion, an electron and an anti-neutrino, wherein the CP symmetry is in fact violated. By the CPT theorem, T symmetry is therefore also violated in these rare decay pathways: the time-reverse event is prohibited and the process is irreversible, providing a glimpse of the arrow of time.

Before we accept the existence of even a single microscopic phenomenon that is aware of the sense of time, we should first question the validity of the CPT theorem itself, the proof of which depends on certain assumptions that are not satisfied, for example, if gravity is included. Nevertheless, if there is a violation of overall CPT symmetry in these rare decays (which has not been seen), it is even smaller than the observed CP violation; thus T violation is well established, albeit indirectly. Roger Penrose of Oxford University commented: 'The tiny effect of an almost completely hidden time-asymmetry seems genuinely to be present in the K^0-decay. It is hard to believe that Nature is not, so to speak, "trying to tell us something" through the results of this delicate and beautiful experiment.'[71]

At the present, however, there exists no satisfactory fundamental explanation of these intrinsically time-asymmetric interactions.[72] It is therefore difficult to decide whether the strange decays of the long-lived kaon are a diversion in the search for time's arrow or whether, as Roger Penrose has maintained, they hold the key to its elucidation. For they do not play any role in the familiar irreversible processes to be described in later chapters of this book.

The seething vacuum

Heisenberg's uncertainty principle has consequences for the measurement of time, as we have seen. There is a limitation to the accuracy with which we can measure energy within a given interval of time. A precise measurement of an atom's energy in a particular quantum state can only be performed at the expense of considerable uncertainty over the time it spends in that state.

In the next section we will see that some cosmologists believe it is possible to fish an entire universe out of this uncertainty. The relationship allows one to violate the cherished notion that energy cannot be created from nothing. In classical physics, energy was neither created nor destroyed but rigorously conserved, merely being converted from one form to another; for example, the chemical energy stored in petroleum is transformed into heat and the motion of cars. An energy balance sheet could account for all the original chemical energy this way. However, over short time intervals the conservation of energy is undermined by Heisenberg's uncertainty principle.

The link between energy and time in Heisenberg's principle says that the shorter the interval of time considered, the more uncertainty there will be in the energy. This allows the law of energy conservation to be suspended over very short time intervals: owing to random quantum mechanical fluctuations, energy may be 'borrowed' at no cost from nowhere at all. Such events occur even within a vacuum in which, classically speaking, there is simply 'nothing' present. Quantum theory thus gives us a quite different view of what a vacuum is. Because of the uncertainty principle, it is in fact seething with activity.

Dirac is largely responsible for the modern picture of a vacuum. He recognised that Maxwell's electromagnetic field must be described in quantum terms if it is to explain correctly the way matter absorbs and emits photons. In his extension of Maxwell's mathematical model, he pictured the electromagnetic field as a collection of a vast number of oscillators, each of whose energy levels was quantised, just like the energy levels of an electron in an atom. But now, by virtue of the uncertainty principle, each oscillator can never have less than a fixed minimum energy – the *zero point energy* – with the result that even a vacuum is always seething with virtual activity: there are endless fluctuations in the energy of the field at all points within space. Fluctuations of sufficient energy enable matter–antimatter pairs of particles, such as electrons and positrons, to be momentarily formed – the more energy borrowed, the more transient the particle-pairs will be.[73]

These vacuum quantum field fluctuations have considerable physical significance. For example, the constant creation and annihilation of photons provides the trigger whereby atoms swollen with energy can spontaneously emit radiation as light.[74] In fact the vacuum fluctuations also resurrect in a certain sense the nebulous æther which Einstein eliminated as excess baggage in 1905. As Christopher Llewellyn Smith of Oxford University remarked: 'Nowadays, we don't even understand the vacuum.'[75]

Dirac's quantisation of the electromagnetic field solved some problems, yet it raises many more. Briefly, the difficulties relate to the fact that there is no limit to the amount of energy that can be carried by the field, which leads to the frequent occurrence of infinities in the theory. Just as in the case of the singularities of general relativity, such wild behaviour is mathematically abhorrent and suggests that the theoretical framework has broken down somewhere along the way. However, in the case of quantum field theory, ingenious theorists such as Freeman Dyson, Richard Feynman, Julian Schwinger and Sin-itiro Tomonaga discovered how to tame these divergences by means of a technique known as *renormalisation*; the infinities are cleverly absorbed by, in effect, cancelling them out with other infinities. The resulting formalism provides meaningful and frequently remarkably successful predictions. But doubts about the propriety of renormalisation methods will always linger; Dirac himself believed that the problem was symptomatic of a truly fundamental weakness within the theory. Even Stephen Hawking admits that renormalisation 'is rather dubious mathematically'.[76]

One of the principal shortcomings of quantum field theory has been its inability to handle the force of gravity. The infinities that then occur cannot even be accommodated by means of the ruse of renormalisation. Nevertheless, it is widely believed by cosmologists that a successful quantum version of gravity would transcend the singularity problems arising in general relativity, which were encountered in Chapter Three. Very recently, considerable interest has arisen in a new method of approach to this difficulty which goes under the name of string (as opposed to field) theory. One can think of string theory as being a kind of 'higher-dimensional' version of field theory, whose smallest units are no longer visualised as points but objects of finite extension, which may be open or closed strings. Among its leading advocates are Michael Green of Queen Mary College, University of London, and John Schwartz of the California Institute of Technology. Since its earliest days, there has been a great deal of optimistic hope that string theory might be capable of handling and indeed unifying all the basic interactions between particles, including gravity, without the embarrassment of uncontrollable infinities.[77] For this reason some highly optimistic physicists, including Stephen Hawking, believe that with the advent of string theory, the end of theoretical physics is in sight.

Although of undeniable æsthetic appeal on mathematical grounds, there are no compelling scientific reasons to regard string theory as a panacea. In the words of John Barrow: 'As yet there are no experimental facts for or against it. In coming years . . . predictions will surely be forthcoming. Only then will we know whether the unique prescription it offers is a theory of everything or a theory of nothing.'[78] In spite of the great claims made on its behalf, from its time-symmetric structure string theory seems unlikely to shed any new light on the nature of time, particularly in regard to its direction, or on the related fundamental – but all too frequently overlooked – problem of measurement.

The cosmic free lunch

Because of the unavoidable presence of abhorrent singularities, we recall that general relativity, in the words of Dennis Sciama, 'contains within itself the seeds of its own destruction.'[79] Notwithstanding the fact that, as we have just noted, there exists as yet no satisfactory theory of quantum gravity, modern cosmologists have in the past ten years or so begun to speculate on how quantum theory might modify the unpleasant features of Einstein's relativity.

The *ad hoc* fusion of the quantum and relativistic approaches in a quantum theory of gravity has led to fascinating ideas about the origin of the universe and thus of time itself. The seething vacuum has, for example, inspired a model of how the universe was born. If gravity is quantised, then random fluctuations in the gravitational field must occur which provide a 'mechanism' for the very act of creation *ex nihilo* – 'out of nothing'. Alan Guth of the Massachusetts Institute of Technology, a leading contemporary cosmologist, remarked: 'I have often heard it said that there is no such thing as a free lunch. It now appears that the universe itself is a free lunch.'[80] This idea – based on the concept of a seething vacuum, the energy of which fluctuates wildly by virtue of Heisenberg's uncertainty principle – was originally put forward by an American physicist, Edward Tryon, in 1973, but his suggestion did not find favour for a further five years.[81]

Then, starting in 1978, a fascinating series of proposals came from a group of cosmologists working at the Free University of Brussels.[82] According to their version of the Bible, in the beginning was an 'empty' vacuum where spacetime was flat. This is quite distinct from the unpleasant Big Bang singularity where everything is initially condensed into a single point. From Einstein's famous relation between mass and energy, a quantum fluctuation in the energy would be capable of producing the equivalent mass of particles.[83] These in turn, through their mutual gravitational attractions, cause the originally flat spacetime to become curved, as we saw in Chapter Three. There is no violation of energy conservation in this process, since the positive energy taken up by the particle masses is offset by the negative energy of their gravitational attraction. The authors maintained that the state of nothingness was unstable with respect to the creation of something: a cascade starts, leading to 'the spontaneous creation of all the matter and energy in the universe'.[84]

Other proposals have followed, each amounting to a modification or development of this model. Alex Vilenkin[85] and later Stephen Hawking together with Jim Hartle[86] have tried to describe the very creation of space and time from absolutely nothing – that is, not even from the 'empty' spacetime vacuum seething with quantum mechanical creation and annihilation of matter. Hawking and Hartle attempt to explain away the singularity at the beginning of time – the Big Bang – by assigning a wavefunction to the entire universe and then calculating the probability of something appearing literally from nothing.

Part of Hawking's approach employs Feynman's sum-over-histories technique together with so-called 'imaginary time'. According to Hawking, imaginary time 'may sound like science fiction but is in fact a well-defined mathematical concept'.[87] Imaginary numbers give negative numbers when multiplied by themselves. This is not possible with the numbers most people are familiar with: plus *and* minus two squared both equal plus four. Using imaginary time, which like ordinary time in Einstein's relativity has no direction, the distinction between space and time completely disappears in the mathematics of spacetime.[88] In the context of a closed universe (in other words, one with a Big Crunch), it is claimed that such a mixed-up spacetime might exist at a very early stage, being finite in extent yet having no boundaries or uncomfortable singularities. Time would cease to be well defined in the early universe, which would be a bit like the surface of the Earth but with two more dimensions.[89] For such a self-contained universe, it is claimed that no initial conditions are required and that the singularities of Einstein's theory can be overcome.

Imaginary time is merely a mathematical subterfuge which does not tell us anything new about reality. However, in Hawking's view: 'If you take a positivist position, as I do, questions about reality do not have any meaning. All one can ask is whether imaginary time is useful in formulating mathematical models that describe what we observe. This it certainly is. Indeed, one could even take the extreme position and say that imaginary time was really the fundamental concept in which the mathematical model should be formulated. Ordinary time would be a derived concept that we invent as part of a mathematical model to describe our subjective impressions of the universe.'[90]

It is important to retain a healthy scepticism about all these speculative models of creation *ex nihilo*. There are many technical objections as well as problems of principle. As an example of the latter, if 'nothing' really means nothing, then we do not appear to have any right to apply our scientific laws prior to the act of creation. Thus one is still left wondering what brought the universe into existence.

But what about the arrow of time? On this fundamental question, all these models remain silent. For both quantum mechanics (excluding the awkward measurement problem) and relativity, as they are formulated today, are strictly reversible. A much more exciting possibility has already been mentioned in Chapter Three. This is Roger Penrose's suggestion that the 'true' form of quantum gravity, hitherto unknown, should be time-asymmetric. In other words, it should contain the arrow of time explicitly. We shall discuss the reasons for Penrose's point of view in Chapter Five when we consider the relationship between cosmology and thermodynamics. Such a radical proposal could ultimately lead to a new revolution in physics for which our existing outlook cannot prepare us.

In this chapter, we have surveyed the perplexing subject of quantum theory and reached the conclusion that it is at best an incomplete theory. The principal difficulties surround the act of measurement, in which the wavefunction containing all possible outcomes is collapsed into one single reality – for instance a dead cat in Schrödinger's famous *Gedankenexperiment*. Because of the intrinsic irreversibility of the measurement process, it cannot be incorporated within the existing reversible theoretical framework.

In the next chapter we shall consider thermodynamics, a scientific theory of irreversible processes which contains the arrow of time, and statistical mechanics, which attempts to build a bridge between the microscopic quantum world and the macroscopic world of thermodynamics. In this way, we shall be able to throw more light on the shortcomings of the existing quantum description.

The arrow of time: thermodynamics

> Our present picture of physical reality, particularly in relation to the
> nature of *time*, is due for a grand shake-up – even greater, perhaps, than
> that which has already been provided by present-day relativity and quan-
> tum mechanics.
>
> Roger Penrose
> *The Emperor's New Mind*

TIME'S arrow becomes clearer as we move away from the microscopic
world of quantum mechanics to the macroscopic world of everyday life.
This is the realm of thermodynamics, a theory of immense power which
locates the passage of time in the same kind of images that haunt poets
and novelists. Robert Graves once movingly described 'counting the
beats, counting the slow heart beats, the bleeding to death of time in
slow heart beats'.[1] Thermodynamics does much the same thing. It
reveals the reality of transience in the same irreversible processes that
give human existence its poignancy and its meaning, from the ageing of
our bodies to the drying of our tears. Not that all its applications are
quite so richly symbolic, of course. Thermodynamics also explains how
steam engines work and why cups of tea go cold.

Time is linked by thermodynamics to ideas about organisation and
randomness. The flow of time becomes apparent because there is an
inexorable tendency in any system left to its own devices for organisation to diminish and randomness to increase. If milk is added to black
tea, the milk molecules mingle and diffuse with those of the tea.
Eventually, both are evenly distributed, giving tea its characteristic
murky brown appearance; when the mixing is complete, no further
change occurs. Within the tea in its final state, the molecular randomness – or to be thermodynamically accurate the entropy – has reached a
maximum value. This is the equilibrium state, when molecules of milk
and tea are to be found uniformly throughout the mixture and the
capacity for more mixing is lost. We never see the reverse process, in
which the uniform brown liquid spontaneously separates into white
milk and black tea. For in order to do that, we would have to travel back
in time.

The arrow of time is made explicit in the so-called *Second Law of
Thermodynamics*, which states that all physical processes are irreversible

because some energy is always dissipated as heat. The writer C. P. Snow argued that it should be part of the intellectual complement of any well-educated person. He described being with highly educated people 'who have with considerable gusto been describing their incredulity at the illiteracy of scientists. Once or twice I have been provoked and have asked the company how many of them could describe the second law of thermodynamics. The response was cold; it was also negative. Yet I was asking something which is about the scientific equivalent of: Have you read a work of Shakespeare's?'[2]

Unfortunately, just as literary critics squabble over the interpretation of Shakespeare's plays, so scientists fiercely dispute the meaning of the Second Law. The basic propositions of thermodynamics are rather vague and lend themselves to a great diversity of opinion. The subject highlights the absurdity of popular ideas about scientists forming a po-faced coterie of like-thinking men in white coats. In fact, differences of opinion between scientists are so strong that they often surface as personal sniping and stormy rows. The American philosopher David Hull has recently pointed out that physicists can readily compile a list of 20 or more different formulations of the Second Law. The appearance of consensus that so impresses outsiders, wrote Hull, is largely due to each scientist's insistence that his own view is correct.[3] And like thermo-dynamics itself, the vexed question of time regularly generates more heat than light.

Some of these conflicts will become evident as we discuss the different approaches to thermodynamic theory. One version which we shall look at, based on thermodynamic equilibrium, is such a special case that it suppresses all the change and flux that is at the theory's heart – yet it is the only part of the subject that most scientists ever encounter. We shall show that equilibrium thermodynamics is quite literally a dead end. Some people make the mistake of thinking that thermodynamics rules out the natural emergence of order, and implies that the whole course of the universe is a straightforward descent into random muddle. We will show that this too is incorrect, and that in reality the theory holds the key to ordered life. But for our present purposes, the most fundamental debate is about how to reconcile thermodynamics with microscopic theories of molecular mechanics which, as we have already seen, are time-symmetric. This apparent clash between two crucial strands of scientific thought is known as the irreversibility paradox (some call it the reversibility paradox). It has led some scientists to conclude that the thermodynamic arrow of time has no reality outside our minds, and is purely subjective.

148

The birth of thermodynamics

The theory of thermodynamics took shape with the advent of steam power in the Industrial Revolution in Britain during the early nineteenth century. The first effective steam engine had been constructed in 1782 by James Watt, a former scientific instrument maker for Glasgow University. Steam engines operated by burning coal to heat water, thus creating steam pressure to pump a piston or turn a turbine blade. But in order to calculate an engine's maximum efficiency, it was vital to understand the full theory behind the mechanism. This subject became known as thermodynamics, from the Greek words meaning the movement of heat.

The flow of energy when a steam engine turns, translated into the motion of all the component molecules, is of mind-blowing complexity. But thermodynamics is not concerned with atoms and molecules (recall that atoms were not widely subscribed to in those days); instead, it focuses on directly perceptible 'macroscopic' quantities like volume, temperature and pressure. Pre-eminent among the early thermodynamicists was the French engineer Sadi Carnot. The elder son of one of the leading figures in the First French Republic, Carnot died in a cholera epidemic in 1832 at the age of 36. Ten years earlier, however, he had provided a penetrating thermodynamic analysis of how an idealised heat engine operated. His idealisation was to consider a perfectly reversible engine, one with no irreversible heat losses. He showed how its efficiency depended on the fact that heat flows out of a hot body and into a cold one. In a steam engine, this meant that the heat flowed from a hot chamber where the steam was formed to a cold chamber where it condensed. And it was the difference in temperature between the two, said Carnot, which determined how well the engine worked. Significantly, even for such a perfect heat engine, the efficiency could never reach 100 per cent.

In practical terms what a steam engine does is to convert heat into work, with work simply meaning a more useful and organised form of energy. The equivalence of heat and work was demonstrated by James Prescott Joule, the son of a Manchester brewing family. He looked at the problem in the opposite direction from a steam engineer, and applied work to produce heat (by stirring water with paddles or compressing air in a container) instead of the other way round. The results showed that a given amount of work, of whatever kind, produced the same amount of heat. In recognition of this discovery, Joule's name is now given to the most basic unit of energy (roughly equal to the work needed to raise an apple through a vertical distance of one yard on the Earth's surface).

The equivalence of heat and work as forms of energy is the basis of the *First Law of Thermodynamics*, which lays down that energy will always be conserved in a physical process, even though it may be converted from one form to another. In other words, if you draw up an energy balance sheet for any physical event, the figures for both 'before' and 'after' will always match. The only difference is that some or all of the energy that you started with will have to be re-labelled afterwards as heat. This is because some energy is always 'burnt up' during a physical process – for example in fighting against friction or air resistance. None of it will actually vanish off the balance sheet, but *dissipation* of energy in the form of waste heat makes it change in form.

Such dissipation of energy as heat occurs in every energy conversion process, from a woman burning up calories of chemical energy as she runs a 100 metres race to an incandescent bulb turning electrical energy into light (electromagnetic energy). The recognition of the deeper ramifications of dissipation owes much to the insight of the German physicist Rudolf Clausius, who was born in 1822, the son of a pastor and schoolmaster. He realised that although heat and work were equivalent in the sense showed by Joule, dissipation created a fundamental asymmetry between them. In principle, any form of work can be converted entirely into heat. But dissipation means that the reverse is not true; heat energy always get frittered away as it is converted back into work. For example, not all of the heat in a steam engine can be used to pump the piston. Some of it is squandered in warming up the machinery, the surrounding air and the operator's hands. Some remains within droplets of water when the engine is turned off. Clausius recognised, dimly at first, that this meant that heat loss was irreversible; once it had happened, the wasted energy could never be put to work again. His breakthrough in 1850 fully justifies naming Clausius as the scientific father of the arrow of time, but at that stage his ideas were rather ill-defined. The Belfast-born mathematician William Thomson (later Lord Kelvin) turned his fumblings into a universal statement, the Second Law of Thermodynamics. According to this principle, there is an inexorable trend toward the degradation of mechanical work into heat, but not *vice versa*.

The implication of the Second Law is that all energy transformations are irreversible. By turning the crankshaft through one complete revolution, an engine may be returned to a position which the most able mechanic could not distinguish from its initial state. But the wasting of energy as heat will have ensured subtle changes that cannot be wiped out. The orderly mechanical motion of the crankshaft (work) will have encountered frictional forces. This will have turned part of its mechanical energy into heat, which we may think of as the random and disorderly motion of the molecules forming the shaft. Energy from this disorderly motion will be carried off from the shaft by air molecules. The net result is the irreversible dissipation of energy as heat.

A key factor to remember here is Carnot's point that heat only flows from a hotter to a colder region. This provides us with an alternative formulation of the Second Law, to the effect that it is impossible to perform work by transferring heat from colder to hotter regions. For a more humorous account of the unshirkable principle in action, we can appeal to the rendition of comic duo Michael Flanders and Donald Swann in their song 'The First and Second Law':

> You can't pass heat from a cooler to a hotter.
> Try if you like, you far better notter,
> 'cause the cold in the cooler will get hotter as a ruler,
> 'cause the hotter body's heat will pass to the cooler.
> Oh you can't pass heat from a cooler to a hotter.
> Try it if you like, you'd only look a fooler.
> Cold in the cooler will get hotter as a ruler
> – That is a physical law![4]

The Second Law was sharpened up in 1865 when Clausius distinguished reversible and irreversible processes by introducing the concept of *entropy*, a quantity that relentlessly grows with dissipation and attains its maximum value when all the potential for further work is spent. According to his version of the Second Law, the entropy change in a reversible process is zero, while entropy always increases during irreversible processes. Entropy was so named by Clausius from the Greek words *en* (in) and *trope* (turning), with the intention of representing the 'transformation content' or 'capacity for change'. It is undoubtedly the most important concept in thermodynamics, and furnishes an explicit arrow of time: increasing entropy coincides with time's forward movement. Eddington was greatly impressed by the scope of entropy; he wrote: 'I wish I could convey to you the amazing power of this concept of entropy in scientific research.'[5] He likened it to beauty and melody because all three are connected with arrangement and organisation.[6]

To gain further insight into the meaning of entropy, it is very helpful to strip away any clutter and confounding factors. Scientists do this by

151

using idealised situations, in which the processes of interest define the system, with the rest of the world constituting the system's surroundings. As an example, thermodynamicists often like to consider the special case of an *isolated system*, that is one without any contact with the external world (*see* Figure 13(a)), like a box with rigid walls which do not permit matter or energy to pass through in either direction. An example would be an idealised Thermos flask of coffee that cannot lose water

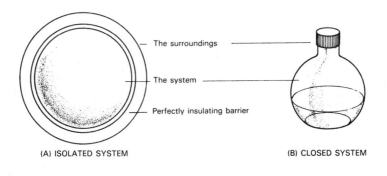

(A) ISOLATED SYSTEM

The surroundings

The system

Perfectly insulating barrier

(B) CLOSED SYSTEM

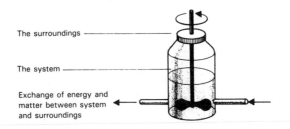

The surroundings

The system

Exchange of energy and matter between system and surroundings

(C) OPEN SYSTEM

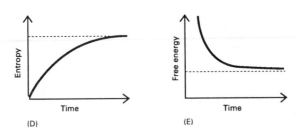

(D)

(E)

Figure 13. The three types of system encountered in thermodynamics: isolated (A), closed (B) and open (C). The two graphs show increasing entropy (D) and decreasing free energy (E).

[Adapted from P. V. Coveney. *La Recherche* **20**, 190 (1989).]

152

vapour or heat energy. Although in reality no system can be perfectly isolated (except perhaps the universe itself), it is nevertheless a very helpful idealisation. Two other general types of system are also frequently employed in thermodynamics: *closed systems* (*see* Figure 13(b)), set-ups which can exchange energy with the surroundings, and *open systems*, which exchange both energy and matter (*see* Figure 13(c)). In this terminology, living things such as people are open systems, since they exchange both energy and matter with their surroundings, ranging from wine and sausages to hot breath and excreta.

Any naturally occurring processes in an isolated system must be accompanied by an increase in the system's entropy, according to the Second Law. Entropy therefore furnishes an arrow of time for all isolated systems. The temporal evolution of an isolated system stops when the entropy attains its maximum value, when the system is at its most random state. The system has then exhausted all its capacity for change – it has reached thermodynamic *equilibrium*.

Clausius recognised that the universe itself is a perfectly isolated system (what else is there outside it?) and in 1865 cast the first two laws of thermodynamics in a cosmological form.[7] The First Law stated that the total energy of the universe is constant and the second, as we have seen, that the total entropy of the universe is remorselessly increasing towards its maximum value. The German physicist Hermann von Helmholtz was the first to infer from the Second Law that the evolution of the entire universe amounts to progressive cosmic degeneration, finally grinding to a halt at thermodynamic equilibrium when all change ceases. A universe at equilibrium is one where entropy and randomness are at their greatest, in which all life has died out. This *Heat Death* of the universe provided another target for the wit of Flanders and Swann:

> Heat is work and work's a curse
> And all the heat in the universe
> Is gonna cool down
> Because it can't increase.
> Then there'll be no more work and there'll be perfect peace.
> Really?
> Yeah. That's entropy, man!
> All because of the Second Law of Thermodynamics![8]

This is quite a fearsome prediction for a scientific theory that spurted out of a cloud of steam, and illustrates perfectly Mark Twain's remark: 'Get your facts first and then you can distort them as much as you please.'[9]

The argument is flawed because it ignores the role of gravity (and black holes): when gravity is included, it turns out that the universe must go further and further away from the uniform distribution of matter envisaged in the Heat Death.[10] Regardless of this, such a gloomy prognosis for the universe in the long run has no relevance to its behaviour in the short and medium terms. We know from astronomical evidence (Chapters Three and Four) that the universe as a whole is expanding, so that it cannot be anywhere near a state of thermodynamic equilibrium. And our tour of thermodynamics will show that when a system is a long way from equilibrium, perhaps because of a local hot spot like a star in the heavens, rather interesting things are possible, such as life.

Here we come to an apparent contradiction between thermodynamics and the Darwinian theory of the evolution of living species. In his theory of natural selection, Darwin showed how nature could preferentially select rare events (mutations) and thereby evolve progressively more complex forms of life. The driving forces for change in his theory are random events. However, Ludwig Boltzmann (who will play an important part in this chapter) showed that in a gas swarming with molecules there was a statistically random disappearance of highly organised configurations with time, in line with the Second Law of Clausius. It has been stated that 'Clausius and Darwin cannot both be right'.[11] We shall return to the reconciliation of Darwin's ideas with thermodynamics in Chapters Six and Seven, only pausing here to point out that Boltzmann held Darwin, his elder contemporary, in high esteem. Writing towards the end of the nineteenth century, Boltzmann commented: 'If you ask me for my innermost conviction whether it will one day be called the century of iron, or steam, or electricity, I answer without any question that it will be named the century of the mechanical view of nature, of Darwin.'[12] He made the same point in his lecture on the Second Law in 1886, during a meeting of the Austrian Academy of Science.[13]

If scientists decided there was indeed an unbridgeable disagreement between Darwin and thermodynamics, then most physicists would surely say it was Darwin who was at fault, such is the universal respect in which the Second Law is held. In the words of Arthur Eddington: 'The law that entropy increases – the Second Law of Thermodynamics – holds, I think, the supreme position among the laws of Nature. If someone points out to you that your pet theory of the universe is in disagreement with Maxwell's equations – then so much the worse for Maxwell's equations. If it is found to be contradicted by observation – well, these experimentalists do bungle things sometimes. But if your

154

theory is found to be against the Second Law of Thermodynamics, I can give you no hope: there is nothing for it but to collapse in deepest humiliation.'[14]

Equilibrium thermodynamics

If you are interested in steam engines, the heat loss through dissipation predicted by the Second Law is obviously a nuisance. Therefore, the early thermodynamicists sought ways of thwarting irreversibility. In the ideal world of *Gedankenexperimenten*, this seemed an easy task: one could simply arrange for the steam engine to work infinitesimally slowly, so that at each instant in time the system and its surroundings would be in thermodynamic equilibrium. In this 'quasi-static' state there would be no change in the overall properties as time passed. The entropy would at all times be at its maximum value, and no irreversible heat losses could occur. This corresponds to a perfectly reversible heat engine.[15] Even then, as we have noted, the efficiency in the conversion of heat into work could not reach 100 per cent – there can be no violation of the Second Law. Of course, there is one slight drawback. To the extent that all this could be achieved, the steam engine would take forever to complete the simplest operation.

These problems might suggest that the equilibrium approach is rather flawed. It is tantamount to suppressing time's essential role in the very notion of process. For all processes occur in a finite period of time and therefore cannot involve an infinite succession of these equilibrium states along the way. Nevertheless, many scientists somewhat paradoxically still try to think about thermodynamic processes in this way. One reason for this is that concentrating on equilibrium allows them to do away with the inconvenience and difficulty of describing irreversible processes, despite the fact that these are what get a system to equilibrium in the first place. In point of fact, the Clausius definition of entropy is only made for equilibrium states and it is still an open question to ask whether entropy can even be defined in general out of equilibrium.[16]

The American physicist Josiah Willard Gibbs is justifiably regarded as one of the founding fathers of modern equilibrium thermodynamics. Gibbs was in large part responsible for broadening thermodynamics from a limited investigation of the relationship between heat and work to the study of the transformation of energy in all its forms. His three key papers appeared between 1873 and 1878, when he was aged 39. They were published by the Connecticut Academy, despite a later admission by its publication committee that none of them understood what Gibbs was talking about. As one put it: 'We knew Gibbs and took his contributions on faith.'[17] Gibbs, and many others who followed him, studiously avoided all reference to non-equilibrium phenomena. Although this work is of vital significance for systems which have reached equilibrium, it is nevertheless a little like confining the subject of medicine to the dead.

By all accounts, Gibbs was a deeply uncharismatic man, whom even senior academics used to confuse with the then prominent US chemist Wolcott Gibbs.[18] It took a long time for his ideas to gather recognition among the wider scientific community. Today, their main interest is the insight they give us into what makes the evolution of a system cease in time. When an isolated system (see Figure 13(a)), like a gas in a well-insulated container, is in its most random state, it has achieved thermodynamic equilibrium. Then a single quantity, the maximum possible value of the entropy, is all that is needed to describe the macroscopic state of equilibrium and the end point of evolution. But for closed and open systems (see Figure 13(b) and (c)), which have an increasingly important dialogue with their surroundings, the state of maximum entropy must take into account the entropy of the surroundings as well. Thus, for a cooling cup of coffee in a kitchen, one has to take into account the tiny heating of the kitchen that results from the energy transfer between the two. This is a nuisance if we wish to investigate the equilibrium properties of the coffee alone – for the sake of simplicity, we wish to avoid bringing the behaviour of the kitchen explicitly into the discussion.

To exclude the kitchen, we can call on a new quantity, called free energy, which assumes its minimum value at equilibrium. The free energy of a system represents the maximum amount of useful work obtainable from it. Although free energy is only a disguised form of the total entropy, its value is that it can be thought of as an intrinsic property of the coffee, thus removing the need to refer to what is happening elsewhere in the kitchen. Free energy plays a central role throughout physics and chemistry in describing the equilibrium properties of systems, whether they be magnetic materials, refrigerators or chemically reacting mixtures.

Entropy and free energy are examples of what are called thermodynamic potentials. By this we mean that their respective extrema – the highest value of entropy and the lowest value of free energy – reveal the position of thermodynamic equilibrium. An analogy can be drawn with the swinging of a pendulum in a clock. At the start of each swing the pendulum has 'potential' energy under the force of gravity, which is transformed into the 'kinetic' energy of movement as it swoops through its arc. The effect of air resistance gradually dissipates the amount of potential energy left at the start of each swing; the pendulum's arc diminishes and ultimately it comes to rest in the vertical position. This is its position of minimum potential energy. Much the same thing happens in a chemical reaction in a closed vessel. Eventually it reaches a state of minimum free energy, when no further chemical change can take place.

A more fashionable way to refer to the extrema of the thermodynamic potentials is to say that they act as *attractors* for the system's evolution through time.[19] In the same way that the final resting place of a ball rolling into a valley (*see* Figure 14) remains unchanged, regardless of the position from which it is started, so the equilibrium state of a chemical reaction depends on the point of maximum entropy or minimum free energy, no matter what the starting conditions.

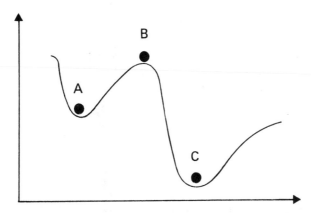

Figure 14. Mechanical analogue of the thermodynamic potentials (entropy and free energy). A mechanical system reaches equilibrium at positions of minimum potential energy. Positions A and C are local minima. The ball at point B is unstable – the slightest movement causes the ball to slide either to A or C, dissipating energy in the process.

Equilibrium attractors mark an important end point, a signpost declaring 'All change stops here!' Just as death is the Great Leveller and will bring us all to the grave, whatever our varying fortunes in life, so these attractors drag their systems inexorably towards equilibrium. Attractors come in a variety of forms, which we shall subsequently encounter. The equilibrium attractor of a chemical reaction can be likened to the bottom of a funnel: whatever the starting position of a ball placed in the funnel it will always roll to a point at the bottom. But not all systems have a simple fixed-point attractor like this – metaphorically speaking, other (non-equilibrium) attractors can trap the moving ball in a one-dimensional loop, rather like the brim of a sombrero, or allow the ball to roll around in a more extensive region of higher spatial dimension, like the toroidal surface of a ring doughnut. Much stranger attractors are also possible, as we shall see.

Shattering equilibrium

Equilibrium thermodynamics is valuable for studying the time-independent properties of macroscopic systems – a test tube where all chemical reaction has ceased, the ashes of a fire, or a cold cup of tea. The concept of an attractor also provides us with a 'target' for the arrow of time. However, in a very real sense, equilibrium is also a dead end.[20] Since it is concerned with the end-state of thermodynamic evolution and thus of time, it cannot describe the very processes by means of which time becomes manifest. As Peter Landsberg put it: '. . . in so far as the time coordinate is absent, nothing *happens* in thermodynamics'.[21] The real world has little to do with the dreary state of equilibrium. 'We breathe, we change! We lose our hair, our teeth! Our bloom! Our ideas!'[22] For living things, thermodynamic equilibrium is only attained with death, when a decaying corpse ultimately crumbles into dust. Life consists of many processes, from cell division and heart beat to digestion and thinking, all of which can only occur because they are out of equilibrium. If one wanted to be provocative, one might say that the predictive power of equilibrium thermodynamics holds as much interest as a fortune teller who says you are going to die some time in the future but omits to give any details of the handsome strangers you are going to meet in the interim. This would seem a deeply disappointing conclusion for a theory based on the Second Law and its powerful insight into transience and change. In order to make further progress in our investigation of time, we must examine what thermodynamics can tell us about irreversible processes which are not in equilibrium.

It turns out that non-equilibrium thermodynamics divides naturally into two parts: the 'linear' version describes the behaviour of systems close to equilibrium, while the 'non-linear' version deals with systems from which equilibrium is still far away. Linearity is a mathematical term for any kind of behaviour which could be charted on a graph in a straight line, based on a directly proportional relationship between two quantities. But while the action of a linear system is the sum of the behaviour of its parts, the behaviour of a non-linear system is more than the sum of its parts. The difference can be illustrated by comparing ordinary household fan-heaters with the fourth reactor at Chernobyl, the source of the world's worst nuclear accident to date. If one had two fan-heaters, one would expect to receive double the heat; if one had three, one would expect triple the heat, and so on. This pattern represents a linear relationship between the heat produced and the number of heaters. But in the case of Chernobyl, the overheating of the reactor's core increased the nuclear chain reaction which in turn boosted heat production. The resulting surge of heat, which was out of all proportion to the original rise in temperature, is known as non-linear positive feedback. Sloppy operating procedures conspired with this deadly characteristic of the Soviet design to produce an explosive release of energy. On 26 April 1986, it devastated the reactor, with worldwide consequences.

Order out of chaos

The natural tendency for macroscopic systems is to wend their way along the arrow of time towards equilibrium. But what happens if the system is stopped in its tracks before it reaches its destination?

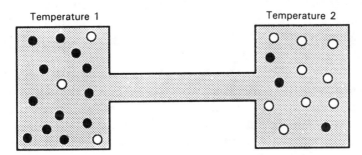

Temperature 1   Temperature 2

Figure 15. Apparatus for thermodiffusion. If a temperature difference is maintained between the two containers, more molecules of one kind are found in one container and fewer in the other. The concentration gradient is proportional to the difference in temperature $(T_1–T_2)$.
[After I. Prigogine and I. Stengers. *Entre le temps et l'éternité*, p. 50.]

The vessel containing a mixture of hydrogen and hydrogen sulphide gas shown in Figure 15 illustrates how maintaining a system out of equilibrium can overturn the simple-minded dogma that 'entropy equals disorder'. One might imagine that by heating the mixture it would become more randomised: the more heat, the more vigorously the gas molecules rush around in the vessel. But now let us prevent it from reaching equilibrium by maintaining a small temperature difference at opposite ends of the vessel. Experiments reveal that a gradual separation of the gases emerges along the vessel: the hotter region is richer in the lighter hydrogen gas, while the cooler end has a higher concentration of the more massive hydrogen sulphide. This effect is known as thermodiffusion, and is observed to lead to analogous concentration profiles of substances within oil reservoirs.

At first sight this phenomenon appears counter-intuitive – heating the vessel increases the entropy yet produces a less random arrangement of molecules, manifested by what is called the concentration gradient of the gas. In spite of the popular interpretation of the Second Law of Thermodynamics as linking entropy in a facile way with 'disorder', thermodiffusion shows how structural organisation can spontaneously emerge from randomness. In this simple example, the organisation is the gradual increase of hydrogen concentration and concomitant decrease of hydrogen sulphide along the vessel's temperature gradient. To be sure, there is randomness in the frantic motions of the gas molecules, yet overall this is clearly less than at equilibrium. Thermodiffusion provides the first indication that irreversible, non-equilibrium processes can give rise to organisation. Thus there is a link between the arrow of time and the possible emergence of structure.

Equilibrium potentials allowed us to predict the final state of a thermodynamic system. They showed that the system was like a ball rolling in a road, which would always end up in the gutter no matter where on the camber it started. Can we make similar statements for *non-equilibrium* behaviour, when a system is prevented by its environment from reaching equilibrium? The answer is generally yes, provided that the system is not pushed too far from thermodynamic equilibrium: mechanically speaking, the ball will again always roll into the gutter. But it may roll elsewhere – for example, into a pot-hole in the road – if it is driven too far from the gutter, making predictions much more difficult. To determine a thermodynamic potential for thermodiffusion, the temperature difference across the vessel – that is, the temperature gradient – must be sufficiently small.

The temperature gradient that provides the 'push' for this system can be described as a thermodynamic force. Such forces produce the flow and flux of heat and matter in the same way that a kick moves a ball. Close to equilibrium, these flows and fluxes behave in a simple way – if the force is doubled so is the flux, and so on. At equilibrium the forces vanish and hence so do the fluxes. The fluxes are directly proportional to the forces

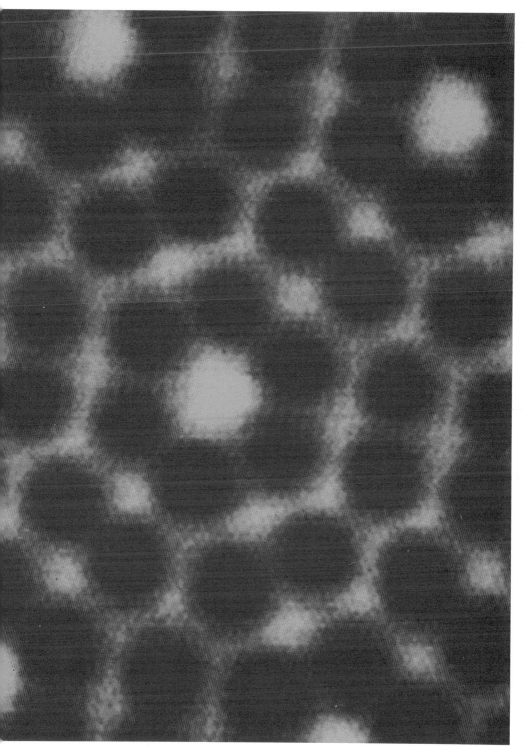

anning tunnelling microscope image of atoms.

Overall view and close-up of the hexagonal convection pattern in a layer of silicone oil 1 millimetre deep.

A strange attractor.

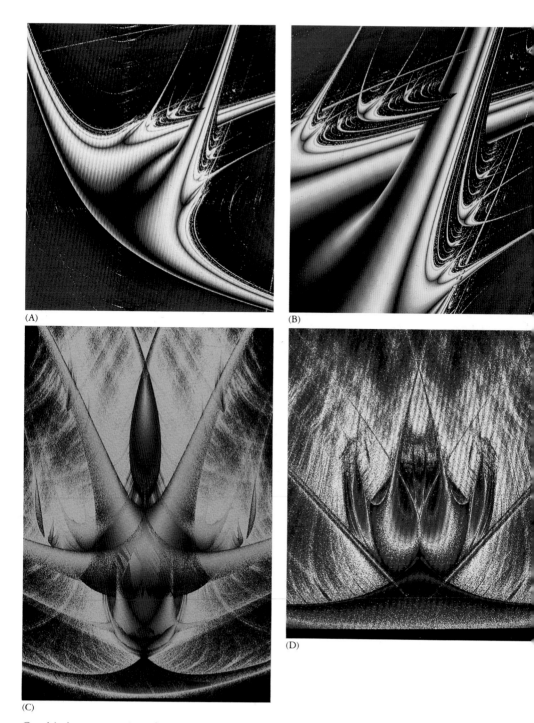

Graphical representation of the logistic equation, a simple model in population dynamics, being put through its paces. A and B show the same parameters at different magnifications; C and D are different parameters at the same magnification. Lighter regions correspond to organisation, and darker regions to chaos.

producing them, which is why we refer to linear thermodynamics in this region.

Linear thermodynamics was put on a firm footing largely thanks to the efforts of one man during the 1930s, Lars Onsager of Yale University, who won the Nobel Prize for Chemistry in 1963.[23] His 'reciprocity relations' show that in linear systems there is an aesthetically pleasing symmetric relationship: forces produce fluxes, while fluxes produce forces. In the case of thermodiffusion, a flow or flux of matter is caused by a force (expressed by the thermal gradient). But in turn it follows from Onsager's reciprocity relations that a concentration gradient will generate a heat flow, an effect that has also been verified experimentally.

In 1945, a new character stepped on to the thermodynamic scene – Ilya Prigogine. Born in Moscow in the revolutionary year of 1917, the young Prigogine came with his family to western Europe ten years later, where they settled in Brussels. He grew up as an apprentice in the so-called Brussels School of thermodynamics founded by the Belgian physical chemist, Théophile de Donder. Most of Prigogine's university career has been spent at the Free University of Brussels, an institution created by the Freemasons as an antidote to what they saw as the oppressive authority of the Catholic Church in purely secular affairs, a view which still informs the official policy of the university.

The 28-year-old Prigogine showed that in this well-behaved linear regime, thermodynamic dissipation moves down into its lowest possible gear. Thus the rate of change of the system's entropy, referred to as the internal entropy production, will decline: in general, the system will evolve to a steady or stationary state in which the dissipation is at a minimum.[24] In the case of thermodiffusion, the total entropy may be increasing but the rate of internal entropy production is at its lowest value when the final concentration gradient of the gases is established.

Prigogine's doctoral thesis was published in 1947 under the title 'A Thermodynamic Study of Irreversible Phenomena'.[25] It contained the minimum entropy production theorem[26] and set the stage for his life's work. Prigogine began a meteoric rise which saw him take over the leadership of the Brussels School. With his elder colleague, Paul Glansdorff, he sought to extend thermodynamic analysis into new realms.

Prigogine's picture of minimised dissipation is more useful to us than the equilibrium notion of maximising entropy because it is more relevant to the real world where nothing is truly in a state of equilibrium. There is always potential for further evolution, for liquids to mix, buildings to crumble, objects to cool. So long as there is some small outside influence to keep a system out of equilibrium (such as the temperature gradient in the diffusion vessel) then it will persist in a *steady-state* rather than collapse into total randomness. Most home-owners are familiar with the steady-state. The effects of wind, rain and weathering will in time reduce a house to a pile of rubble, its equilibrium state. But a house usually occupies a steady-state for many years, where the rate of repair equals the rate of decay. Only if the home-owner stops interfering will the house start to fall apart (although at some point – perhaps far into the future – this is bound to happen in any case).

In the case of equilibrium thermodynamics, the 'target' of the arrow of time was described by a fixed-point attractor with quantities like minimum free energy or maximum entropy drawing the system to equilibrium. In systems which are kept out of equilibrium, but remain close to it because the outside influences on them are relatively small, we can see that much the same thing still applies. The end-state – such as the concentration gradient – still does not vary in time; it stays steady. By extending the analysis a little away from equilibrium, Prigogine had shown that a system will generally evolve to a point where the entropy production is at a minimum. Now he hoped to take the bold step of developing this still further for more complicated situations – specifically, for non-linear systems held far from equilibrium – so as to furnish a broad picture of the way things evolve through time. However, the similarities with equilibrium behaviour no longer hold. The new regime may be harder to understand, but it is much more exciting because it has a great deal more to do with the world we see around us, and offers a much more sophisticated picture of time and change.

Far, far from equilibrium

Prigogine's minimum entropy production theorem is an important result. However, its proof depends (among other things) on the well-behaved linear relationship between fluxes and forces described by Onsager. Together with his colleagues in Brussels, Prigogine set out to explore systems maintained even further away from equilibrium, where the linear law breaks down, to see whether it was possible to generalise his theorem into a criterion of evolution that would work for non-linear systems, in which the simple relationship between flux and force fell apart. It was a path which was to lead to some fascinating consequences, tempered by bitter controversies.

We have mentioned that scientists, like everyone else, have fierce disagreements and rivalries. The controversies surrounding the Brussels School continue unabated, unfortunately fuelled by the occasional exaggerated claims made by the group in reporting its findings to the outside world. The root problem is largely one of presentation but it has led to a rather unsatisfactory situation: the polarisation of opinion along almost party political lines. Sadly, this tribal behaviour often undermines rational debate on what the Brussels School has to offer.[27]

Over a period of some twenty years, the Brussels School elaborated a theory which has come to be known as 'generalised thermodynamics', whose aim was to apply thermodynamic principles to far-from-equilibrium situtations.[28] Essentially Glansdorff and Prigogine used an approximation which makes systems far from equilibrium look and behave locally as a good-natured patchwork of equilibrium systems; conceptually, this is rather like visualising curved spacetime in general relativity as consisting of local regions of flat spacetime stitched together.[29]

Glansdorff and Prigogine used this 'local equilibrium' approximation as a means to move a long way away from the cosy world of equilibrium thermodynamics. They formulated what they termed a 'universal criterion of evolution', which they thought would provide a much more sophisticated view of the way things evolve through time. This work achieved its apotheosis in their book, *Thermodynamic Theory of Structure, Stability and Fluctuations*, published in 1971.[30] Unfortunately, the language in which they couched this criterion – in particular their choice of the word 'universal' – has been at the centre of many subsequent controversies. All the authors meant was that they had obtained a purely thermodynamic criterion of evolution, but by appearing to claim more than they had actually established, they left themselves open to vituperative polemics from those who were only too eager to launch a broadside on the work of the Brussels School.

The Glansdorff–Prigogine criterion deals with the most common situation one is likely to encounter – open systems far from equilibrium through which both energy and matter can flow, where very complicated non-linear relationships govern behaviour. Their criterion makes a general statement about the stability of steady-states when they are pushed further out of balance. It says that when they are far from equilibrium, such stationary states may become unstable. There may arise a 'crisis point', technically referred to as a 'bifurcation point', at which the system prefers to leave the steady-state, evolving instead into some other state (*see* Figure 16).

The important new possibility is that beyond the first crisis point one can have highly organised behaviour in time and space. In some chemical reactions pushed far from equilibrium, for example, one can witness the appearance of regular colour changes or beautiful scrolls of colour. These are stable states to which the system evolves in the course of time, but ones that are no longer associated with minimal internal entropy production.

Perhaps we can give a clearer view of the relationship between entropy and organised behaviour by drawing an analogy with the choices open to a man who wants to save money (a woman would, of course, do equally well). If he is completely broke, the change in his expenditure – corresponding to the change in entropy at equilibrium – is zero. However, the real world is not at equilibrium. Our would-be saver most probably has at least a small amount of money, and must spend some of it to eat and drink. The best he can expect to do in this situation is to minimise his expenditure at a level which allows him to survive. In a similar way, many thermodynamic systems pushed to a stable state a little away from equilibrium will minimise the production of entropy. Now let us look at what happens far from equilibrium. This is equivalent to the predicament of the would-be saver who is a wealthy man. In these circumstances, it is no longer enough for him to minimise his rate of expenditure. There are now many more options available to him than plumping for the cheapest food and accommodation. He could invest his money in a high-interest account. He could spend vast quantities on speculative ventures that promise high returns. All of this expenditure in the short term is quite compatible with his urge to be wise with his money. There is no longer a universal criterion for saving money (not spending it) as there was when he was poor and his lifestyle was simpler. A huge range of possibilities of organised behaviour now unfolds.

The remarkable fact is that while far from equilibrium the global entropy production increases at a furious rate, consistent with the Second Law, we can nevertheless observe exquisitely ordered behaviour. Thus we must revise the received wisdom of associating the arrow of time with uniform degeneration into randomness. True, at the 'end' of time, when no further change can occur, randomness may win. But over shorter timescales we can see the emergence of evanescent structures which survive for as long as the flow of matter and energy continues. Indeed it is important to appreciate that a system can only be held away from equilibrium if it is open to its environment: this enables the entropy produced by the system to be exported to the surroundings, thereby permitting the maintenance of organisation while allowing an overall increase in the entropy of the system *and* the environment.

164

The Glansdorff–Prigogine statement is 'weak' in that it only suggests the possibility of a thermodynamic crisis occurring, not its inevitability.[31] It does not govern the way things evolve in time in the same way as the iron rule set by the Second Law. It depends on the Second Law and some additional assumptions. Far from equilibrium there is in general no thermodynamic potential – no single attractor – which can act as the target for the arrow of time. This lack of universality was pointed out by Joel Keizer of the University of California at Davis and Ronald Fox of the Georgia Institute of Technology in Atlanta.[32]

Something stirs, far from equilibrium

In the same way as the dog that did not bark furnished vital evidence for Sherlock Holmes in *The Adventure of Silver Blaze*,[33] the weakness of the Glansdorff–Prigogine criterion can be turned to considerable advantage. The reason why the criterion is not a universal rule is that there is an enormous range of possible behaviours available far from equilibrium. And this sheer complexity – which makes the drawing of straightforward causal connections impossible – helps to suggest why the taking of one route, among many, confers a special status on the system's path through time. This can be represented with the help of what is called a bifurcation diagram, a simple pictorial representation which describes the various possibilities on offer (*bifurcation* means that there is a fork at the crisis point). In the case of our thrifty saver, the diagram would show only one possibility near equilibrium, when he is on the bread line (the thermodynamic branch). Far from equilibrium, many more possibilities or 'branches' are available. The situation is shown in Figure 16(a): the bifurcation represents the dilemma faced by a system when it has reached its thermodynamic crisis point.

Let us apply this kind of treatment to chemical reactions, which form an excellent example of irreversible processes. Chemical substances combine and are thereby transformed into chemically distinct products as time goes on; for example, oxygen and hydrogen together form water, while iron reacts with oxygen in the atmosphere to produce rust (an iron oxide). The bifurcation diagram reveals the behaviour of a special kind of chemical reaction as it is pushed further and further away from equilibrium. The vertical axis in Figure 16 represents the concentration of a chemical, called A, in a chemical reaction mixture. The horizontal axis represents the 'distance' (denoted by the Greek letter λ) of the chemical reaction from thermodynamic equilibrium, at which the chemical reaction stops. The point A_{eq} (the equilibrium concentration

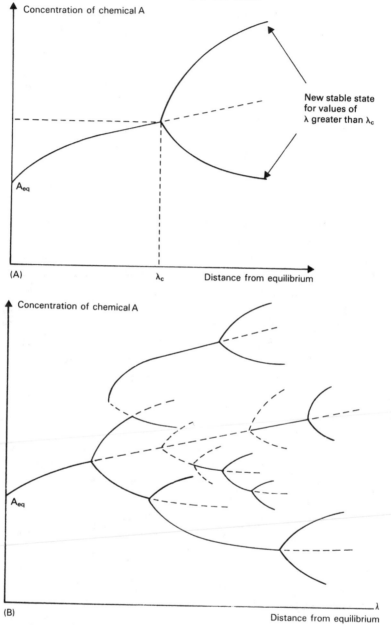

Figure 16. Bifurcation diagrams for chemical reactions far from equilibrium. (A) Primary bifurcation. λ_c is the distance from equilibrium for which the thermodynamic branch of minimum entropy production becomes unstable. The bifurcation or crisis point corresponds to the concentration λ_c. (B) Full bifurcation diagram. As the nonlinear reaction is driven further from equilibrium, the number of possible stable states available increases dramatically.

of A, where λ is equal to zero) is the equilibrium concentration of A in the system. In a system prevented from reaching equilibrium (for example, an open system in which the supply of substances is constantly replenished as the products are eliminated), the concentration of A will differ from A_{eq}. In this case, the reactant A is being poured in as fast as it is consumed by the chemical reaction so that the concentration of A is held constant. The difference between this concentration and A_{eq} is a measure of the 'distance' that the system is held away from thermodynamic equilibrium.

What does the diagram reveal? At equilibrium and nearby, λ lies in the well-behaved linear regime (the thermodynamic branch) and the system obeys Prigogine's theorem of minimum entropy production; the reaction mixture remains in a stable steady-state. But beyond a critical threshold distance from equilibrium, a choice emerges, shown by the sudden emergence of two new lines at the thermodynamic crisis point, the bifurcation point λ_c. The dotted line shows the thermodynamic branch, which has now become unstable.

This crisis point has special significance, for beyond it (depending on the details of the chemical reaction under study) we can sometimes see organised behaviour emerge. This region is also special because the chemical reaction is presented with a choice. In this particular instance, the reaction has two new stable states on offer: the concentration of A assumes radically different values in each state as the distance increases away from the linear region. The crisis point in effect introduces a sense of history into the bubbling reaction mixture. This arises because, among many choices, one has to be made that subsequently takes the system along a given route in the bifurcation tree: in effect, it has 'memory' of the choice it made. Another mixture in a separate vessel may have made a different decision. The chemical reaction can only exist in one stable state at any time: which of the two new branches it actually follows can only be predicted by the toss of a coin since there is a 50–50 chance of the system being found in one or other of the new states. We shall see in Chapters Six and Seven how situations can arise where there is so much choice available to the system – in the form of vast numbers of adjacent stable states – that very unpredictable dynamical behaviour occurs, known as deterministic chaos. This kind of behaviour is indicated by the more extensive bifurcation diagram shown in Figure 16(b), in which a whole succession of further crisis points appear beyond the primary one.

Only a vague picture of what can happen beyond the first crisis point is given by Glansdorff and Prigogine's non-equilibrium thermodynamics.

167

It hints that something very important can happen but it says nothing about what it means in real terms. For a chemist, these choices could mean a chemical clock that changes colour periodically, or the emergence of coloured patterns in a test tube. For an ecologist, the possibilities could mean new steady-states or oscillations in animal populations. For a doctor, it might be the moment when a heart attack occurs. To handle these scenarios adequately, we must employ analyses based on a detailed mathematical study of irreversible kinetic equations, which describe, for instance, the behaviour of a specific chemical reaction as time passes. The remarkable properties of such kinetic equations and the wealth of complex effects which they describe form the theme of Chapter Six.

The new states which arise far from equilibrium can possess an extraordinary degree of order wherein trillions of molecules coordinate their actions in time and space. Prigogine coined the term *dissipative structures* to describe them, since they result from the exchange of matter and energy between system and environment, together with the production of entropy (dissipation) by the system. The complex and mutually dependent processes leading to the formation of these structures are collectively called *self-organisation*.

We can now see that the spontaneous creation of order is not forbidden by thermodynamics, although the Second Law is commonly – and erroneously – regarded as being a watchword for the uniform degeneration into randomness. That is the real message of Glansdorff and Prigogine's work. In 1977, Ilya Prigogine was awarded the Nobel Prize for chemistry; the citation mentioned his contributions to non-equilibrium thermodynamics, with particular reference to the theory of dissipative structures.

How do dissipative structures fit in with the notion of the Heat Death? Our cosmological view, resulting from the rash application of equilibrium thermodynamics but neglecting the role of gravity, would imply that the evolution of the universe is synonymous with an inexorable growth of randomness, finally terminating at thermodynamic equilibrium when all change ceases. But the study of non-equilibrium processes has shown that a universe evolving far from equilibrium is not capable of being described in these simplistic terms. We get a much more optimistic picture than this. In such a universe, irreversible non-equilibrium thermodynamics allows for the possibility of spontaneous self-organisation leading to structures ranging from planets and galaxies to cells and organisms.

The irreversibility paradox

Many scientists only believe that they have understood a phenomenon when they can relate it to the actions of atoms and molecules. This is called a *reductionist* approach because it is claimed that everything else can be reduced to this supposedly more fundamental level. But the approach runs into problems with thermodynamics and its sophisticated arrow of time. For as we have seen, no such arrow seems to be present in the microscopic world, whether it is described by Newtonian, relativistic or quantum mechanics. One consequence of this is that there can be no state of microscopic equilibrium: time-symmetric laws do not single out a special end-state where all potential for change is spent, since all instants in time are equivalent. The apparent time reversibility of motion within the atomic and molecular regimes, in direct contradiction to the irreversibility of thermodynamic processes, constitutes the celebrated *irreversibility paradox* which was put forward by Loschmidt among others in 1876.[34] The paradox suggests that the two great edifices – thermodynamics and mechanics – are at best incomplete. It represents a very clear problem in need of an explanation which should not be swept under the carpet. Our search through science may finally have brought us a deep insight into transience, but we cannot rest until it is reconciled with the theories we discussed in our earlier chapters – for as Lord Kelvin once pointed out, a microscopic world that can defy time's arrow has the power to stand the macroscopic world on its head. 'If the motion of every particle of matter in the universe were precisely reversed at any instant, the course of nature would be simply reversed for ever after. The bursting bubble of foam at the foot of a waterfall would reunite and descend into the water; the thermal motions would reconcentrate their energy, and throw the mass up the fall in drops reforming in a close column of ascending water . . . living creatures would grow backwards, with conscious knowledge of the future, but no memory of the past, and would become again unborn.'[35]

Bridging the gap

Our attempts to bridge the gap between the two levels of reality will draw upon the discipline of *statistical mechanics*, which has the job of contracting the vast amount of information needed in quantum or classical mechanics to the handful of parameters used in thermodynamics. For example, it shows us how to move away from describing all the separate molecules in a litre of a gas and to talk in more general terms about the gas's pressure and temperature.

At the microscopic level, the gas consists of an enormous number of molecules in motion – of the order of ten followed by 23 zeros. A phenomenal amount of information would be required in order to put together a 'full' description of their behaviour. One would need to know the velocities and positions of all the molecules at some given initial time. But on the macroscopic level, we know that a tiny amount of information is needed to describe the gas's general properties. At equilibrium, for example, only three quantities are required – the pressure of the gas, the volume which it occupies and its temperature.[36] Clearly, this involves a tremendous compression in information. It is the business of statistical mechanics to show how this contraction can be achieved.

The key step is to supplement the laws of mechanics with probability theory, which allows us to base our calculations on averages. There are two pillars of support for this approach. The first is that a probabilistic description, even for a Newtonian model, has to be used since we never have precise information on the speeds and whereabouts of all the molecules (a deeper insight into this uncertainty is given in Chapter Eight). This makes all our subsequent statements statistical, rather as in quantum mechanics (see Chapter Four), although for quite different reasons, and means we can only talk about average properties. For example, we can possess information about the average energy of the molecules.

The second pillar concerns the very large size of macroscopic systems compared with the atomic and molecular building blocks from which they are made. Suppose that we wanted to test whether there is a 50 per cent chance of turning up heads (or tails) when tossing a coin. This probability does not mean that, if tossed twice, a head and a tail must turn up. Average behaviour only emerges if we toss a very large number of identical coins simultaneously or toss the same coin many times – the more flips or the more coins, the closer we get to the 50 per cent probability. Returning to the use of averages for describing the behaviour of molecules in an everyday object, we are fortunate in that the number of molecules such objects contain runs to a figure of one million million million million. The averages calculated in statistical mechanics generally provide an excellent description – for the same reason that if we tossed a coin one million million million million times, any deviation away from the average behaviour of half heads and half tails would then be negligible.

170

One big problem with statistical mechanics is that it is a lot simpler to implement for equilibrium phenomena, when all change has ceased, than for non-equilibrium phenomena, which are still evolving in time. We will start with statistical mechanics in equilibrium, which provides a very clear link between molecular and thermodynamic properties, for instance between the average speed of gas molecules and the temperature of the gas. The key object is a complicated mathematical device known as the *partition function*, which permits the calculation of all the macroscopic properties of a system – be it solid, liquid or gas – at equilibrium, such as its entropy and pressure. For our purposes, all we need to know is that, in principle, the partition function can be calculated exactly from a knowledge of the energy levels available to the system. These energies are found by solving the Schrödinger equation we encountered in Chapter Four. From the practical point of view, this is a formidable and daunting task which can only be overcome with the mathematical equivalent of fancy footwork, amounting either to the use of clever approximations or to the construction of idealised models.[37] However, once we have the partition function we can use it to unlock all the equilibrium properties of matter. For example, one can calculate the entropy or free energy of one litre of hydrogen gas at equilibrium and determine its pressure at a given temperature without performing a single experiment.

Life gets harder when we try to do the same for non-equilibrium processes, such as describing the behaviour of the hydrogen gas when the vessel is opened and the gas escapes. Suddenly the partition function becomes irrelevant as it is strictly meaningful only for thermodynamic equilibrium.[38] This neat route to entropy and other thermodynamic quantities turns out to be a dead end.

Sadly, a return to first principles suggests that there is no other easy route. Let us take a closer look at one of those pillars of statistical mechanics – the use of averages rather than the precise whereabouts and speeds of every molecule in the gas. The manner in which this averaging is done depends on whether we choose to describe the microscopic level in classical or quantum mechanical terms. In reality we do not have a choice, since the microscopic world is best described by quantum theory. However, in spite of the differences between the two, the basic ingredients of the macroscopic description arising from either is not usually substantially different, save in explaining unusual phenomena like superconductivity, when at low temperatures a material loses all resistance to the flow of electric current.[39] (The similarities arise from a very close formal mathematical parallel between classical and quantum statistical mechanics. In essence, the bizarre quantum effects are smudged out by the large number of particles that are under consideration.)

To avoid the hopeless task of trying to determine the position and velocity of every molecule in a large system which would be needed to start off Newton's equations, we resort instead to a statistical entity known as the *probability distribution function*. In the same way that an exit poll conducted on Election Day provides an idea of how an election will turn out, the probability distribution function simply tells us the probability of finding the system in a particular state, in which all the molecules have exactly known positions and velocities which evolve in time according to Newton's equations.

If quantum mechanics is preferred, we use wavefunctions, the quantum 'equivalent' to the trajectories of classical mechanics. Of course, like the Newtonian model, the quantum description also suffers from a massive overload of information at the level of atoms and molecules: there are a great many microscopic quantum states (each described by a separate wavefunction) compatible with the observed macroscopic behaviour, so we cannot hope to know what state the macroscopic system is actually in. For a typical gas consisting of millions upon millions of molecules, the number of quantum states is unimaginably large. Hence we must turn to the quantum mechanical analogue of the exit poll or classical probability distribution function. Called the (probability) density matrix, this quantity simply tells us the probability of finding the system in one or another quantum state, each described by a different wavefunction.

So far, so good. Now that we have managed to find a way to describe the statistical behaviour of a large number of molecules by using either quantum or classical mechanics, we have to represent the way that a system evolves irreversibly away from equilibrium. What we need are some equations which describe how the classical probability distribution function and the quantum density matrix vary in time. These happen to be formally identical, and are called the Liouville–von Neumann equations. But, alas, the Liouville–von Neumann equations are traitors to the cause. They are directly based on classical and quantum mechanics which we know do not distinguish the directions of time. Hence these equations are also time-symmetric: while they can be used to compute the entropy of a system at equilibrium, as was done with the methods based on the partition function, they cannot alone explain the increase in entropy for systems evolving away from equilibrium which furnishes us with the arrow of time.

172

This is the fundamental problem of non-equilibrium statistical mechanics.

Boltzmann's arrow of time

The great Ludwig Boltzmann, whom we first met in our Prologue, is worthy of special attention for the way he tried to resolve this embarrassing paradox. The son of a Viennese tax official, he was born on 20 February 1844, the evening between Shrove Tuesday and Ash Wednesday, as a Mardi Gras ball came to an end. He used to joke in later life that this was the reason why his temperament would swing between great joy and grief.[40] Boltzmann was the first person to give a fundamental physical law – the Second Law – a statistical interpretation, but his ideas met with violent objections from other theoretical physicists and mathematicians. He killed himself before his ideas were widely accepted. Yet he also had powerful allies. In 1900, just six years before Boltzmann's suicide, the 21-year-old Einstein wrote to his girlfriend, Mileva Marić, that Boltzmann 'is a masterly expounder. I am firmly convinced that the principles of [his] theory are right, which means that I am convinced that in the case of gases we are really dealing with discrete mass points of definite finite size, which are moving according to certain conditions.'[41]

Boltzmann was brought up as a devout Catholic in the Austrian towns of Salzburg and Linz during the time of the Austro-Hungarian Empire. A gifted mathematician, he studied for his doctorate in mathematical physics at the University of Vienna under the physicist Joseph Stefan and the eccentric mathematician Josef Petzval. It was Stefan's doctoral supervisor who had introduced James Clerk Maxwell's kinetic theory of gases to continental Europe. Boltzmann extended this work, using an approach to statistical mechanics pioneered by another student of Stefan's, Josef Loschmidt (1821–95). In 1872, many years before quantum mechanics appeared on the scene, Boltzmann reported on the application of the Liouville equation to a large assembly of molecules in a gas and analysed the approach to thermodynamic equilibrium. His work at last seemed to give time a sense of direction at the microscopic level. He obtained a time-asymmetric equation of evolution, now known as the Boltzmann equation, which is satisfied by a single-particle distribution function, a fancy term for the statistical description of the motion of a single molecule in the gas. From this equation, he constructed a new mathematical function, the so-called $\mathcal{H}$ function, which decreases as time passes. It provided a complementary arrow to that of

173

entropy, which increases as time passes *en route* to thermodynamic equilibrium. (The $\mathscr{H}$ function is equal to the entropy, but of opposite sign.) Boltzmann thus claimed to have solved the irreversibility paradox on the molecular level.

It was a vital achievement for Boltzmann, since he held the opinion that: 'Mechanics is the foundation on which the whole edifice of theoretical physics is built, the root from which all other branches of science spring.'[42] However, in order to arrive at his irreversible equation Boltzmann had made one crucial approximation – the assumption of 'molecular chaos'. Put another way, he assumed that molecules which are about to collide are uncorrelated but that following collision they are correlated (because their trajectories are changed by the collision). Since the molecular chaos assumption is time-asymmetric, it explains why Boltzmann's equation describes irreversible time evolution.

Naturally enough, therefore, molecular chaos violates Newton's time-symmetric laws. For these and related reasons, Boltzmann was severely criticised by Ernst Zermelo, the young Berlin assistant of Max Planck, and by his friend Josef Loschmidt. Put most straightforwardly, their arguments are that one cannot hope to build a unique arrow of time into equations which are oblivious to any such direction. Loschmidt's weapon in 1876 had been the time-symmetric nature of mechanics.[43] Twenty years later, Zermelo attacked Boltzmann's work armed with Poincaré's recurrence theorem, which, it will be recalled, says that every isolated system will eventually return to its original state.

Boltzmann was so shaken by these attacks that he stepped back from his original attempt to find the arrow of time through the use of mechanics; he decided that the Second Law was based on probability theory and not mechanics alone, eschewing mechanistic reductionism for a form of atomism in which atoms had to be treated statistically.[44] He even came to favour the view that the universe was already in a state of Heat Death, or overall thermodynamic equilibrium. We just happen to be in a region in which a fluctuation has taken us away from equilibrium and is now returning us there with an accompanying increase in entropy. There will be just as many patches of the universe in which the fluctuations away from equilibrium are accompanied by a decrease in entropy. Thus Boltzmann maintained that, on a cosmic scale, there is no arrow of time: there would be as many regions with the reverse arrow of time as there would be with our own.[45]

It is an ingenious argument. Unfortunately, it is undermined because we never do observe other portions of the universe possessing an inverted arrow of time, where bulls grow younger and have the power to clear up demolished china shops. Indeed, as we noted in Chapter Three, modern astronomy and cosmology show that the entire universe is expanding and so it cannot be in thermodynamic equilibrium. There has to be a cosmological arrow of time, for at earlier times in the universe galaxies were closer together. All known phenomena are in accord with a uni-directional time for which the state of thermodynamic equilibrium lies in the future, not in the past. Boltzmann's original aim to reconcile thermodynamic irreversibility with the reversibility of quantum and classical mechanics was correct. He was ahead of his day but lacked the knowledge with which to overcome his critics, a point we shall return to in Chapter Eight.

In his later years, Boltzmann continued to argue fervently that thermodynamic behaviour is a manifestation of atomic and molecular phenomena. As we have mentioned before (*see* Chapter One), this induced his opponents Mach and Ostwald into intense polemics, fuelled no doubt by Boltzmann's prolixity and occasional tendency to change his views without saying so explicitly. His critics were men of considerable authority, and Boltzmann − who was handicapped in debate by a rather high-pitched voice − was thrown into depression, making a suicide attempt while at Leipzig. Not that his opponents always had things their own way. The German mathematical physicist Arnold Sommerfeld wrote about one discussion at the Lübeck Scientific Conference in 1885: 'The fight between Boltzmann and Ostwald was like the fight of a bull with a supple fencer. But this time the bull overcame the *torero* in spite of all his art of fencing. The arguments of Boltzmann won. We younger mathematicians were all on Boltzmann's side . . .'[46] Mostly, however, it was a lonely and difficult battle. Following a renewed surge of opposition to his ideas, Boltzmann wrote in the foreword to his classic work, *Lectures in Gas Theory:* 'I am conscious of being only an individual struggling weakly against the stream of time.'[47]

In 1902 Boltzmann returned from Leipzig to his personal chair in Vienna.[48] He was the only survivor of the triumvirate of theoreticians who had created the kinetic theory − Clausius and Maxwell were already dead. Boltzmann's intellectual isolation was compounded by a deterioration in his health − worsening eyesight, frequent asthma attacks, angina and headaches. In 1903 his wife Henriette (who called him her 'sweet fat darling') wrote to their daughter, Ida: 'Father gets worse every day. I have lost my confidence in the future. I had imagined a better life in Vienna.'[49]

Boltzmann fell into a deep depression following a banquet to celebrate his sixtieth birthday in 1904. One student described the depth of his torment after his last lecture on theoretical physics in the winter of 1905–6. 'A nervous complaint [headaches] prevented him from continuing his teaching activity. Together with another student I took and passed my oral examination in his villa in Währing. On leaving after the examination was over we heard from the front hall his heart-rending groans.'[50] The combination of polemical debates, an innate fear of loss of creativity and worsening attacks of *angina pectoris* culminated in his suicide on 5 September 1906. Only a few years afterwards his atomistic ideas were to prevail, in considerable measure due to Einstein's work on Brownian motion and the counting of molecules and its rapid experimental confirmation by Jean Perrin in Zürich.[51] In his wide-ranging study, *The Austrian Mind*, William Johnston furnishes an account of how it became almost fashionable for Austrian intellectuals of Boltzmann's period to resort to death by their own hand as a refuge from the vicissitudes of life in a turbulent era. Indeed, Boltzmann once spoke of 'the idea that fills my thought and my action: the development of theory', adding, 'no sacrifice for it is too much for me: since theory is the content of my whole life'.[52]

It is all in the mind: 'coarse-graining'

The Boltzmann equation lives on. It has found wide use for describing irreversible 'transport processes' such as diffusion and viscosity in dilute mixtures of fluids, where the assumptions Boltzmann invoked are actually good approximations. There have also been several attempts to obtain an improved version of the $\mathscr{H}$ theorem. One of the initially most promising approaches is known as coarse-graining. Unfortunately, it leads to a conclusion that is utterly alien to all Boltzmann championed. For it suggests that the arrow of time is purely subjective, existing only outside the microscopic world through our own approximations.

Coarse-graining is a technique used to describe events at smaller scales than we can observe. It involves calculating the average motions of molecules in arbitrarily defined sub-compartments of a system, for example a sample of gas. Using this approach, one can effectively rescue Boltzmann's $\mathscr{H}$ theorem (the equivalent of increasing entropy) and an associated 'irreversibility' in the system's evolution. A link is thereby forged between microscopic reversible equations and the arrow of time.

There is, however, nothing to tell us how fine the coarse-graining should be. Entropies calculated in this way depend on the size-scale decided upon, in direct contradiction with thermodynamics in which entropy changes are fully objective. Moreover, if the coarse-graining is done in an *ad hoc* manner, there is no guarantee that the entropy will increase with time – it may even decrease. Even if the coarse-graining happens to work, it only does so because we are introducing an approximation at a certain level and ignoring what goes on below it. This leaves us in the odd situation where there is an arbitrary cut-off point at which the reversible microscopic world ends and the irreversible macroscopic world begins. In other words, time only has a sense of direction above this subjective limit: coarse-graining relegates the whole question of irreversibility and the arrow of time to the realms of the illusory.

At a stroke, time's arrow becomes once again a subjective notion. The approximations we make in following the behaviour of a thermodynamic system are supposedly due to the fact that we cannot possibly follow the motions of all the millions of molecules comprising it. As the physicist Ed Jaynes remarked: 'it is not the physical process that is irreversible, but rather our ability to follow it'.[53] The suggestion is that if only our senses were sharp enough, we would be able to watch the individual motions of molecules (ignoring the problems this raises in quantum theory) and thereby testify to the true reversibility of all processes on this microscopic level. This quickly leads to the notion of entropy as a measure of our ignorance of the precise details of a process, an idea that has been developed successfully in the subject called information theory.

Information theory deals with the problems of encoding and sending messages. Every communication system – from a hi-fi to a computer or telephone – is beset by random interference, or noise. The aim of information theory is to show how a message is selected at the receiving end from amid all the noise. The foundations of the theory were laid down in 1949 by Claude Shannon and Warren Weaver. As far as they were concerned, information could consist of an utterly meaningless string of nonsense words. Its only technical importance was that it could be encoded, transmitted, selected and decoded. Shannon proposed a purely mathematical definition of information for any probability distribution within a system. One could use this definition to calculate the probability of the information being found in the muddle of interference.

Shannon's mathematical formula looks rather like the one for entropy in statistical mechanics. Many people have concluded that this means there is a direct relationship between the two concepts, one consequence of which is the technique known as the 'maximum entropy formalism'. 'Maxent', as it has been dubbed, is an extremely powerful technique for finding a needle of useful information in a haystack of noise or interference. The very essence of noise is randomness, contrasting with the organised nature of information. It seems but a short step to say that the greater the information in a message, the lower its entropy. In subjective interpretations advocated by Jaynes and his acolytes, the information 'entropy' is taken as a measure of our ignorance of more detailed processes below the scale of observation – the equivalent of the cut-off point between the reversible and the irreversible. Information itself is defined as the negative of the Shannon entropy, sometimes called 'negentropy'.

Information theory is replete with fertile ideas which are of tremendous use in theoretical computer science and for analysing the engineering problems of communication systems. However, the close analogies which exist between entropy and information in no sense make either concept necessarily subjective.[54] In fact, Shannon, the pioneer of information theory, was only persuaded to introduce the word 'entropy' into his discussions by the mathematician John von Neumann who is reported to have told him: 'It will give you a great edge in debates because nobody really knows what entropy is anyway!'[55]

What should not be assumed is that information theory is a vindication of coarse-graining and the retreat into subjectivism. In actual fact, there is no coarse-graining in maxent – rather than making an arbitrary partitioning, the entire system is taken to be a 'black box' of whose details we are ignorant. But whatever scraps of support we find elsewhere for it, coarse-graining is irretrievably flawed. As Prigogine and Stengers wrote: 'Irreversibility is either true on all levels or on none: it cannot emerge as if out of nothing, on going from one level to another.'[56] We have already mentioned in this chapter, and will discuss in more detail in the next, how irreversible processes occurring far from equilibrium are known to generate macroscopic organisation. In particular, they play an essential role in many key biological processes necessary for the existence and maintenance of life itself. Should the coarse-graining or subjective information theory argument be correct, we would be forced to accept that all this is illusory. Indeed, it would have the paradoxical result that macroscopic processes as manifestly irreversible as the functioning of our brains are really only the result of our own approximations.

The cosmological arrow of time

We have seen how the Second Law appeared to imply the Heat Death of the universe, the 'winding down' into a final state of utter disorder. How does this view compare with that of the cosmologists?

The origins of time and the universe have been touched on in previous chapters. We know the universe is expanding, and we can predict two possible extreme fates for it: continued expansion until Heat Death is attained[57] (although some speculate, as we will see in Chapter Eight, that even under these conditions a revitalised universe can emerge); or a Big Crunch, where the ubiquitous gravitational force eventually halts the expansion and draws together all matter ineluctably into one huge final singularity. Which – if either – of these scenarios is correct remains unknown, depending as it does on the actual amount of matter present in the universe, a point already made in Chapter Three.[58]

Suppose that the universe is closed and it collapses. Given the connection between increasing entropy and the arrow of time, does the Big Crunch mean that time would run backwards as soon as collapse began? Some have speculated that it would. Rivers would run uphill.[59] Mike Berry of the University of Bristol quipped: 'Light would be absorbed by stars and emitted by eyes.'[60] Underlying these strange images is the contention that, during expansion, the arrow of time points from a highly organised Big Bang singularity to some maximally disorganised intermediate state, and then points back again as the universe starts contracting to a Big Crunch singularity that looks as well organised as the original Big Bang. Roger Penrose argues cogently against this, maintaining that even during the Big Crunch entropy would be increasing, the Second Law would hold good, and the direction of time would be preserved (although he still regards the Second Law as being more a 'secondary' than a 'primary' law of Nature). This is because of the structure of the Big Bang and Big Crunch singularities, which are *not* equivalent. Many cosmologists have advocated a symmetrical model, where the Big Bang is indistinguishable from the Big Crunch, in that both consist of a fireball of infinitely compressed matter. However, Penrose maintains that all initial spacetime singularities are subject to a constraint which does not hold for black-hole or final singularities: the Big Bang singularity is much more ordered, and thus of much lower entropy, than the Big Crunch.[61] This surprising result is a consequence of the geometry of spacetime in the vicinity of the singularities, which is different in either case (*see* Figure 17). Observational evidence indicates that the Big Bang singularity was isotropic – like a blancmange, if it were cut in half it would show no structure – with a high degree of organisation and thus a low entropy. But *en route* to the Big Crunch, imperfections like black holes are created. These congeal into a great mess during the Big Crunch, which boasts a structure as disordered as a fruit cake and has a correspondingly high entropy. Were the initial singularity not so constrained, there would be no Second Law, and we would expect to find white holes as well as black holes (*see* Chapter Three).[62]

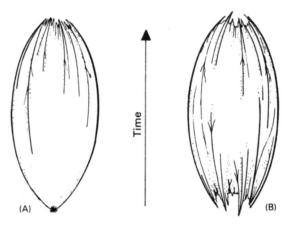

Figure 17. (A) The history of a closed universe which begins with a highly constrained, low-entropy Big Bang and ends with a messy, high-entropy Big Crunch. (B) Without the special initial constraint, the Big Bang is also of high entropy. In the Penrose scenario, only A can account for the Second Law of Thermodynamics on the cosmological scale.
[Adapted from R. Penrose. *The Emperor's New Mind*, pp. 339, 341.]

If Penrose's conjecture is correct, then we need to know why there is such a time asymmetry in the structure of the singularities that ensures a low-entropy bang and a high-entropy crunch. Many physicists might be content simply to accept that the special, low-entropy state of the Big Bang was just the result of an 'initial condition' (due to the hand of God?) and leave it at that. Penrose, however, believes that the distinct nature of the singularities at the 'beginning' and at the 'end' of time provides evidence that a full theory of quantum gravity must be time-asymmetric. In his view, a fully satisfactory theory should explain both the time evolution *and* the initial conditions.[63] At the present time, we still seem to be a long way from uncovering it.

Nevertheless, the Penrose scenario seems to have the right conditions for an arrow of time. One problem remains: how to arrange a flow of time from the state of low entropy – the bang – to the state of high entropy – the crunch. Drawing the arrow between these extremes is tricky with the present laws of physics, which have scant regard for the direction of time: why not start at the Big Crunch and go to the Big Bang? Roger Penrose's arguments are based on the use of coarse-graining to compute entropies, with all its attendant problems of subjectivity. While agreeing that coarse-graining can give different results, depending on how it is carried out, he maintains that it does not make much difference in practice because the two entropies concerned at the start and end of time are 'so ridiculously different'.[64]

There is another interesting corollary to Penrose's speculations. He thinks that a complete theory of quantum gravity equipped with an arrow of time would be able to mop up one of the central problems of contemporary quantum theory discussed in Chapter Four – how to make sense of the measurement process. A quantum gravity theory including the arrow of time might be able to account for irreversible wavefunction collapse whenever a 'significant' amount of spacetime curvature is present. In effect, gravitational interactions would pop the wavefunction, providing an explanation for why collapse only takes place at the macroscopic level (owing to the large number of particles and thus appreciable gravitational effects present under such conditions).[65] As Penrose admits: 'So far, this is only a germ of an idea for what I believe to be a much needed new theory. Any totally satisfactory scheme would, I believe, have to involve some very radical new ideas about the nature of spacetime geometry.'[66]

Finally, we note that the notion of a highly ordered Big Bang makes the anthropic principle (see Chapter Three) look somewhat insecure. The odds against mankind emerging in the universe may have been enormous, but the anthropic principle argues that the fact that we are here to ask the question shows that fortune smiled upon our creation. Yet the odds against Roger Penrose's low-entropy Big Bang scenario – the initial conditions – are truly astronomical in comparison with those for our creation. Certainly the more extreme expressions of the anthropic principle – that the universe was created for our benefit – are undermined because the improbability of mankind's existence is dwarfed by that for the existence of the universe itself.

Although we now have a rather sophisticated idea of the meaning of entropy, we have still not solved the conflict it engenders between the microscopic and macroscopic worlds.[67] This clash between thermodynamics and mechanics will not be resolved until Chapter Eight. Before that, we want to show why the insights of thermodynamics are too important and wide-ranging to be lightly dismissed as subjective aberrations superimposed on a mechanistic world. This involves a short tour of irreversibility from physics through to biology.

Creative evolution

God has put a secret art into the forces of Nature so as to enable it to fashion
itself out of chaos into a perfect world system.

Immanuel Kant
Universal Natural History and Theory of the Heavens

CHEMICAL reactions form a broad class of processes that display the arrow of time: they are irreversible. Some exhibit such precise and regular changes that they become a kind of chemical clock. By understanding more about these processes we can sort out how and why their time ticks away. In turn, since the chemistry of living cells is also the essence of life, we will find that these clocks are the microscopic 'cogs' that turn the wheels of our own human bodies. The arrow of time will emerge as the arrow of life. And it will throw up such exquisite patterns along the way that it becomes almost impossible to believe that the arrow is the illusion that reductionists pretend.

The arrow of time inherent within the Second Law does not express itself simply as blind destruction in a steady and uniform descent into disorder; on the contrary, a study of the principles governing irreversible processes far from equilibrium, begun in Chapter Five, can help us to understand how the boundaries are drawn between slow, inexorable decay, the intricate patterns of life, and turbulent conflagration.

The rhythms we are interested in start to form when a system is driven far from equilibrium. A refreshing example is a bottle of beer.[1] While it is standing upright, the bottle's contents lie in a state of mechanical equilibrium. A polite drinker would push the contents into a near equilibrium steady-state by gently tilting the bottle so that the beer flows out smoothly into a mug. But someone desperate to slake a thirst is likely to push the system even further from equilibrium. Seizing the bottle to swig from, he would notice that if it is tilted beyond a certain point the beer gurgles out in regular glugs. There is a 'magic angle' at which the flow of beer starts to oscillate. It is a crisis point – a springboard from disorder to organisation.

This, of course, is what we started to look at in the last chapter when we examined the basic thermodynamic recipe for self-organisation. Take the Second Law and apply it to any open system – that is, one

182

which will take in and/or give out matter and energy. Push the system far enough from equilibrium to a crisis point and organisation may emerge. We have already encountered the possibility of a chemical reaction in which the crisis point ushers in regular oscillations in colour: a simple chemical clock. Our problem now is how to describe a clock like that as it leaves the first crisis point far behind.

Thermodynamics is not enough. It describes the arrow of time through the tendency of entropy to increase and also provides milestones *en route* to equilibrium which indicate when things may change. But it does not give any clues as to what those changes will be. There is no universal thermodynamic criterion which describes how any system evolves through time. We are forced to leave thermodynamics and resort instead to completely new techniques.

Order and chaos

We could attempt to describe chemical clocks by drawing on quantum or classical physics. Apart from the fact that any such attempt would be ridiculously complicated, this approach must be ruled out because neither makes a distinction between the two possible directions of time. A different tack must be taken. To understand a mundane phenomenon like the comings and goings of a train, one can call on a timetable rather than having to understand the form of locomotion, be it steam, electric or diesel. In a similar fashion, a chemical clock can be described by constructing a purely empirical account that is not derived from a complete understanding of what is going on at the molecular level.

This is known as taking a phenomenological approach, and is particularly useful when irreversible processes are pushed to extremes. Take our bottle of beer, for example. If we turn it upside down we get even more remarkable behaviour than the regular glugs when it was tilted. The beer becomes turbulent and forms eddies, where millions of milling molecules are organised into a vortex. We shall see, too, how irreversible processes such as chemical clock reactions, pushed beyond a certain limit, become chaotic. This is deterministic chaos, mentioned on p. 167, when precise laws lead to apparently random behaviour which is in fact minutely organised. Some scientists believe it underlies all sorts of complex events, including the random flutterings of a sick human heart, the seemingly erratic fluctuations in populations of wild animals, and temporal variations in climate.[2]

Chaos may seem far removed from the organised behaviour of a chemical clock, but both have the same physical (and mathematical)

183

genealogy. This is a very important point, and has its basis in time's arrow. Assuming that time is continuous, rather than a series of discrete moments, all dissipative systems can be modelled by means of *differential equations*.[3] These are mathematical descriptions of the instantaneous change of properties with time. They can deal just as happily with the rate of change in the population numbers of a North American gipsy moth as with the varying concentrations of ingredients in a chemical clock. Unlike Newton's, Einstein's and Schrödinger's equations, they have the arrow of time built into them. They permit a vast range of possibilities, from self-organisation to chaos, and can account in many ways for the richness of the world we inhabit. The implications for biology will be investigated in this chapter and the next.

The factor that allows these time-asymmetric differential equations to give birth to both order and chaos is non-linearity. As we saw in thermodynamics, this basically means 'getting more than you bargained for'. A linear relationship implies a proportional one: ten oranges cost ten times more than one orange. Non-linearity implies a disproportionate effect for bulk buying: crates of oranges will cost less than the number of oranges bought multiplied by the cost of a single orange. An important element of this interdependence is feedback – the size of the discount will affect the number of crates one decides to buy.

Although it may sound simple, this interdependence can lead to quite unexpected behaviour. It can cause an instability or crisis point to be reached, in much the same way as positive feedback from a loud-speaker to a microphone turns a whisper into a deafening howl by an amplification loop. In a chemical clock, feedback occurs when the production of one chemical affects the rate at which more is formed. This amplifies fluctuations in the chemical mixture into the coordinated colour changes of a chemical clock. Feedback also leads to the untimely demise of lemming colonies: the lemmings may multiply so fast that they eat all the available food. In biochemistry we come across analogous situations where, for example, a reaction produces an enzyme whose presence then encourages its own production, eventually exhausting the reactants. In biology, positive and negative feedback loops abound, conspiring to turn the genetic blueprint imprinted within the nucleic acid *DNA* into a complex living organism.[4]

The full scope and power of dissipative non-linear equations have only begun to be exploited within the past twenty years. While linear equations can be explored and solved by existing analytical mathematical techniques, non-linear differential equations, except for a few rather special situations, are much harder to solve in this way. The most unsubtle method of all has to be used – numbers are fed into the equations and the solutions calculated numerically. This explains in large part why such a rich field of investigation was overlooked for so long by many scientists, for before the advent of computers the method was impractical to implement. But now it is possible for the non-linear jungle to be explored in great detail.[5]

Non-linear mathematics appears to be devious. Until the computer became an everyday accessory, people sought rough and ready descriptions of non-linear systems by making approximations – they linearised them. Take an estimate of tax to be paid after a salary rise. Although taxation in general follows non-linear rules, being linked in a complicated way with the amount of one's earnings, an estimate of the amount of tax due can be calculated assuming one pays the base rate only. Inevitably, this linear approach is of limited use, not just for dealing with annual tax returns but also for non-linear dynamics. The anaemic linear approximations miss all the novel non-linear features that lead to the very organising and chaotic features which we seek to explain. We shall see later in the chapter how, by employing non-linear dynamics together with a kind of portrait of their behaviour, called bifurcation analysis (see p. 165), we can make a much more detailed investigation of the time evolution far from equilibrium than was possible by thermodynamics alone.

Cooking up self-organisation

Unexpected non-linear effects can be seen in the simplest processes. We had a glimpse in the last chapter, in the guise of thermodiffusion (Figure 15), when a mixture of two gases was pushed away from equilibrium by heating: organisation emerged as a simple gradient of gas concentrations. There we saw how only small excursions away from equilibrium could lead to macroscopic order. But this hardly prepares us for the spectacular organisation which spontaneously wells up beyond the first crisis point. Heating a thin layer of liquid wedged between two glass plates can cause organisation to appear in the form of a honeycomb pattern consisting of hexagonal cells of convecting liquid. It is a shocking thought for anyone who adheres to a traditional equilibrium-based view of the world. The more heat is applied, the more frenetically the molecules should rush about in the liquid. Why should structure arise?

The honeycomb pattern of self-organisation was first discovered by the French investigator Henri Bénard in 1900. An attempt to explain it was made by Lord Rayleigh in 1916. Now it is known to arise from what has been dubbed the Rayleigh–Bénard hydrodynamic instability (*see* colour plate section).[6]

In the experiment, a liquid is poured into a clear dish and placed on a heat source like an electric cooking hob. The heat rises from the bottom to the top of the dish by conduction, convection or both. Before the heat is turned on, the fluid appears quiescent, even though the microscopic picture consists of molecules in more or less random motion. When heat is applied, a temperature difference is established vertically through the liquid. Macroscopically, the fluid remains at rest, however, until a certain threshold difference in temperature between the top and bottom is reached. Below that temperature difference, the heat is conveyed by conduction alone. Above this value, convection sets in as the warmer regions at the bottom of the vessel rise through the cooler, denser fluid. At the same time, the honeycomb pattern arises from the coupling of the buoyancy, heat diffusion and viscous forces.

The equilibrium-based image of the fluid would suggest that the more heat is applied the more the countless molecules should swarm willy-nilly around the dish. But the honeycomb state is much more organised than before heat was applied, as a glance at the colour plate shows. The distance over which this hexagonal pattern forms is 100 million times greater than the distance between individual molecules. Vast numbers of molecules mill in unison over these colossal distances to make the hexagonal 'convection cells'. The honeycomb remains visible to the naked eye for as long as the temperature difference is maintained.[7] This structure is sustained by the dissipation of heat, which exports entropy out of the system.

In a thermodynamic description, the temperature at which the hexagonal cells appear is the crisis or bifurcation point we encountered in the last chapter. There the system can decide to go down one of two branches. In the case of the Rayleigh–Bénard instability, neighbouring cells have fluid convection currents turning in opposite directions. It is a certainty that as soon as the temperature exceeds the critical value, these cells will appear. But the direction of rotation of the cells is unpredictable, depending on the cranking up of uncontrollable microscopic fluctuations to the macroscopic level each time an experiment is carried out. Gregoire Nicolis, a colleague of Prigogine, has described this effect as being due to a 'remarkable cooperation between chance and determinism'.[8]

Fluctuations are vital to seed self-organisation. Near equilibrium, the convection currents in the fluid are small and controlled, exerting a negligible effect. Like a dying whisper, the fluctuations quickly fade away. But a whisper can become a shout if there is feedback. Far from equilibrium the non-linear properties of the system build up a microscopic convection current into an organised state that invades the entire dish and turns it into a liquid honeycomb. Some people try to explain this in terms borrowed from thermodynamic equilibrium. But while these ideas can cope with the monotonous regularity of an ice crystal, for example, they are utterly inadequate to describe *dynamic* dissipative structures like Rayleigh–Bénard cells, which only exist for as long as the fluid is heated.[9]

In their world view, most chemists and molecular biologists like to put the emphasis on the activities of individual molecules, an approach which works rather well for many systems at equilibrium. But it conveys no idea of the 'communication' between molecules in a self-organising medium. The forces which allow one water molecule to influence another in ice at equilibrium usually only extend over distances as little as one-hundredth of a millionth of a metre. The organisation which arises in a dissipative system stretches over relatively enormous distances as great as centimetres – this is a dynamic structure analogous to thousands of fridges making ice cubes at the same rate, or everyone in New York City performing a gymnastics exercise in unison.

Yet self-organisation is not a mystery; as we shall show, it is a consequence (albeit unexpected) of far-from-equilibrium physical laws which contain the arrow of time. We are not so much interested in where a particular chemical reaction is going 'at the end of time', when equilibrium thermodynamics governs behaviour, as with the fact that the motions of countless billions of molecules are synchronised to create macroscopic patterns in space and oscillations in time *en route* to equilibrium.

To realise how unexpected this is, imagine a truck filled with a uniform mixture of white and black tennis balls. The analogue of heating the fluid would be to drive the lorry over rugged terrain so that the balls are violently bounced around. Imagine patterns of balls forming in this seething activity, with, for example, all the black balls at one end of the lorry and all the white ones at the other. The macroscopic order we find in a Rayleigh–Bénard cell is equally remarkable: it implies the cooperative behaviour of vast numbers of individual molecules in both time and space. If anything, it is rather more remarkable, since the organisation of the molecules stretches over a relatively much greater distance than that of the tennis balls. It is as if, sufficiently far from equilibrium, the molecules become aware of a single, synchronised time with which they all move in step. The system takes on a life of its own: it can no longer be viewed as a haphazard agglomeration of randomly moving molecules. The molecules in the fluid have spontaneously self-organised.

Self-organisation in chemistry

It is simple to arrange for a mixture of reacting chemicals to be maintained far from equilibrium. They need only be fed through a continuously stirred flow reactor, a set-up typical in any chemical factory. Chemicals enter the reactor, are stirred to induce reaction, and the products drawn off downstream.[10] Non-linearities are spawned if a product of a reaction (say X) catalyses its own production, a kind of feedback process known as *autocatalysis*. The amount of X formed at any instant depends on how much X was there in the first place – the signature of a non-linearity that is analogous to positive feedback between a microphone and a loudspeaker. Thus, without too much trouble, we have all the ingredients necessary for self-organisation and phenomena like chemical clocks.[11]

It was the mathematician Alan Turing, one of the greatest figures of British science in the twentieth century, who was the first to envisage this possibility. His ideas were laid down in a remarkable paper published in the *Philosophical Transactions of the Royal Society Part B* in 1952, when he was 40 years old and working at the University of Manchester. They were the last great flowering of a career dotted with intellectual achievements which, like some human chemical clock, seemed to follow a five-year oscillation: he conceived the 'universal Turing machine' in 1935, combining a simple mechanistic picture of the mind with pure mathematics to show how machines could emulate thought; in 1940 he was a cryptanalyst at Bletchley Park in Buckinghamshire where his expertise in computation was put to work deciphering German naval messages encoded on an Enigma machine; in 1945 he worked on ACE, the Automatic Computing Engine, the practical construction of his universal machine or electronic brain; finally, in 1950, Turing, who had spent much of his time cracking enemy codes, turned his attention to the codes used by Nature to create patterns.

Turing was interested in furnishing a chemical basis for the means by which shape, structure and function arise in living things, a process known in biology as *morphogenesis*. Turing asked himself a simple question. How does an organism marshal a chemical soup into a biological structure, or turn a spherical bundle of identical cells into an organism? It is one of the greatest puzzles of life. Yet Turing wrote in the abstract of his paper that 'the theory does not make any new hypotheses; it merely suggests that certain well-known physical laws are sufficient to account for many of the facts. The full understanding of this paper requires a good knowledge of mathematics, some biology, and some elementary chemistry.'[12]

Take the process of gastrulation, in which a mammalian embryo – a sphere of cells – loses its symmetry as some cells begin to develop into the head and others into the tail. Starting off with such a perfect sphere of cells, one might expect that uniform irreversible diffusion of the biochemical reactions which control development of the sphere would retain this symmetry, and we should all be spherical blobs. But in his paper, Turing showed that the breakdown of symmetry necessary for a fertilised egg to develop into the complicated form of a creature can indeed arise. Qualitatively, it is the same concept as one we have already encountered: near equilibrium, the homogeneous state of maximum symmetry is stable, but it can become unstable at greater distances from equilibrium owing to omnipresent fluctuations. Turing illustrated this with a mechanical analogy: 'if a rod is hanging from a point a little above its centre of gravity it will be in stable equilibrium. If, however, a mouse climbs up the rod the equilibrium eventually becomes unstable and the rod starts to swing.'[13]

Few eggs are spherically symmetric and there are other factors such as gravity which break this symmetry. Nonetheless, Turing's ideas have led to some impressive descriptions of pattern formation in nature, from snail shells to snake skins, which we will investigate in the next chapter. The underlying processes are, of course, irreversible and express the arrow of time. Tragically, Turing was not to develop his ideas much further: two years after his morphogenesis paper appeared, he killed himself.

Time ran out for Turing because of prevailing public moral attitudes in Britain during the 1950s. The beginning of the end came in 1952, when he went on trial for 'Gross Indecency, contrary to section II of the Criminal Law Amendment Act 1885'. As a result of police investigations into a burglary at his home, he was led to confess that he was a homosexual.[14] He was put on probation and sent for medical treatment – 'organo-therapy' – where hormones were administered to curb his sexual urge.[15] But on the Whit Monday of 1954, the coldest and wettest for 50 years, he ate an apple he had dipped in cyanide. In his biography of Turing, Andrew Hodges described how in 1938 Turing had gone to see *Snow White and the Seven Dwarfs* in Cambridge: 'He was very taken with the scene where the Wicked Witch dangled an apple on a string into a boiling brew of poison, muttering: "Dip the apple in the brew. Let the Sleeping Death seep through." He liked to chant the prophetic couplet over and over again.'[16]

Just like the suicide of Ludwig Boltzmann, Turing's death was a hammer blow to science. Nevertheless, he had already made the important discovery that chemicals can vary their concentrations to form spatial patterns if several coloured substances with different diffusion rates react with each other in a liquid. The theory is counter-intuitive, because one would expect any type of irreversible mixing process to wash out pre-existing patterns or structures, just as the patterns made by milk added to black coffee eventually disappear. Way ahead of his time, Turing formulated a mathematical recipe for stationary patterns, which do not change with time, and oscillatory patterns, which can be seen in chemical clocks as wavelike ripples of colour. The point where the system is driven far enough from equilibrium for the first of these patterns to emerge has come to be known as the Turing instability, an example of the crisis points we have been referring to. However, the pattern formation Turing proposed at this instability, although theoretically possible, has not yet been unambiguously observed for any real chemical system, and certain details of his model have been criticised on other grounds. Nevertheless, self-organisation is widespread in biology, some examples of which can be understood on the basis of Turing's reaction–diffusion theory. But for now we will concentrate on simpler chemical phenomena.

For the next fifteen to twenty years, Turing's work went largely unnoticed by chemists and biologists. There were several reasons for this hiatus. In order to make the mathematics tractable, Turing had to tame these non-linear equations: he took the approach of assuming that the mathematics behaved in a linear, predictable fashion, over a limited range of circumstances. This made his analysis of what it had to say somewhat myopic – he could not stray beyond the first crisis point when equilibrium is left behind. In other words, Turing was able to work out when a pattern might be formed, but not any subsequent changes in that pattern if the system was pushed still further from equilibrium. Turing realised that what was needed to make further progress was a fast computer, yet none was then in existence. Moreover, no one knew of a chemical reaction for which any aspects of his theory would work.

Birth of the Brusselator

Many laboratories are today pursuing research on self-organisation along lines broadly similar to those originally outlined by Turing. Two

190

key developments in particular have led to an explosion of interest in the subject within the past twenty years. One of these was a symposium held in Prague in 1968, where Western scientists first learned about the magical Belousov–Zhabotinsky chemical reaction (shortly to be described) and began to make a comparison with oscillations that take place in biology to help organisms harness energy, like glycolysis and photosynthesis.[17] The second also occurred in 1968 with work published by Ilya Prigogine and René Lefever.[18] Citing Turing's seminal paper, they formulated and analysed a model of a chemically reacting system possessing some of the necessary ingredients for spatial self-organisation. This model was dubbed the *Brusselator* by John Tyson of Virginia Polytechnic University in 1973, since it was born in the Belgian capital. In their paper, Prigogine and Lefever showed that the self-organising features of the Brusselator appear in a manner consistent with the Glansdorff–Prigogine criterion for thermodynamic evolution which was discussed in Chapter Five. (This criterion, it will be recalled, is based on the Second Law of Thermodynamics as it applies to far-from-equilibrium situations.) Their model was the simplest they could define which both fulfilled the conditions of the evolution criterion for possible instability of the thermodynamic branch, and which was easily tractable. It thus has a firm thermodynamic foundation. The Brusselator was later shown by Lefever and Nicolis to be capable of exhibiting sustained regular oscillations in the concentrations of certain chemicals.[19]

It is easy to see how in a 'chemical clock' like the Brusselator feedback and non-linearity are essential for self-organisation, where enormous numbers of molecules seem to 'communicate' with each other in creating a pattern. The Brusselator is an idealised model involving two chemical substances, A and B, which are converted into two products, C and D. Instead of the conversion occurring in a single chemical reaction, in order to generate interesting non-linear features four simple steps are included, involving two chemicals X and Y as intermediates. The details are straightforward: a molecule of A is first converted into one of X, which reacts in the second step with a molecule of B to form Y and product C; in the third reaction, two molecules of X combine with one of Y to produce three molecules of X. The final reaction is the direct conversion of X into product D. The springboard to self-organisation – non-linear feedback – is held in the third step of the Brusselator, where three molecules of X are formed from two by a reaction with intermediate Y. The feedback occurs because one molecule – X – is involved in its own production, an autocatalytic step. The non-linear nature arises because for every two molecules of X reacting another one is produced, making three in all.

Now let us hold the Brusselator far from equilibrium by constantly replenishing chemicals before they are used up. To achieve this, all we have to do is carry out the reaction in a well-stirred open reactor. There we can maintain the concentrations of A and B at the appropriate values by controlling their flow rates into the reactor, and similarly for the products C and D which are removed; only the concentrations of X and Y can vary with time. To find out how they vary, we must write down and solve the mathematical description of the Brusselator – a series of coupled differential equations describing X and Y.

If the Brusselator is put through its paces, one gets a feeling for the elaborate behaviour locked up in these coupled differential equations. The detailed mathematical analysis is complex but the results are colourful if we assume that the chemical X is red and Y blue.

Let us start with the end of the chemical reaction. The ingredients are mixed together, they react, and the reaction achieves equilibrium, when all chemical change ceases. At equilibrium, we have an un-interesting purple soup – a mixture of blue and red molecules. Little changes if the concentrations of A and B are maintained close to these equilibrium values, which is the steady-state region in which Prigo-gine's minimum entropy production theorem holds good. The inter-esting behaviour begins when the flow of chemicals A and B into the reactor is cranked up to levels beyond a certain threshold from the equilibrium concentrations. (In the case of beer pouring out of a bottle, the flow broke up into 'glugs' beyond this critical point.) For the chemical clock, no matter what the starting concentrations of X and Y, oscillations occur beyond the critical point. The reaction mixture turns red, then blue, and so on, at regular intervals. This – the equivalent of the oscillating flow of beer – is known as the Hopf instability, after the mathematician who discovered it.

Simple terms are used to describe these colour changes. The chemical clock can be represented as a closed loop or cycle, called a *limit-cycle*. All one has to keep in mind is that, as the chemical reaction changes from blue to red, it trundles around this loop like a ball bearing in the brim of a sombrero (Figure 18). You can think of the reaction as moving around the cycle, alternating from red to blue to red again as it passes between the poles. Even if the amounts of added ingredients are changed slightly, the point representing the chemical reaction will always return to this regular cycle of colour change. It always rolls back into the loop just as a ball bearing will always roll to the bottom of the sombrero's brim. Because of this behaviour, the cycle is an attractor, a concept which we encountered in the last chapter.[20] When no more chemicals are added, the chemical reaction runs its course, the colour changes stop and the purple mixture reappears. In this case – thermodynamic equilibrium – the attractor is a single fixed point, a funnel into which, metaphorically speaking, the reaction rolls. There the entropy is maximised and X and Y have constant values through time.

It is important to pause here to appreciate fully what has been done. The Brusselator highlights how to create order from disorder through the phenomenon of self-organisation. By holding the system sufficiently far from equilibrium – by constantly adding chemical ingredients – the chemical soup in the reactor turns periodically from blue to red to blue and so on, rather than remaining a featureless purple mixture. The Brusselator is, of course, just a model. Nonetheless, the behaviour which it describes is not only theoretically possible; it offers many insights into the oscillations that are seen in the dramatic Belousov–Zhabotinsky reaction as well as more widely in biology, as we will see.

What is the difference between an oscillating chemical reaction and a more conventional one? One can compare an ordinary (linear) chemical reaction to car manufacture, where the number of components in a warehouse falls as cars build up in the parking lot outside the factory. The intermediates – partly built cars – remain fairly constant during the process. In a test tube, the concentrations of reacting chemicals drop constantly as they are consumed to make products. All the while, as the reactant concentrations fall and product concentrations rise, there is a low and constant number of intermediate chemicals present.

During a non-linear chemical clock reaction, the concentrations of reactants still decrease, those of products still rise; however, as long as the concentrations of reactants remain above a threshold level, the concentrations of the intermediates (and therefore the colour of the mixture) oscillate through the regular limit-cycle as the conversion of reactants to products takes the reaction to equilibrium. It is a process akin to the glugging beer that we encountered earlier. Only if the flow of reactants into and out of the Brusselator is maintained will the colour changes continue.

The behaviour is as startling as the honeycomb patterns of the Rayleigh–Bénard instability. All the molecules in the Brusselator are able to communicate with each other over vast distances: they all know when to turn blue, and when to turn red. The 'tick' of this clock as it rolls round the limit-cycle is a function only of certain physical properties of the Brusselator. It is completely independent of the initial conditions. This is, of course, another example of a dissipative structure

193

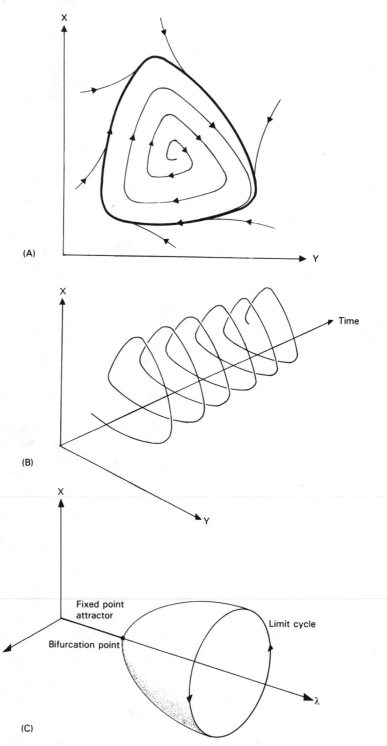

Figure 18. Limit-cycle behaviour. (A) Two-dimensional projection: X and Y are concentrations of chemical intermediates (blue and red, say). (B) Three-dimensional view, showing time going on. (C) Limit-cycle attractor emerging from a point attractor.

(see p. 168), a term which emphasises its origin in an irreversible (far-from-equilibrium) thermodynamic process dependent on the arrow of time.

As is the case so often, the mathematicians had been there first, exploring some of these ideas more for aesthetic than for scientific reasons. The French mathematician Henri Poincaré, and later Andronov's school in the Soviet Union, had studied these kinds of non-linear coupled differential equations and proved that limit-cycle behaviour results (the name dates from Poincaré's era); the Brusselator brought this abstract mathematics to life in a very real sense.

The concept of dissipative structures struck a chord in many diverse areas. It helped to foster scientific, as opposed to purely mathematical, interest in non-linear differential equations. The chemistry of clock reactions was taken up because they were controllable and relatively easy to model. In turn, this effort paved the way for attempts to develop mathematical models of biological processes in single cells and multi-cellular ensembles, suggesting that biochemical cousins of the clock reactions, which can also be described by limit-cycles, have major significance for life. Biochemical clocks appear to be part and parcel of the regulatory processes found in living organisms and have given great impetus to the study of non-linear phenomena.

Once mathematical methods for studying non-linear systems became widely available, they invaded physics, chemistry and biology, leading to impressive interdisciplinary studies. Indeed, systems that evolve usually require for their mathematical description the use of differential equations that are capable of following the way things change instantaneously with time.[21] Because feedback is such an important part of the recipe for non-linear behaviour – as in the chemical clock where a molecule plays a role in its own destiny, whether it be through self-enhancement, self-destruction or competition – similar analyses have been applied to the 'softer' areas of science, such as sociobiology, sociology, socio-economics and economics, where feedback also occurs. Such is their power that they have also inspired a fashionable new buzz-phrase – artificial life – described by Christopher Langton of Los Alamos as the study of man-made systems which exhibit behaviours characteristic of living systems and which may even give insights into life as it could be rather than as it is.

Chemical patterns and waves

Non-linear models of chemical clocks hold many secrets. We have dealt with patterns in time. What of patterns in space? In the analysis above

195

that led to the prediction of limit-cycles in the Brusselator, the possible role of another irreversible process – diffusion – was overlooked. We assumed that the chemical cocktail in the reactor was strongly stirred so that all the species (A, B, C, D, X, and Y) were uniformly distributed in the liquid. If we return to the car factory analogy, it is a little like assuming that cars are sprouting everywhere in the factory. Clearly, this is unrealistic: in a similar way, the various chemicals milling around in the Brusselator reaction can take time to meet. Certainly, when the reactor is not stirred we cannot suppose that intermediates X and Y and products C and D are automatically formed in equal amounts in all regions of the vessel. Thus we should think about what happens when we have to take into account the fact that little pools of reacting chemicals may form in one part of the reactor vessel and then have to migrate to other regions to take part in the feedback reaction.

It is a case of molecular logistics: how do we model the way these reacting molecules wander around in the reactor before they bump into each other and react? The answer is simple. The effects of diffusion are readily incorporated into the analysis by stirring in a term from Fick's law, named after Adolf Fick, which relates the concentration of a substance at a given point in space with how it changes in time, one for each of the different chemicals in the soup. The conversion factor in Fick's law that links the two is the diffusion coefficient, a property of a given chemical: it reflects the fact that bulky molecules diffuse slowly, slim ones diffuse quickly, and also takes into account the viscosity of the solution they have to move through.

With the help of Fick's law, we can link the behaviour of the chemical mixture in time with the emergence of patterns in space. Technically speaking, the system is now modelled by partial (rather than ordinary) differential reaction–diffusion equations, an additional feature that we shall not worry about, save to note that as the mathematics becomes more complex the physical chemistry becomes concomitantly richer. For instance, the limit-cycle resulting from a Hopf instability can turn in space as well as time. One familiar object which changes in space and time is a wave – think of one breaking on a pebble beach. And indeed, in a chemical clock where the Hopf instability rules, we should expect to see ripples of red and blue passing through the reactor rather than the entire solution in the reactor simultaneously changing colour to red or blue instantaneously.[22]

There is also the possibility of evolution to a fixed spatial pattern that does not change with time, a process which mathematical biologists have employed to show, for instance, how a zebra may get its stripes, butterflies their wing patterns and how gradients of chemicals may direct the development of a bundle of identical cells within an egg into an embryo.[23] In the case of the Brusselator model, you would add some chemicals, allow for diffusion to occur, and spots or stripes could appear in the test tube. This is the Turing instability which we have already encountered and will discuss again in Chapter Seven.

Courtesy of the Brusselator, we have seen various permutations of self-organising behaviour. The message is that the non-equilibrium Brusselator can become self-organising in time, in space or in space and time. These ideas have profound biological significance. For within your body, as you read this sentence, a vast range of processes organised in both time and space are taking place, from the movement of your eyes and heart to the nerve cells firing in your brain.

For the moment, we will continue to restrict our discussion to chemistry. It should not be forgotten that the Brusselator we have talked about is only a model, chosen by dint of its mathematical simplicity.[24] Nevertheless, its study helped to pave the way for the understanding and acceptance of all kinds of self-organising phenomena to which we now turn, and which initially appeared to contravene the tenets of the Second Law of Thermodynamics.

The enchanted Belousov–Zhabotinsky reaction

The Belousov–Zhabotinsky reaction has a history as intriguing as its name suggests. Art Winfree, a leading American authority on the ins-and-outs of the reaction, has described how Boris Pavlovitch Belousov performed the key work on this chemical reaction while head of a laboratory of biophysics attached to the Soviet Ministry of Health in the early 1950s.[25] During his research he concocted a strange mixture of chemicals meant to resemble and so throw further light on aspects of the Krebs cycle, a crucial metabolic pathway by which living cells break down organic foodstuffs into energy (in the form of molecules called adenosine triphosphate) and carbon dioxide gas.[26]

Belousov's caricature of the cycle contained citric acid, an ingredient of the actual Krebs cycle; potassium bromate to mimic the biological equivalent of burning the citric acid (oxidation); sulphuric acid; and a catalyst of ceric ions, which he thought would bear a passing resemblance to the business end of many enzymes. (Enzymes frequently possess a charged metal atom at an 'active site' where chemical reactions are carried out.) To his amazement, the solution started to oscillate between being colourless and of a yellow hue – corresponding to two distinct forms of the charged ceric ion – with clockwork regularity. There is evidence that, during subsequent investigations, he also observed the formation of spatial patterns.[27] Thus Belousov provided the first real chemical reaction which supported the notion of self-organisation through the twin irreversible processes of reaction and diffusion, as had been anticipated theoretically by Turing at around the same time. However, as Winfree has recently written: 'its antics turn out to resemble nothing foreseen in the thirty years devoted to the subject by theoretical chemists and biologists'.[28]

Unfortunately for Belousov, the reaction was so peculiar that he had great trouble in convincing the scientific establishment of its veracity. In 1951 a manuscript of his work was rejected. The editor told him that his 'supposedly discovered discovery' was quite impossible. Belousov submitted a more comprehensive analysis six years later but the editor would only offer to publish a savagely cut version in the form of a brief communication. Belousov's work eventually appeared as an obscure contribution in the proceedings of a symposium on radiation medicine. It consisted of two pages before another of his papers.[29]

The scientific establishment was so besotted with the simplistic interpretation of the Second Law – order decaying uniformly to disorder – that no one was prepared to accept Belousov's reports of the spontaneous emergence of self-organising features in a chemical system. People thought the Second Law said that a chemical reaction always heads for degenerate equilibrium. A chemical clock which switches between two colours implies that the reaction is somehow turning back on itself, a travesty of the Second Law.[30] (In fact, Belousov was not the first to suffer from this misinterpretation. The discovery of an oscillating chemical reaction in the conversion of hydrogen peroxide to water by William Bray of the University of California at Berkeley in 1921 was dismissed as an artefact caused by poor experimental procedure.[31])

It was only when Anatoly Zhabotinsky studied Belousov's oscillating recipe that the real interest in the reaction began, albeit slowly at first, confined as it then was behind the Iron Curtain. During the 1960s, as a graduate student of biochemistry at Moscow State University, Zhabotinsky tinkered with Belousov's basic reaction in a number of ways, such as replacing the cerium ion with an iron reagent that gave a much more distinctive colour change from red to blue.[32] In this way he finally managed to capture the imagination of his conservative peers. Other people began to take up the study of this amazing system; over the past twenty years, research on self-organising chemical reactions has grown into a fashionable field of investigation. Testimonials to the importance of the work were sought from scientists worldwide in 1979[33] and in 1980 both Belousov and Zhabotinsky were awarded the Lenin Prize, along with Valentin Israelovitch Krinsky, Genrik Ivanitsky and Albert Zaikin. Unfortunately, Belousov died in 1970, before receiving belated international recognition for his seminal contribution.

The discovery of Belousov and the many variants subsequently developed have together come to be known as the Belousov–Zhabotinsky (BZ) reaction. It is simple to prepare and works very reliably (the would-be experimenter should consult the Winfree references to this chapter for instructions). All kinds of beautiful phenomena can be associated with this apparently magical mixture. A selection of figures in the black-and-white plate section shows the BZ reaction being put through its paces.

The chemistry of this remarkable and complex reaction has been studied in great depth by many people and entire books have been devoted to it.[34] It is thought that some thirty distinct chemical species participate in the overall reaction, including short-lived intermediates which act as stepping stones in the various interlocking cycles of chemical reactions. These were summarised by a University of Oregon team – Richard Field, Endre Körös and Richard Noyes – in an eleven-step chemical reaction mechanism, a much more complicated affair than the four steps of the Brusselator. If you rummage around among these eleven steps you can find the tell-tale signs of a process where one chemical influences its own manufacture. Thus we have autocatalysis, a key ingredient for feedback and non-linearity. From the plethora of complicated intermediates, the Oregon team then suggested a rather simplified yet important model scheme comprising just five separate steps which has been baptised the 'Oregonator' by the scientific community. The Oregonator model is a theoretical representation of the state of affairs in the evolving BZ reaction, capable of describing in many respects the behaviour of the experimentalists' clock solutions, such as the limit-cycle attractors which generate the chemical oscillations.

If one insists on the tendency to compartmentalise knowledge, one would probably have to say that the enchanting BZ reaction falls within an area of chemical science known as 'inorganic' chemistry. Detailed understanding of the reaction has inspired chemists to probe further into the inorganic world. For instance, oscillations have been discovered by Thomas Briggs and Warren Rauscher at the Galileo High School Lux Laboratory in mixtures of hydrogen peroxide, malonic acid, potassium iodate, manganese sulphate and perchloric acid which change periodically from blue to yellow.[35] The number of such oscillating reactions is constantly growing, but the general principles governing them are now well understood. Other models of such chemical clocks have subsequently emerged – there is a 'K model' (K for Kyoto), an 'IUator' (for Indiana University oscillator) and a 'Bubbleator', a model that describes a chemical reaction producing surges of gassy foam.[36]

An essential aspect of the BZ reaction is its so-called 'excitability' – under the influence of stimuli, patterns develop in what would otherwise be a perfectly quiescent medium. Some clock reactions, such as Briggs–Rauscher and BZ using the chemical ruthenium bipyridyl as catalyst, can be excited into self-organising activity through the influence of light.[37] This property of excitability, which describes how the BZ reaction can be pushed into action,[38] was quite unknown to Turing, and is still frequently overlooked by theoreticians today. Indeed, excitability remains a somewhat vaguely defined concept.

The subtleties of the BZ reaction are still being pursued by mathematicians, physicists and biologists, and with good reason. For, as we shall soon begin to see, it is impossible to overlook its relationship with much of the organisation we are familiar with in the living world. Spiral waves formed in chemical clock reactions bear more than a passing resemblance to those formed in heart attacks, primitive slime moulds (see black-and-white plate section), spiral galaxies and hurricanes. Indeed, Winfree maintains: 'Though it lacks anything like a genetic system through which it could mutate and evolve, the [BZ reaction] shares many of the features that make living systems interesting: chemical metabolism (oxidation of organic acids to carbon dioxide), self-organising structure, rhythmic activity, dynamic stability within limits, irreversible dissolution beyond those limits, and a natural lifespan.'[39] Thus the study of chemical clocks has in a real sense helped to breathe life into inorganic chemistry, a subject that has often been dangerously short on understanding and long on the chemists' equivalent of stamp-collecting – the amassing of vast quantities of purely factual information.[40]

Inorganic self-organising systems consist of a lot of simple chemical species. But the actual chemistry is somewhat haphazard – there is little specificity, since the molecules milling around in the soup may all react with one another to some extent. As we shall see, living organisms tend to be at the other end of the spectrum of possibilities. There the (bio)chemistry is both complex and finely tuned: every reaction is highly specific and proceeds with astonishing efficiency. Prigogine and

200

Stengers remarked: 'This can hardly be an accident. Here we encounter an initial element marking the difference between physics and biology. Biological systems *have a past*. Their constituent molecules are the result of an evolution; they have been selected to take part in the autocatalytic mechanisms to generate very specific forms of self-organisation.'[41] It is chemistry with a purpose. It is the miracle of life.

Of fractals, strange attractors and chaos

In a chemical clock, non-linear complexity translates itself into regular behaviour in time: the colour of the reacting mixture of chemicals undergoes rhythmic colour changes. This behaviour, as we have seen, is described by a limit-cycle attractor in which the chemical reaction rolls around, like the ball bearing in the rim of a sombrero. This behaviour should be contrasted with the fixed-point attractor of thermodynamic equilibrium (*see* Figure 19(a)) which we have likened to the bottom of a funnel.

But there is another attractor owing its existence to irreversible processes for which the system displays a completely different – chaotic – behaviour in time. The degeneration into chaos is best illustrated with the bifurcation diagram shown in Figure 16(b). The diagram shows the possible behaviour of a system like a chemical clock as it is pushed away from equilibrium: at the first crisis point it branches and two possibilities are available. In turn these branches sprout twigs and so on, corresponding mathematically to the fact that non-linear systems can display several different kinds of behaviour in exactly the same circumstances. Eventually, however, these crisis points or bifurcations occur so often that they overlap (on the right-hand side of the diagram) into a dense clot of possibilities. Recalling what is plotted along the horizontal axis, this requires the system to be driven a long way from thermodynamic equilibrium.

A dazzling range of behaviour is on offer in this overlap because of the enormous number of possible states densely packed together. The system is no longer confined to a limited number of branches, but is free to sample an infinitude of possibilities. To reach such a chaotic state starting from equilibrium (the origin of the horizontal axis), the system may have to pass through an infinite number of crisis points as it is driven further and further (yet still a finite distance) from equilibrium. Naïvely, one might think that the further one goes from equilibrium, the more chaos occurs in this bifurcation tree. But the level of complexity is greater than this, for the bifurcation diagram is more akin to a

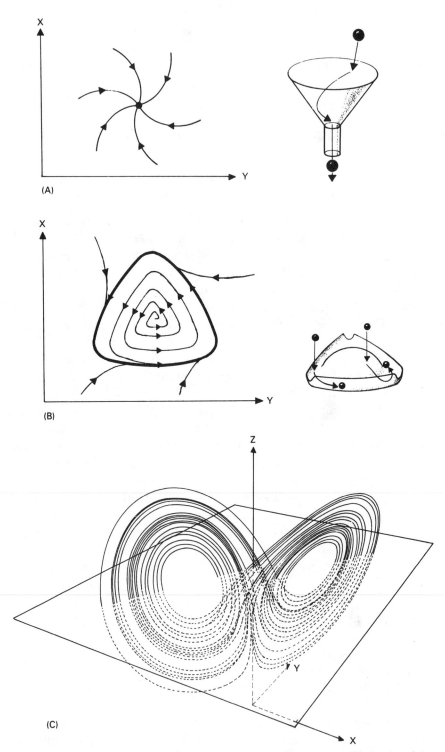

Figure 19. Attractors. (A) Fixed-point attractor (steady-state; equilibrium) and its mechanical analogue. (B) Limit-cycle (periodic) attractor and its mechanical analogue. (C) Lorenz strange attractor.

plane tree, where there are gaps between layers of foliage. Thus one can encounter 'islands' or 'windows' of regularity between regimes of chaos, inside which are windows within windows *ad infinitum* and *vice versa*. Various routes to chaos will be revisited in more detail later on in this chapter.

Chaotic evolution is apparently the antithesis of all that has been said up to now: it denies any long-term regularity or predictability in evolution through time.[42] The regular tick of the colour changes in a chemical clock is lost if the concentrations of its ingredients are changed and it is driven too far from equilibrium: it turns into a chaotic mixture. In such a state, the chemical oscillator changes from red to blue at random: there is no predicting when the next colour change will occur. And the measurements recorded in one experiment can never be repeated. The next time around, another random set of time intervals between colour changes will appear.

In spite of such delinquent behaviour, chaos can still be understood in terms of the concept of an attractor, as David Ruelle, a Belgian-born mathematical physicist at the Institut des Hautes Etudes Scientifiques at Bures-sur-Yvettes near Paris, and Floris Takens, of the University of Gröningen in the Netherlands, showed in 1971.[43] Their paper has the title 'On the Nature of Turbulence'. Notwithstanding a deceptively short abstract – 'A mechanism for the generation of turbulence and related phenomena in dissipative systems is proposed' – it is dense with advanced mathematics. The authors hoped to understand what happens, for instance when you turn a tap full on, and the initially smoothly flowing water undergoes the transition to a turbulent and inherently complex flow. But the scope of their conclusions has proved far wider, revealing a fascinating beast called a *strange attractor*.[44]

What relevance does it have to the real world? Ruelle illustrated its use in the context of smoke rising in still air from a cigarette: 'Oscillations appear at a certain height in the smoke column, and they are so complicated as apparently to defy understanding. Although the time evolution obeys strict deterministic laws, the system seems to behave according to its own free will. Physicists, chemists, biologists and also mathematicians have tried to understand this situation. [In so doing] they have been helped by the concept of the strange attractor and by the use of modern computers.'[45]

The origins of the strange attractor were described by Ruelle: 'I asked Floris Takens if he had created this remarkably successful expression. Here is his answer: "Did you ever ask God whether he created this damned universe? . . . I don't remember anything . . . I often create without remembering it." The creation of strange attractors thus seems to be surrounded by clouds and thunder. Anyway, the name is beautiful, and well suited to these astonishing objects, of which we understand so little.'[46] On the other hand, the British mathematician Christopher Zeeman believes that 'perhaps *chaotic attractor* would be a better name than strange attractor, since by now many of them are quite familiar'.[47] Both terms are used by scientists.

A strange attractor (*see* colour plate section) is quite unlike the attractors we have previously encountered – fixed-points and limit-cycles – although it too is stable and represents a possible state in which such a system can reside like a target for the arrow of time. It has two distinct features. Unlike a limit-cycle, it displays an immense sensitivity to the initial conditions: the long-term behaviour of a system trapped on a strange attractor depends on the minutest details of how it was started off. Second, unlike a limit-cycle, it is a *fractal* object.

The word fractal has been in existence since 1975, coined by Benoit Mandelbrot[48] to describe the peculiar geometry of irregular shapes which look the same on all scales of length. In the same way, regardless of how much any region of a strange attractor is magnified, it contains essentially the entire structure of the attractor. This property of showing a motif within a motif within a motif *ad infinitum* is known as self-similarity. The motif is mirrored at every scale of length: the edges of a clover leaf will be bristling with smaller clover shapes which will bristle with still smaller clover shapes (*see* black-and-white plate section). This is called invariance under scaling, for the form of the pattern is the same no matter on what scale the object is viewed.

Mandelbrot's work has shaken the way we think of dimensions. We are all familiar with the idea that a line has a dimension of one, while the area enclosed by a rectangle is two-dimensional. But in fact these are almost always idealisations: objects can be one-and-a-bit dimensional. The 'and-a-bit' means it is a fractional or fractal dimension. Mandelbrot illustrated this in a paper in which he posed the question: 'How long is the coast of Britain?'[49] A little thought shows that the answer depends on the length scale chosen to measure the coastline. The linear distance ('as the crow flies') between adjacent seaside towns would give one crude estimate. But if you strolled along the coast you would find that the country's coastline had grown, having traversed every cove and inlet. To an ant, mere pebbles could add considerably to the journey, while for a wriggling bacterium the length of Britain's coast would be interminable. Clearly, the answer depends on the scale of measurement used, owing to the fact that there is structure on essentially *all* scales of length. Indeed, if we could shrink the length scale to the infinitesimal, then the coastline would have an infinite length. The apparently paradoxical result is that the coast is in fact a 'line' of infinite length contained quite happily within a finite area (draw a circle round Britain).

A real coastline has the fractal property of self-similarity, although in an average, statistical sense. A mathematically defined object which has a strong resemblance to a coastline is the Koch curve (introduced by Helge von Koch in 1904) constructed from successively reduced triangles, as shown in Figure 20. The Koch curve has a dimension somewhere between that of a one-dimensional Euclidean line and a two-dimensional plane – it has a fractal dimension of approximate value 1.2818.[50]

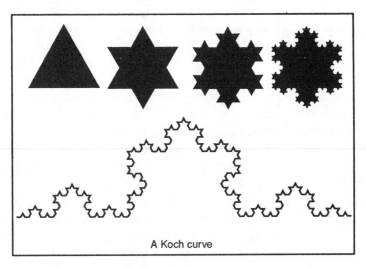

A Koch curve

Figure 20. A Koch curve. To construct one, start with a triangle. Add a new, smaller triangle at the middle of each side. Continue to build up the curve shown at bottom.

The discovery of fractal patterns has revealed a new way of looking at the beautiful and endless levels of intricacy in nature. Mandelbrot's work – as well as that of the mathematicians who preceded him – lends itself to describing the natural shapes around and within us. Clouds and coastlines are fractal objects.[51] Nor are fractals limited to inanimate objects. A two-dimensional projection of the root system of a tree or a nerve looks remarkably like the view of a river delta photographed by a passing satellite. They can be thought of as fractals; they are similar because the large-scale features of their growth can be generated by repeatedly iterating a simple mathematical rule. Fractal organisation controls a number of structures within our bodies. Mandelbrot wrote:

'Tissue . . . contains no piece, however small, that is not crisscrossed by both artery and vein. It is a fractal surface.'[52] And on the way folds and contours form within the brain he remarked that 'a quantitative study of such folding is beyond standard geometry but fits beautifully in fractal geometry'.[53] Indeed, it is amusing to speculate about the tiniest length scales at which nature is fractal – it may render the quest for the 'ultimate' building blocks of matter a futile one.

What do strange attractors have to do with time? Part of the answer is that they describe chaotic evolution, and chaotic evolution, as we will see in Chapter Eight, blows apart time-symmetric determinism. The first idea to grasp is that a point representing a chemical reaction on a strange attractor can explore an unending series of points because of the attractor's fractal properties (*see* colour plate section). Indeed, whereas fixed-point and limit-cycle attractors have dimensions of respectively zero, one, two, three and so on, a strange attractor can be defined as one of fractal dimension. Ruelle wrote: 'These systems of curves, these clouds of points, suggest sometimes fireworks or galaxies, sometimes strange and disquieting vegetal proliferations. A realm lies there of forms to explore, and harmonies to discover.'[54] The fact that the dimensionality of a strange attractor is fractal should prepare us for its second property, that of chaos. The attractor contains an infinity of possibilities, albeit confined to a finite region: the system samples different configurations as time passes, never repeating itself. One can imagine the system endlessly tracing out patterns within patterns within patterns. At first sight this is difficult to envisage. Yet armed with the notion of fractal forms, it becomes easier to see how a system, though restricted to a finite region – the strange attractor – can nevertheless discover unlimited opportunity.

Once an irreversible dynamical system has been sucked into a strange attractor, it is totally impossible to predict its long-term future behaviour. This is because, as we have seen, strange attractors show incredible sensitivity to the initial conditions: unless the system is started out with initial conditions of literally infinite precision, it will end up being completely unpredictable. Although the differential equations governing the way these irreversible systems evolve through time are deterministic, in the sense that knowledge of the initial conditions suffices in principle to predict the entire future behaviour, their exquisite sensitivity smashes the dream of a clockwork and predictable universe.

We can highlight this extraordinary behaviour by comparing it with that of a chemical clock trapped on a limit-cycle. No matter how the ball representing a chemical clock was tossed into that sombrero, it would end up rolling round the bottom of the rim. But within the confines of a chaotic strange attractor, something completely different occurs. Suppose that the ball rolled into a strange attractor and you wished to repeat the complicated path it followed. You would find that any neighbouring starting point you chose subsequently, which would always be different no matter how close it was to the first, had a trajectory which rapidly diverged from the original, leading to a completely different motion on the attractor – a different path through the fractal's infinite patterns within patterns.[55]

Dissipative chaos is born in the nested worlds of the strange attractor. Only if an observer knew with infinite accuracy what the starting conditions were in an experimental study of such a chaotic system, would he or she be able to make a cast-iron prediction. But the slightest uncertainty – which will always be the case in the real world – denies this, since no matter how small the imprecision, it will be amplified exponentially as time passes. This profound connection between chaos and sensitivity to starting conditions is crucial, for it is instrumental in formulating a consistent scientific description of the arrow of time.

There are, nevertheless, certain regularities in deterministic chaos, as it is called, because it results from deterministic non-linear dynamical equations. This kind of chaos is internally generated by a system – it is an intrinsic feature. Conceptually, therefore, it is clearly distinguished from the uncontrollable effects of random or 'stochastic' fluctuations in the external environment ('noise'). However, such stochastic processes can generate random, chaotic-looking behaviour in a system which is not trapped in a strange attractor. Differentiating between deterministic and stochastic chaos is one of the principal hurdles confronting scientists. It will return to obstruct us in the next chapter, when we look at a variety of complicated biological phenomena.

Deterministic chaos blurs the ideas of order and disorder. There has been a recent trend of using the word 'chaos' (meaning deterministic chaos) as an explanation of everything, not only of things that are unpredictable or unstable, but even when self-organisation would be more appropriate.[56] However, no one should be blinded by the buzzword 'chaos'. Order and deterministic chaos spring from the same source – dissipative dynamical systems described by non-linear differential equations. But, as we shall see in the next chapter, the ordered regimes are very often more important for biology and life itself than the chaotic ones. One should also retain a fair amount of scepticism when confronted with the many claims scientists make in the trendy name of chaos. Every case must be evaluated separately.

Chemical chaos was first proposed by Ruelle as long ago as 1973.[57] As we saw in our chemical clock example, the presence of a strange attractor is betrayed by the absence of any kind of clock-like regularity in the colour changes from red to blue. Ruelle recounts why deterministic chaos was an anathema to classical science.[58] Traditionally, a scientist looks for patterns and regularities in data because there is a good chance of making sense of them. In 1971 Ruelle asked a chemist who specialised in oscillating reactions if they were ever found to have a chaotic time dependence. 'He answered that if an experimentalist obtained a chaotic record in the study of a chemical reaction, he would throw away the record, saying that the experiment was unsuccessful. Things fortunately have changed and we now have several examples of non-periodic chemical reactions.'[59]

Chaos can be generated in chemistry in various ways. One recipe involves setting up an oscillating reaction in the usual way, as an open system maintained far from equilibrium in a continuously stirred tank reactor: with a constant flow of reactants into the tank, the reaction may settle into a stable sequence of colour changes. Suppose we increase the flow of reactants into the tank, varying their concentrations with time at a frequency different from that of the chemical clock. Think of the flow of chemicals as representing the distance from thermodynamic equilibrium: the smaller its value, the closer the reaction is to equilibrium; the larger its value, the further away. As the flow rate is increased, therefore, the reaction is pushed through more and more crisis points: beyond a threshold value, chaotic chemistry may rear its head (see black-and-white plate section).[60]

Strong evidence for the existence of a strange attractor underlying the chaotic regimes in the BZ reaction has been found by Harry Swinney and collaborators at the University of Texas at Austin from a detailed study of the reaction dynamics.[61] Deterministic chaos is something of an academic oddity in chemistry itself, although better understanding of it would be valuable for chemical engineers because of the non-equilibrium character of many industrial chemical processes. However, in the context of living systems, chaos may have an important – some would say an essential – role to play.[62]

The idea of strange attractors, although only explicitly articulated in 1971 by Ruelle and Takens, was implicit in the work of Edward Lorenz, a professor of meteorology at the Massachusetts Institute of Technology, in 1963.[63] He was trying to make sense of the all-too-frequent discrepancies between what weather forecasters say and what actually happens. A British television forecaster, Michael Fish, would have found Lorenz's ideas appealing. Fish told viewers on 15 October 1987: 'A woman rang to say she'd heard there was a hurricane on the way – well, don't worry, there isn't.' There was.

Lorenz's work provides a neat excuse for such erroneous forecasts. Armed with a computer (a rarity in those days) and an unusually strong mathematical background for researchers in that field, he sought to make as simple a mathematical model of atmospheric weather flow as possible, while retaining the essential physics. Lorenz's equations give an approximate description of a horizontal fluid layer heated from below. The warmer fluid is lighter, tending to rise and stir convection currents. If the heating is sufficiently intense the currents are irregular and turbulent.[64] He was led to a set of three coupled non-linear differential equations – the minimum required for a strange attractor to emerge; studying these, he gradually realised that however slight the variation in the initial weather conditions fed into the computer which solved the equations, the resulting solutions (the weather forecast) changed totally in a very short time. It would have been tempting to blame this on some problem with his computer but it was undoubtedly his meteorological training that led him to be so receptive to the unexpected result – and in this respect he was years ahead of his time. His strange attractor (now named after him) was not recognised as such for more than a decade. Even now it has not been rigorously proved to be a strange attractor in the mathematical sense, although all its physical properties are what one would expect.

Because of chaos, the aim to produce ever more accurate weather forecasts with ever more sophisticated computer calculations faces a huge drawback: the sensitivity of Lorenz's system of differential equations to the initial input data, a facet which led him to coin the phrase the 'butterfly effect'. This vivid image captures the idea that through chaos the smallest of events can lead to the most massive of consequences. The slightest movement on the strange attractor wreaks a quite different outcome: the beating of a butterfly's wings in the Amazon could spark off a hurricane in the West Indies, and so on. However, in the face of all the hyperbole, it is usually forgotten that if one adds further variables to Lorenz's equations in an attempt to make the picture more realistic, chaos becomes harder, not easier, to find.

The cascade to chaos

When can one expect to observe chaos? It is a question of considerable importance – chaos may be good or bad news, depending on whether we are talking about epileptic fits or heart attacks (*see* Chapter Seven). But the full answer lies beyond the frontiers of today's knowledge about the extraordinarily complex behaviour of irreversible, non-linear systems. A complete theory of all the scenarios under which chaos can emerge remains a major enterprise. Many scientists seem content to search for chaos in model problems, and to compute the fractal dimension of the resulting strange attractors, for all the good that does (primarily it swells the research literature). Cynics who are irritated by the exaggerated emphasis currently accorded to non-linear chaos, as opposed to its sister subject, self-organisation, acknowledge that if you work with such systems, whatever the field, at some point you are almost bound to come across the phenomenon; yet whether it has deep intrinsic significance is debatable.

A fair amount is known about certain scenarios which do generate chaos. Strange attractors can be spun out of the breakdown of a regularly oscillating state, ruled by periodic limit-cycles, or from time-independent steady-states, governed by fixed-point attractors. The former transition proves to be of great physiological significance, as we shall see later on when we describe biological abnormalities that arise when the regulatory properties of a limit-cycle break down into chaotic dynamics.

We have already described the Ruelle–Takens path to chaos, which requires a system to be driven through three or more limit-cycle bifurcations; this is often referred to as the 'quasi-periodic route'. The emergence of a chaotic attractor from the remnants of a limit-cycle can arise in two other ways: via the equally mysterious-sounding pathways of 'subharmonic cascades' and 'intermittency'. Both are somewhat technical to describe in detail and we shall not attempt to describe the latter, which was discovered by the French scientists Yves Pomeau and P. Manneville in 1980.[65]

The best way to illustrate the subharmonic or Feigenbaum cascade (named after Mitchell Feigenbaum of The Rockefeller University) to chaos in a chemical clock is, as we have already mentioned, with a simple bifurcation 'tree'. It shows the possibilities available, and what happens when a system is pushed from a region with a few possible states, near the trunk, to the blurred band of chaos high in the foliage. Although the representation as a bifurcation tree looks similar, the mathematical and physical details of the subharmonic and Ruelle–Takens paths to chaos are quite different. Moreover, in the subharmonic route the clock is driven by regular changes of the concentration of chemicals within it over a regular cycle. Suppose we have a chemical clock which oscillates with a regular period of time T seconds. At this stage, we have passed the first crisis point in Figure 21 where the trunk first divides. Now suppose that the incoming flow of reactant concentrations varies twice as often: in effect, that the clock is driven by an external 'force' delivered at every half-period (T/2).

Let us see what happens when we climb the bifurcation tree. As the clock is pushed further from equilibrium by increasing the reactant flow rate, we move to the right in Figure 21. Beyond a certain threshold or crisis point, the first clock oscillation becomes unstable, and flips to a new period of (approximately) twice the duration, 2T. This new behaviour, where the colour changes twice as often per period, is represented in Figure 21 by the pair of lines beyond the first crisis point. Further increase of the flow rate sends the clock through successive crisis points: after each bifurcation, the clock's period doubles again and again, to 4T, 8T, 16T, and so on. This process is called *period-doubling* and is the most well-trodden route to chaos. Ultimately, at a finite value of the flow rate – as a result of an infinity of such cascading bifurcations – the clock breaks down. It reaches a completely aperiodic state, that is, the period is infinite and the clock never repeats itself. Now it is lost in a strange attractor where it will never trace the same path twice. This, the limit of the repetition of period-doubling, is synonymous with chaos. It is as though chaos emerges when there is an overload of temporal organisation on offer.[66]

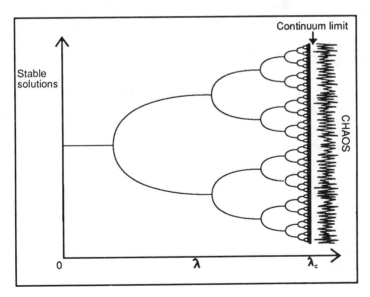

Figure 21. The birfurcating cascade to chaos in a simple non-linear system (period-doubling). Note the regularly repeated bifurcations; there are infinitely many branches in a finite distance λ_c from the origin (equilibrium).

211

A physico-chemical view of the living world

We have reached a remarkable turning point. This chapter and the last have dealt with the time evolution of processes like chemical reactions based ultimately on the Second Law of Thermodynamics. From the arrow of time contained within the Second Law, we have seen that non-equilibrium processes can be both a source of self-organisation in 'inanimate' matter and of (deterministic) chaos.

In Chapter Five we saw that living beings exist under far-from-equilibrium conditions; consider how a nerve impulse is generated, a myriad of which, acting in unison, are responsible for the thought processes by means of which this very sentence can be read. At root, the firing of a nerve cell depends on maintaining a separation of potassium and sodium ions across the cell membrane, high concentrations of the former being found within, and of the latter outside the cell membrane, furnished by a series of molecular 'pumps'. Special ion-channels in the membrane flip open to allow the ions to flow down these concentration gradients; and the rules of behaviour of these channels are such as to allow currents of ions to flow in a series of sudden pulses into the cell, just as beer can flow out of a bottle in a series of violent glugs. This is very plainly a non-equilibrium state of affairs: at equilibrium, one would find a uniform and equal mixture of the two kinds of ions either side of the cell membrane. One can see why equilibrium equals death – with no concentration gradient there is no way to make a nerve cell fire or, equivalently, for a thought to be had.

It is tempting to regard the unsurpassed complexity that is the hallmark of life as the consequence of self-organising processes. With the concept of self-organisation at hand, it is possible to conceive of ordered structures in time and space and, where appropriate, deterministic chaos. There are now many examples of striking biological order that can be understood on the basis of such self-organisation: fundamental processes of life can be explained through non-linear differential equations which incorporate the irreversible nature of time.

A fascinating feature of these developments in the biological sciences is the way in which they have grown from a fruitful cross-fertilisation of ideas from mathematicians and physical scientists working on non-linear problems. Until the late 1960s, there was a serious impediment to progress: the 'language barrier', a fog of jargon obscuring work in one field from the scrutiny of scientists in another. This was compounded by the physical scientists' dislike for the dizzying complexity of biological systems, which handicapped the development of theories and experiments to test them; and by the biologists' traditional wariness of mathematical reasoning because of the oversimplifications that such models usually entail. However, in seeking to pin down the fundamental generic properties that govern self-organisation in biological systems, most biologists now respect the need for 'caricature' descriptions. We could attempt to include every detail in our models of reality, but in practice this is too complicated, a Herculean endeavour that would in any case almost certainly obscure the principal aspects of what is going on.

Molecular evolution, replication and the origin of life

On the basis of painstaking empirical research carried out over many years, Charles Darwin came to the conclusion that all contemporary species have a common ancestry dating back over a few thousand million years. The ultimate ancestor from which all living things are descended must have been one or a few single-celled organisms. Darwin's theory of evolution, which relies on variation and selection through competition, has been continually buttressed by the ever-increasing wealth of data accumulated by biologists since that time. But from where did the simplest organism appear? Darwin himself did not provide an answer and, although God has been proposed, modern science gives short shrift to the notion of divine intervention.

Why should there not be a natural process by means of which a single cell could be generated from inorganic matter? One of the principal early protagonists of this point of view was the Jesuit priest Pierre Teilhard de Chardin (1881–1955), who aimed to fuse science and religion into a single consistent framework. He advocated that both living and inanimate matter are gradually organised into more complex forms with the passage of time. It must rate as one of the earliest ideas about self-organisation. Unfortunately, he paid a high price for his views. In 1924 he was forbidden by the Jesuits from lecturing at the Institut Catholique in Paris and left France in 1926 for China, where he lived as a virtual exile, eventually dying in New York City.[70]

One of the most influential early ideas about where this evolution might have taken place came from Darwin himself. In a letter sent in 1871 he wrote: 'But if (and oh what a big if) we could conceive in some warm little pond with all sorts of ammonia and phosphoric salts, light, heat, electricity and etc, present, that a protein compound was chemically formed, ready to undergo still more complex changes . . .'[71].

From the modern standpoint, we can understand in principle how the organisation of matter in space and time, which is such a striking feature of living systems, could conceivably have arisen through irreversible processes occurring far enough from equilibrium. Each cell is a well-organised factory, where remarkable chemical transformations take place in a very uneven distribution of chemicals. An account of beauty that can be seen at this level, where the style more than compensates for the biological jargon (and some obsolete science), was written by Santiago Ramón y Cajal (1852–1934), who did pioneering work on the nervous system. In his autobiography published in 1937, one of the classic books in experimental biology, Cajal describes the world he observed under the microscope:

> The tracheal and laryngeal fields, sown with vibrating cilia which wave, in virtue of hidden stimuli, like a field of grain before the wintry blast; the tireless lashing of the spermatozoön as it hastens breathlessly towards the ovum, the lodestone of its affections; the nerve cell, the highest caste of organic elements, with its giant arms stretched out, like the tentacles of an octopus, to the provinces on the frontiers of the external world, to watch for the constant ambushes of physico-chemical forces; the ovum, with its simple and severe architecture guarding the secret of the organic form, its protoplasm resembling the nebula, where there whirl about in embryo innumerable worlds which will emerge in future cycles; the geometrical architecture of the muscle fibre (a sort of highly complicated Voltaic battery) where, as in a locomotive engine, heat is transformed into mechanical energy; the gland cell which, in a simple way, fabricates the ferments of the living chemical laboratory, generously consuming its own substance for the benefit of the other elements, its brothers; the fat cells, models of domestic economy, which, in preparation for future famines, store up the surplus foodstuffs from the feast of life to utilise them when the organs go on strike and in the great nutritive conflicts. All these phenomena, so varied, so marvellously coordinated, draw us with an irresistible attraction and the contemplation of them inundates our spirits with the purest and most lofty of satisfactions.[72]

Who could doubt that life exists far from equilibrium, where all is change, after this breathless description of the microscopic events within our bodies? The architecture which Cajal describes could have arisen, provided that there was the correct recipe for self-organisation present, in the earliest days after the Earth's formation some 4.6 billion years ago. What happened to turn a barren and lifeless planet into the one we see around us today? Very little is certain, but what follows, although in places speculative, contains more than a grain of truth.

Then, the atmosphere of the Earth consisted of hydrogen, nitrogen, carbon dioxide, methane, ammonia, hydrogen sulphide and water, but was starved of oxygen.[73] The conventional view, in line with other ideas of chemical evolution, considers how these simple molecules could have been marshalled to form more complex ones. One of the classic experiments in 'pre-biotic synthesis' was reported in 1953 at the University of Chicago by Stanley Miller,[74] a student of Harold Urey. By passing simulated lightning through a stew of simple substances thought to be similar to the primeval atmosphere, he found that certain amino acids were formed, the basic building blocks of proteins which are essential to life as we know it. Since then, a large body of evidence has been accumulated, principally by Cyril Ponnamperuma of the University of Maryland, Leslie Orgel of the Salk Institute in San Diego, Sidney Fox of the University of Miami and others, showing that a range of biologically important molecules can be made in a similar way, including the basic genetic units, nucleic acids, enzymes and energy-storing biomolecules like adenosine triphosphate.

Not surprisingly, there are other theories for how these organic compounds formed in the absence of life. One suggested alternative is that these simple molecules were made in space, in clouds of gas and dust called dark nebulae, and were imported by a meteorite dropping on some favourable spot, such as a muddy patch.[75] Moreover, clays have been put forward not only as a means of catalysing the manufacture of these simple building blocks but also as part and parcel of the early forms of life that paved the way to life as we know it, which is controlled by the genetic materials DNA (deoxyribonucleic acid) and RNA (ribonucleic acid).[76] No matter which route was taken to the first complex molecules, the problem remains of how these simple molecules assembled into cells. At least three components must evolve for this to occur: a boundary to separate them from their surroundings; a metabolism consisting of a coordinated set of (bio)chemical reactions; and genes that orchestrate the whole affair.

The traditional view is that it all started when the interacting molecules were sequestered in structures separated by semi-permeable boundaries from one another to allow the evolution of more complex molecules in space and time. There is the 'coacervate model' of Alexander Oparin, in which a water droplet forms round a charged particle;[77] the process whereby amino acids can be persuaded to self-organise and form microscopic spheres, put forward by Sidney Fox;[78] and the 'lipid bilayer' model of Richard Goldacre, where molecules of fat join forces to produce simple membrane-like structures.[79] Current views emphasise the autocatalytic properties of RNA polymers as the prime mover, believed to have preceded the formation of a cell membrane. The discovery of these properties earned Sidney Altman of Yale University and Thomas Cech of the University of Colorado the 1989 Nobel Prize for Chemistry; until their work, biological catalysts were all considered to be proteins.[80]

Drawing on the principles of self-organisation, we can take a different, but complementary view of what might have happened. If there was a suitable feedback mechanism operating in the pre-biotic soup which could account for non-linearities, then the general conditions were ripe for self-organisation. For instance, if a molecule in that soup catalysed its own production, then non-linear feedback, the hallmark of self-organisation, could emerge. This could break the spatial homogeneity of the medium and trigger patterns and rhythms (possibly along lines broadly similar to those proposed by Turing in 1952), just as a chemical clock can show patterns in space and time. This suggests that we should seek a mechanism coupling diffusion with suitable non-linear (bio)chemical reactions.

A key ingredient of the pre-biotic soup then becomes a molecule (or molecules) which catalyses its own production: this autocatalysis provides the requisite positive feedback for non-linearity, although other possibilities can occur, such as the more complex cross-catalysis that provides feedback through a more indirect sequence of interlocking reactions. The precise nature of the pre-biotic soup is still a matter of heated debate, being safely out of the reach of direct observation. But only the principles are important here. In this respect, Leslie Orgel and his collaborators at the Salk Institute in California performed experiments of immense significance. They showed that the nucleic acids we have already mentioned possess the all-important property of self-replication: in a purely chemical mixture of the building blocks of nucleic acids more nucleic acid molecules are formed.

The genetic clock

Nucleic acids hold the designs of life. In DNA and RNA are the coded characters, the genes, which spell out the specific instructions for building proteins for life on our planet. This chemical form of information technology employs a four-letter alphabet. It sounds restrictive, but one must remember that computers use the two-letter language of binary arithmetic. In the case of a single human cell there is enough information-storing capacity to hold the 30 volumes of the *Encyclopaedia Britannica* three or four times over.[81]

It is possible to use evolutionary change in DNA and RNA as a kind of molecular clock. By comparing the genetic material of living and extinct species, molecular biologists have found that DNA and RNA mutate at a fairly steady rate over long periods. Mutations can arise owing to exposure to energetic radiation and because there are copying errors during reproduction. These mutations give rise to species as diverse as leeches and lichen. They also lead to a kind of evolutionary clock, in which the 'tick' corresponds to the rate of mutation. The clock can be calibrated with precisely dated fossils and is then put to work in estimating the moment in time when species diverged from one another.[82] It has also been used to show that the genetic code cannot be more than about 3.8 billion years old.[83]

In all forms of life, genetic sentences spell out proteins by instructing the cellular machinery to string together the protein building blocks, amino acids. The proteins are another key group of biological molecules which we have already encountered in their role as biological catalysts or enzymes. Unlike nucleic acids, proteins do not have the ability to self-replicate, but their fantastic specificity as catalysts has ensured a symbiotic relationship with their progenitors, the nucleic acids. The proteins are thus involved in massive feedback loops, for they catalyse nucleic acid replication, which is itself a prerequisite for their own production. Manfred Eigen, Nobel Prize winner in 1967 for his work on chemical reaction kinetics, and director of the department of biochemical kinetics at the Max Planck Institute for Biophysical Chemistry at Göttingen, together with Peter Schuster in Vienna and other collaborators, has sought to elaborate a theoretical framework for such molecular evolution from a pre-biotic mixture of sugars and amino acids in terms of so-called hypercycles – an interconnecting cycle of autocatalytic reactions[84] – an approach which has led to certain predictions which could reasonably be subjected to experimental test in the near future.[85]

Another model of how life began has been developed by Stuart Kauffman of the University of Pennsylvania and the Santa Fe Institute, New Mexico, and is now being explored with Doyne Farmer, Richard Bagley and Norman Packard. It envisages a set of genetic or protein polymers that can catalyse chemical reactions in which other molecules are split and/or combined. Simple chemical 'nutrients' are fed in and converted into more complex molecules. Kauffman and his colleagues have shown that such a system can become self-reproducing.[86]

From this discussion we can begin to recognise the conditions appropriate for the development of self-replicating chemical reactions. And the property of self-replication, it will be recalled, is one of the chief characteristics of life. Provided that the molecular assembly of nucleic acids plus proteins is subjected to non-equilibrium constraints, all forms of dissipative structures can in principle emerge: spatial, temporal and spatio-temporal structures, as well as chaotic behaviour, a veritable *pot-pourri* which might describe the inexhaustible richness of the forms of living creatures appearing in the world about us. This theme will be developed in the following chapter.

Time and creation

We began this chapter with a reminder that in the view of some scientists the arrow of time is an illusion. Like Immanuel Kant and many other philosophers before them, they believe its appearance in the Second Law of Thermodynamics, like our impression of the flow of time, has to do with subjective phenomena or processes in the brain, rather than with nature itself.

Ironically, it is by taking a closer look at the Second Law that we can show that this attempt to dismiss the arrow of time as merely subjective leads us into very real difficulties. Processes which are necessarily irreversible appear to play a vital role in the emergence of life. In the next chapter we shall develop this idea, looking more closely at the application of non-linear dynamics in biology. Echoes of the Belousov–Zhabotinsky reaction will be found in the behaviour of colonies of single-celled creatures called slime moulds and the way muscle behaves during heart attacks. And chaos will also appear in the rise and fall of insect populations and ideas about the origins of sex.

For the adherents of the 'irreversibility-is-an-illusion' school of thought, it is difficult to escape a profoundly paradoxical situation. We have shown how a keen description of living processes can be made using equations that contain the arrow of time. If this arrow is an illusion, one is obliged to conclude that the patterns of life – including ourselves – are a result of our own approximations. It may be that the arrow of time is so much a part of our experience that we overlook its central position. But a scientific theory which cannot accommodate this facet of time must inevitably be barren when it comes to describing whole parts of the real world.

Arrow of time, arrow of life

There could be no self-consciousness and human creativity without living organisation, and there could be no such living dissipative systems unless the entropic stream followed its general irreversible course in time.

Arthur Peacocke
God and the New Biology

ELECTRICAL activity crackles through your brain as you read this sentence. Insect populations in the lush rainforest multiply and fall. Somewhere inside an alligator egg, a pattern of stripes is laid down that the reptile will wear all its days. And within each of these images of life, time ticks steadily on.

Just as there are chemical clocks, so there are biological ones. Their rhythms vary dramatically, but they are all indispensable to life. Some nerve cells fire thousands of times in the twinkling of an eye. Within a single cell, concentrations of substances can rise and fall over periods of a few seconds. The mass of cells which constitutes a human heart beats around 70 times per minute. And the packages of cells that make up plants and animals have built-in cycles of development and reproduction that can last for years. Ultimately, all these rhythms are controlled by molecular, biochemical processes which can be understood along the same lines as the clock-like oscillations in the Belousov–Zhabotinsky (BZ) reaction, which we encountered in the last chapter. The only important difference is that the systems we are looking at now are alive.

In the BZ reaction, self-organisation appeared as whirling spirals of chemical activity where millions of molecules adopted coherent macroscopic structures in time and space. In biology, the same organising processes can be found when single cells club together to make a multicellular organism. Self-organisation also embraces ordered phenomena like the swings in insect populations or the beating of the human heart. Indeed, the whole human body can be regarded as a complex unit, self-organised in time and space.

It is not surprising, then, that biological systems have generically similar internal feedback processes to the BZ reaction. When an enzyme is manufactured in the body, it in turn participates in subsequent processes that affect its manufacture. For example, the enzyme may

220

encourage or suppress the cell's machinery. Such non-linear processes are tricky to predict, for as the quantities of the enzyme change so do the rules governing its manufacture. But life itself is an inherently highly non-linear process. The very genes that contain the blueprint for these feedback processes are responsible for regulating the way they themselves are read and interpreted in our bodies.

Feedback processes abound in biology and can spawn self-organisation in three qualitatively different ways, just as we saw in chemistry with the BZ reaction and the Brusselator model. There is temporal organisation, corresponding to oscillations; spatial organisation, corresponding to patterns; and a combination of the two, when waves of activity propagate through space. Taken together, these three brands of organisation have the power to provide much insight into what makes life tick.

It is important to remember here that there are essentially two different types of feedback: positive and negative. Positive feedback increases output in a system – such as a chemical in a reactor catalysing its own production. Negative feedback decreases output – like a thermostat controlling a central heating system. Once the temperature in a room falls below some set point, the central heating is switched on to warm it to the required level. Once this is reached, the heating switches off again. Invariably the room will carry on warming up after the heater is switched off, causing a slight overshoot in temperature. Equally, when the room starts to cool, the temperature will fall to slightly below the set 'minimum' point while the heater is starting up again. In this way, negative feedback can set up cycles where the temperature gently rises, falls, rises and so on. This is inherently stabilising and it is thought that similar feedback plays a crucial role in the human body, for instance in controlling blood pressure.[1] Positive feedback by itself is not thought to be so important in the body because it can lead to unstable behaviour. However, it does play an important role in the way animal populations vary in time, and it is common to find examples where positive and negative feedback operate together, for instance in the production of white blood cells in the body.[2]

Biological chaos

Of course, alongside chemical clocks we discovered chemical chaos. The same seems to apply when we look at the clocks of life. Although research on biological chaos is still at a relatively early stage, it appears to be responsible for important and often rather unpleasant effects. Chaos seems to occur when organisation in the body breaks down and is replaced by abnormal dynamics. It has been claimed that chaos rears up during heart attacks, as well as in the irregular waxing and waning of some diseases. Others speculate that an understanding of chaos may help in the prediction of epileptic seizures through the analysis of the brain's electrical activity. It might even be a crucial factor for evolution, although it is wise to exercise some scepticism with regard to all these claims. There has undoubtedly been a bandwagon effect where chaos is concerned: science journalists have hyped it while researchers have leapt upon it to gain publicity and financial support. Chaos, its advocates claim, is everywhere. However, the jury is still out on many of the examples cited in biology. As in Chapter Six, we must distinguish between deterministic chaos – intrinsic to the system – and randomness caused by a cacophony of external influences. There are good tests to differentiate between them but they are not straightforward to put into practice.[3] For without detailed data, existing methods for detecting the hallmark of chaos – such as the fractal dimension of purported strange attractors – cannot be relied upon.[4] No one has proved that even the most well-documented examples of these attractors fulfil rigorous mathematical definitions.

Some element of chaos does, in spite of these caveats, seem to stand beside organisation at all levels of biological processes – from events within cells to events between cells, and from events within organisms to events between organisms. All of these are irreversible processes, and irreversibility contains the recipe for chaos as well as for organisation – just like the Indian god, the dancing Shiva, who holds in one hand a fire which destroys and in another a drum which creates. In the course of this chapter we shall examine different examples of these two facets of life, in an attempt to drive home the importance to nature of irreversible, dissipative systems.

All living systems are dissipative, in that entropy is increasing. They are also dynamical, because the processes going on within them possess an immense power to evolve. Technically speaking, we are studying irreversible time evolution in non-linear dissipative systems far from equilibrium. An equivalent phraseology is 'dynamical systems theory'. Some biologists and mathematical-modellers are irritated by epithets such as 'dissipation', 'far from equilibrium', and even 'non-linearity', seeing them as irrelevant trademarks tagged on to existing fields by politically ambitious outsiders. This response, while perhaps understandable, is unfortunate. Some of the most dazzling insights

222

into biology arise from a cross-disciplinary approach. Based on the mathematics of irreversible processes, every example we shall consider underlines the importance of the arrow of time. Of course, it is impossible to prove that a theory is correct simply by applying it successfully. But we believe that the accumulated evidence for the importance of irreversibility in biology, which the following examples are designed to show, is too impressive to be ignored.

The sugar clock

Cells in our bodies must find a means of concentrating energy resources. These provide the cells with the ammunition to fight the way time's arrow — in its simplest manifestation — seeks to drag them to thermo-dynamic equilibrium and death. A complex web of chemical reactions converts the energy of food into the intricate mechanisms of life. Only by remaining far from equilibrium can plants and animals generate the necessary physiological order to live. Their cells need energy to help in digestion and in the synthesis of biochemicals. This is put to work to generate concentration gradients, to cause muscular contraction, body heat, and so on.

We run our homes on gas or electricity. The immediate fuel for the body's economy is a key, energy-rich biomolecule, called adenosine triphosphate (ATP). It is this molecule which must be synthesised if life is to carry on. ATP carries its cargo of energy in the form of a high-energy chemical bond, like a compressed spring, involving a chemical entity called the phosphate group, a cluster of four oxygen atoms around a phosphorus atom. When the phosphate group is lost, ATP is converted into a depleted form known as adenosine diphosphate (ADP). In turn, ADP can be reactivated to ATP by a chemical reaction called phosphorylation.

In green plants, cells harness sunbeams to synthesise ATP from ADP (and carbohydrates) through a process known as photosynthesis. Animals generally obtain ATP by respiration. They consume carbo-hydrates and fats and then burn these substances within their cells in special processing units called mitochondria, through breathing in sufficient quantities of oxygen from the atmosphere. Combustion takes place through an interlocking series of chemical reactions, like cogs in a delicate watch. The waste products that result are similar to those of incineration — water and carbon dioxide. This metabolic pathway is called the respiratory chain.

Not all cells, however, can exploit sunlight or oxygen for energy.

Instead, they draw on glycolysis, the poor man's respiratory chain, which produces far smaller quantities of ATP by fermentation of glucose, a process in which this sugar molecule is snipped in two. Pasteur, experimenting with yeasts in 1861, showed that these oxygen-starved – anaerobic – processes are less efficient than aerobic ones. In the words of one wag: 'These experiments proved that, as in New York City today, life without air is possible, but expensive.'[5] Primitive unicellular organisms, such as yeasts, which appear in yoghurt and food poisons, draw on glycolysis to survive even when deprived of air. So do creatures such as oysters and green sea turtles, who spend much of their time under water. Even within the human body, glycolysis has a role to play, particularly in areas where there is a limited blood supply – for example in muscles that are engaged in frenetic activity.

Biochemical clocks tick in all three sources of energy – photosynthesis, the respiratory chain and glycolysis.[6] The one that we have the most detailed understanding of is glycolysis in yeasts. Yeasts are fungi that exist as single free-living cells, some of which live on grape skins and are responsible for turning grape juice into wine. During fermentation, not only alcohol (ethanol) but up to about five hundred other substances are produced, the precise nature and balance of which determine the quality of the wine produced. This process is so widespread that it was known to ancient civilisation. According to Agnessa Babloyantz of the Free University of Brussels, 'the first written account of glycolysis in the form of a narrative series of drawings was found in the grave of the keeper of the wine cellar of Thoutmosis III, an Egyptian pharaoh who lived from 1505 to 1450 BC'.[7]

The study of how brewer's yeast converts sugar into alcohol, which has fascinated and intoxicated scientists throughout the ages, is one of the antecedents of modern biochemistry. Because it has been so thoroughly studied, it is not surprising that our understanding of rhythmic biochemical patterns is at its apotheosis in this case. Indeed, by about 1940, the entire metabolic pathway of glycolysis had been mapped out. In 1957, L. N. Duysens and J. Amesz noted for the first time that energy is not always produced in steady quantities during glycolysis: it occasionally oscillates in a regular rhythm, as do the concentrations of various chemical intermediates in the process, among which of paramount importance is our energy-rich friend ATP.

Whether the ATP concentration fluctuates with time is critically dependent on how much sugar and ADP are around. This is the key to the role the oscillations play in physiological regulation. When there is only a little ATP in the cell (and therefore more ADP), glycolysis switches on to generate the needed ATP molecules, the cell perhaps drawing on its reserves of starch or glycogen; if there is an abundance of ATP molecules, for instance if the respiratory chain has been working well, the glycolytic pathway cuts off. This regulatory process, known as the Pasteur effect, is essentially controlled by a single enzyme, a large and intricate biological protein molecule which has the ability to accelerate highly specific chemical reactions.

The enzyme in question is phosphofructokinase (PFK for short). Specifically tailored for its job over millions of years of evolution, PFK is switched on by high concentrations of ADP, and turned off by high concentrations of ATP. But PFK is a phosphorylating agent: it uses ATP to hitch a phosphate group to a sugar molecule, thereby converting the ATP into ADP. And it is, of course, the presence of ADP which activates the enzyme to operate more quickly. This feedback is precisely the kind of autocatalytic, non-linear step that is necessary for self-organisation.

To describe the tick of the sugar clock, a number of theoretical models have been put forward. The most successful in general terms is due to Albert Goldbeter and René Lefever of the Free University of Brussels (1972), subsequently elaborated by Goldbeter and other collaborators. They stripped down the problem to its bare essentials. It started out being described by twelve coupled non-linear differential equations. It ended with just two, focusing only on the enzyme PFK, plus our biochemical power-source molecule, ATP.

This massive simplification is repaid by the fact that its rhythms are then described by equations similar to those used for the Brusselator. As we saw before, its chemical clock properties are described by a limit-cycle. For appropriate concentrations, the amounts of ATP and ADP in the sugar clock waltz round a repetitive cycle; they vary over a period of about a minute, in good agreement with the experimental values (see Figure 22). Thus, glycolytic rhythms became the first confirmed example of a biological dissipative structure, a self-organising pattern in time.[8]

Rhythmic biochemistry has been found in a number of processes involving a single enzyme, such as the autocatalysts horseradish peroxidase and lactoperoxidase, and biochemical processes employing a range of enzymes.[9] As well as in biological clocks, these oscillators have been found to play a role in signal transmission, both inside and outside cells, and in the process of cell differentiation, by which cells in a developing embryo turn into, for example, brain and liver cells. As before, where

225

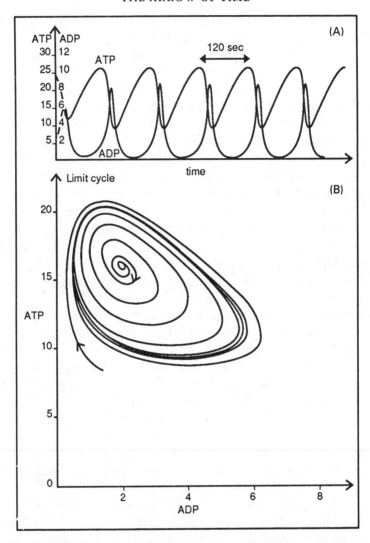

Figure 22. (A) Periodic oscillations in the concentrations of ADP and ATP. (B) Trajectories in the phase plane ATP–ADP evolve towards a limit-cycle.

there is order in terms of predictable cycles there can also be chaos, in which the oscillations occur with ever-changing and unpredictable frequencies and amplitudes.[10] Both regular and chaotic oscillations are of interest in the process of photosynthesis used by plants to turn sunlight into energy and the respiration of the mitochondria, the 'power houses' of cells.

226

A universal clock for cell division

Self-organisation also seems to rule the individual clocks that regulate cell division, the fundamental process of growth in living things. When one of the 10 million million cells in our bodies divides, new genetic material is manufactured, segregated, and walled-off by cell fission. If regulated, cell division can furnish the difference between a nose or an ear; unregulated, it can mean cancerous growth. Within the body, cell division is carried out at a variety of rates. Mature brain cells, for example, do not divide at all. Liver cells divide every year or two, while cells in the lining of the gut divide as often as twice a day. In cell division we meet, as before, both irreversible and cyclic time. To all intents and purposes, cell division is a cycle because freshly divided cells look similar to their precursors. Cell division repeats at fixed intervals, the cell tapping out a beat each time it multiplies. In turn, linear time is spun out by a series of these cycles, for most of our cells are programmed to divide and reproduce for a limited number of times before dying.

As we know only too well, our cells do not divide and proliferate forever. Taken from a human foetus, they will go through about 50 cycles of division before they eventually die. Similar cells from a 40-year-old stop dividing after 40 cycles and cells from an 80-year-old grind to a halt after about 30 cycles. [11] This seems to correspond to the overall ageing of the body, because cells from animals with shorter life spans stop dividing after a smaller number of cycles. And sufferers of Werner's syndrome, who grow old prematurely, have cells that perform unusually few cycles before dying. [12]

Deep insights into the fundamental cyclic clock of the body that marks out cell division have recently emerged. Our bodies grow from a single egg-cell by a succession of divisions, cleaving two-by-two. The process leaves almost nothing to chance: the chain of events, which is programmed very precisely, is controlled by numerous factors. Some are intrinsic to the cell – for instance, it usually divides when it has grown to a certain size. Others are controlled by the environment of the cell, such as its location within the overall organism. But what is or are the molecules which cause cellular division? Can we say anything about the genetic programming governing this vital moment?

In the latter half of the 1980s scientists made great strides towards answering these questions using two approaches. [13] In one, the molecule responsible for triggering cellular division was identified by a group led by James Maller in Denver, Colorado [14] and a group led by André Picard, Jean-Claude Labbé and Marcel Dorée at the Centre for Research

227

in Molecular Biology at Montpellier in France. In the second, a team of geneticists led by Paul Nurse, now at the Imperial Cancer Research Fund's Cell Cycle Group in the University of Oxford, has shown that this important molecule is practically the same from yeast to Man.[15]

Cells can be thought of as tiny ticking clocks (*see* Figure 23) which pass between two states. In one, the cell grows while division is inhibited; in the other, although growth continues, the cell divides. Cell division is an oscillation between the two: one phase is the 'tick' of the cell clock and the other the 'tock'. The blueprint of this living timepiece in the fundamental genetic programming of yeast was found by Paul Nurse and his colleagues. The first step was to identify some one hundred genes that are necessary for cell division. Each gene corresponds to a protein that plays a role in the cell clock mechanism. However this by itself says nothing about how the mechanism works.

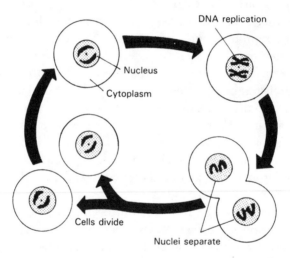

Figure 23. The cell clock: phases of cell growth and cell division. [Courtesy of the *Daily Telegraph*.]

The next stage was to find which proteins were important. Nurse knew that if he found a gene that could boost cell division, it would be responsible for an essential control protein. He explained this by analogy with the clock. It is easy to slow or stop a clock by removing a cog, mainspring, or a host of other components. But only adjustments that make the clock speed up reveal components crucial for controlling speed. Hunting down mutants which divided more quickly was simple: the faster a cell grows the smaller it is on division. At Edinburgh University, Nurse and his team found four such yeasts, appropriately dubbed *wee* mutants. Of this group, a mutant with a defective gene denoted *cdc* (cell division control) *2* was found to be the most interesting – 'the gene of the year', as Nurse put it. It was the genetic blueprint for a member of a large family of proteins called kinases: they switch other proteins on or off by attaching a large negatively charged phosphate group on them, a process that changes their shape and hence their catalytic activity. It seems that, in the run-up to cell division, the *cdc 2* protein kinase switches on many proteins required for the act.

This work dovetailed with a parallel effort to find the factors that push immature eggs to divide. This can be done by adding to them the contents of cells already undergoing cell division. Astonishingly, they can originate from any other creature; taken from starfish, human, yeasts or *paramecium*, the soup can trigger cell division in immature eggs of frogs and starfish. Clearly the key molecular triggers controlling cell division vary little from the most primitive to the most highly evolved organisms. In 1988, one was identified by James Maller in Denver. Nurse asked him if he wanted to check his purified protein against Nurse's yeast protein kinase. It turned out to be the same. Nurse summed up the result as 'Eureka time!'

The human cell clock was also shown to be governed by the same protein as the yeast in an elegant experiment conducted by Melanie Lee of the Oxford group.[16] Although the detailed mechanism of the cell clock remains mysterious, it is clear that the genetic program governing cell division has not substantially changed during the millions of years that life has existed on Earth: yeasts and humans share a common ancestor that lived on Earth 1000 million years ago and used the same control of cell division as we do today, a powerful vindication of Darwin's thesis. Nature's cell-division timepieces, whether the equivalent of a stopwatch or a grandfather clock, have the same control mechanism, a molecular oscillator that regulates the way in which the cell swings from one state to another. This 'one protein suits all' feature appears over and again in nature: when she finds an efficient way to perform a vital task she tends to use the same molecular machinery wherever she can. The cause of this conservatism is easy to understand: a mutation with a slightly different form of this essential enzyme would probably perish. It would not fit into the complex biochemical feedback processes required for self-organisation.

Mathematical modelling will be the next stage of the enterprise to understand cell-cycle regulation. As Nurse put it: 'We have now identified the actors in the play and we know what they do. We now have to find out why they come on the stage when they do.' First several problems must be overcome. With existing technology, it is difficult to measure how the concentrations of biochemicals in a cell vary over time. Instead most methods of analysis used by molecular biologists in the search for the key molecular participants destroy spatial and temporal organisation. 'We try to interpret data which has lost much of what is required for a proper dynamic description of the living cell,' according to Nurse.

When more details of this cell clock have been gathered, it seems almost certain that the secrets of this most fundamental of life's patterns will eventually succumb to the language of dynamical systems theory. It will involve piecing together all the necessary molecular steps, and setting up an appropriate non-linear dynamical model. The American theoretical biologists John Tyson and Stuart Kauffman tried this at an early stage (in 1975), and claimed to have found evidence of limit-cycles at a time when these were the 'in' concept, much as chaos is today. But more detailed experimental studies showed that their conclusions, based on a particular model, were mistaken. In the view of Art Winfree of the University of Arizona, the moral of this tale is clear enough. The use of these non-linear mathematical models 'needs case by case examination, not *a priori* implication of fashionable concepts'.[17]

Socialising slime moulds

Let us now jump to the next level of biological order where we are no longer concerned with organised behaviour within cells but between cells. Advanced organisms such as mammals are comprised of many billions of cells which are organised into enormously elaborate structures during the process of development from egg to offspring. There are scarcely any mechanisms in this development that we yet understand well enough to be able to give a decent mathematical description. Nevertheless, it seems certain that morphogenesis can be interpreted in terms of non-linear dynamics. It is one of the most astonishing of the creative processes which rely on the arrow of time.

A strange creature called a slime mould falls halfway between a collection of single cells and an organism. *Dictyostelium discoideum* (*see* black-and-white plate section) as it is called, is intriguing in that it is at times multicellular (with around 100,000 cells), while at others, as Winfree puts it: 'Its cells wander independently, like the individual workers of an ant colony. Like the ant hive, *Dictyostelium* is a "superorganism", a genetically homogeneous being composed of autonomous

230

individuals, nevertheless organised altruistically for the collective good.'[18]

Cells in the mould feast on bacteria. When food is plentiful, individual cells feed happily and voraciously, multiplying by direct cell division. Needless to say, utopia cannot last and eventually the colony runs short of food. Until this point, the cells have ignored one another's existence, behaving like solitary wanderers. Now they 'notice' each other. For reasons not yet fully understood, certain cells in the colony become active and act as pacemakers, 'ringleaders' that send out rhythmic pulses of a chemical called cyclic adenosine monophosphate (cAMP), a ubiquitous molecule in biology which acts as a molecular messenger between neighbouring cells. This clarion call to close ranks and organise travels at a few microns (millionths of a metre) per second.

Cells slither towards the pacemaker cells, in the direction of increasing cAMP concentration, on receiving a cAMP message. They then amplify and pass on the message, a form of feedback mechanism providing the non-linearity, which induces still more cells to home in on the pacemaker centres. The cells converge in pulsatile waves.

Just by looking at these slime moulds we can see that similar processes may be at work as in chemical clocks. The photograph in the plate section shows concentric and spiral waves of aggregating cell populations that bear a compelling resemblance to the spiral waves occurring in the BZ reaction. These kinds of waves also appear in rhythmic motions of heart muscle which we will encounter later in this chapter, as well as in infectious microbes and the waves of star formation in spiral galaxies.

Once the cells have formed a slimy mass they begin to differentiate, and a tip forms which secretes cAMP continuously. The whole mass becomes organised into a glistening multicellular 'slug', with a head and a tail, which wriggles in search of light and water. All in all, it takes several hours for these cells to form this simple organism. Between one and two millimetres long, it crawls along under the leadership of the pulsating source at its tip.[19] It then rights itself to form a hard stalk above which perches a small head containing spores; eventually, the head breaks open and the spores are cast far and wide by the wind. If they settle in a suitable place, they can germinate and begin the cycle of this strange organism's life anew.

Remarkable biochemistry underlies this behaviour, reminiscent of the glycolytic reactions previously described in the sugar clock. The messenger molecule that provides the clarion call for this wriggling mass, cAMP, is formed from ATP by the good offices of an enzyme called adenylate cyclase.[20] Feedback occurs, just as in glycolysis: cAMP already present in the medium surrounding the cells switches on adenylate cyclase to produce more cAMP from ATP.[21] In this way autocatalysis arises, an essential ingredient of self-organisation. By employing largely the same non-linear analysis as he used to model glycolytic oscillations in yeast cells, Goldbeter was able to show in a detailed way on the basis of limit-cycles how oscillations of cAMP could be produced every few minutes.[22] This is an excellent example of self-organised behaviour; moreover, chaotic cAMP oscillations are also now known. In the case of a mutant form of *D. discoideum*, temporal chaos in the form of cAMP oscillations and spatial disorder manifested in aberrant stalks and fruiting bodies have been observed, all of which can be returned to ordered behaviour by addition of an enzyme called phosphodiesterase.[23]

The creation of biological shapes

Nature has mechanisms to organise cells into a dazzling range of forms and shapes. If they are grouped on the simplest basis – according to size – Man figures in the top 0.001 per cent. Birds, mice and bees may seem small, yet they appear in the top 1 per cent, while the remainder consist mainly of insects and mites with an average body length of just 3 millimetres, 'the size at which we cease to notice things even when they land in our soup'.[24] A detailed description of the recipe of diversity is a long way off. Nonetheless, the very fact that Man is governed to a large extent by the same genetic programming as other creatures like insects, dandelions and chimpanzees fuels the hope that we can make sense of development in general terms.

Some scientists simply want to throw a little light on the processes at work. Others dream of formulating biology's equivalent of Newton's laws. Earlier this century D'Arcy Thompson, professor of zoology at St Andrew's University, Scotland, wrote in his study of the growth of multicellular organisms that the development of living creatures must lie within a framework set by geometry. Many biologists would say that this approach led nowhere: the leaps and bounds in the field have not been made with pen and paper but rather by observations in the laboratory.

Nonetheless, in parallel to the explosion of effort in experimental biological research, theoreticians have quietly worked in the background to understand how these living patterns are woven in time. As well as the temporal patterns encountered in glycolysis and the cell clock, it is theoretically possible to see spatial patterns such as stripes, spots or chevrons, which can be thought of as standing waves: although the coloured molecules that make up the design may be involved in furious activity, the overall pattern they make stands still. This possibility was recognised early on by Turing in his 1952 paper. His ideas on morphogenesis – the development of organic form – have latterly been investigated as a means to account for markings ranging from the zebra's stripes to the leopard's spots, and for the initial differentiation of the parts of the body.

The ingredients needed to persuade a pattern to emerge from a milling mixture of millions of molecules were discussed in Chapter Six. Diffusion is one; this is, after all, the way that different regions of a soup of chemicals 'communicate'. Chemical reaction is another. The feedback that occurs as molecules diffuse and react in a mixture sets up patterns. To model these spatial structures mathematically, the non-linear dynamics must describe the rate of the irreversible chemical reactions, also taking into account the different rates of diffusion of the participating chemical substances. Thus Turing's basic approach is today known as reaction–diffusion theory. The 'excitability' of the medium, mentioned in Chapter Six, may also be important.

The Hydra was variously reputed to have 100, 50 or 9 heads. Although Hercules eventually dispatched it with the help of his charioteer,[25] a much humbler hydra is still alive and well and contributing to research into morphogenesis. One or two millimetres long, pieces of this freshwater polyp can regenerate an entire body. This remarkable feat was witnessed as long ago as 1744 by Abraham Trembley (1710–84), the Geneva-born zoologist. It offers a convenient, though atypical, system for the study of morphogenesis. If a small piece of tissue is taken from near the head of a hydra and put elsewhere on the body, a new head will grow within 48 hours; the head can even be removed and it will spontaneously regrow. 'Something has infected the local tissue,' wrote Hans Meinhardt of the Max Planck Institute for Virus Research in Tübingen, who modelled this pattern-forming ability on the basis of Turing-style reaction–diffusion theory.[26] According to Meinhardt: 'The hydra has a good chance to provide us in the near future with a rather complete picture about how the development of a relatively simple organism is controlled.'

Pattern formation in Hydra seems to depend on two components: short-range chemical activation (by autocatalysis) and long-range inhibition. The resulting non-linearities give rise to patterns with features common to many organisms. Typically, a small patch of tissue becomes

233

slightly different from its surroundings and exudes a tiny amount of an 'activator substance' which rapidly builds up in concentration because it catalyses its own production. The high concentrations in this region trigger the manufacture of an inhibitor signal – another biomolecule – which diffuses into surrounding tissue to prevent other regions from making the activator.[27] The concentration profiles of these so-called morphogens in effect tell cells where they are with respect to this landmark of special tissue, essential information for deciding if they are to evolve into a head or a body cell. For instance, depending on the body segment of an insect, segments of legs or antennae can form. Activation and inhibition not only model the initial pattern development, but are also believed to play an essential role in the spacing of repetitive structures such as bristles, hairs, feathers and leaves. Reaction–diffusion theory has been applied, with debatable success, to cartilage patterns in limbs, feather and scale distributions, animal coat markings and the complex patterns on butterflies' wings.[28] In fact, such reaction–diffusion equations can generate an astonishing variety of different patterns. Turing's mother wrote: 'He showed me some of these [patterns] and asked whether they resembled the blotches of colour on cows, which indeed they did to such an extent that the sight of cows always calls to mind his mathematical patterns.'[29]

In animals, such patterns are generally laid down on the embryo inside a shell or womb. The precise moment at which the patterns form and the size of the embryo at that instant are critical factors in determining the patterns that will adorn the adult animal. Mathematical models suggest why mice and elephants, at the two extremes of the size spectrum for mammals, tend to have a uniform colouration. They can also show why the patterns of animals of intermediate size such as alligators, leopards and zebras can be very exotic. Such models indicate that a leopard's tail is too thin to support spots, which coalesce into stripes; indeed, according to Professor James Murray of Oxford University's Centre for Mathematical Biology, the approach explains why 'you can have a spotted animal with a striped tail but never the other way round'.[30]

There is, however, a major problem with using the Turing instability to describe pattern formation: no known experimental system has been shown to undergo it. Substances that appear to control development have been put forward as morphogens: a husband-and-wife team at the Harvard Medical School found good evidence that retinoic acid triggers the development of a cluster of cells in a chick embryo into a leg or wing;[31] and a team at the Max Planck Institute for Developmental

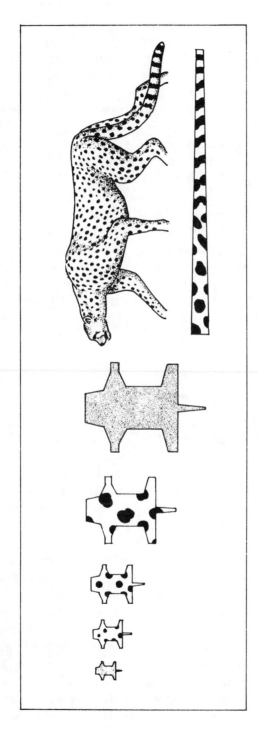

Figure 24. Patterning on animals established at an embryo stage relates to their size. Most mice, being too small, have no spots. Large animals, such as elephants, also tend to have a uniform colour. Mathematics predicts that spots give way to stripes when there is not enough area: compare a cheetah's body with its tail, as illustrated by the wedge shape.

Biology in Tübingen found evidence that a gradient of bicoid protein influences development in the embryo of a fruit fly.[32] But these examples do not fit readily into Turing's picture of pattern formation. The reason has been alluded to already in Chapter Six: Turing's reaction–diffusion theory is unable to describe all possible mechanisms for self-organisation. A related but qualitatively different approach has been proposed by Jim Murray and George Oster at Berkeley, California and others. They constructed a similar model which, however, uses directly measurable biochemical quantities and cell densities, called a 'mechanochemical' approach, since it combines both chemical and mechanical influences on cells. This model can be related to quantities which are either already known or directly measurable by an experimenter, making it easier to refute or verify. Murray has used a mechanochemical model to make predictions about the precise moment at which alligator stripes are laid down during gestation, a concrete result which can be directly tested. Wider application of this approach may furnish insights into wound-healing, how limb cartilage architecture is laid down during normal development and what factors lead to birth defects and deformities.[33]

The throbbing heart

Morphogenesis is a powerful example of irreversible evolution, but it cannot match the symbolic impact of that ultimate biological clock, the heart. A human heart beats about 70 times each minute, 40 million times each year and some 3000 million times during a lifetime. From irreversible processes spill all the regular beats and rhythms of this, the most vital pacemaker. Not only is the heart a symbol of life, it is also a common limit of life: mammals seem to clock up a similar number of heartbeats (2000 million) during their lifetimes, from shrew to whale.[34]

The evolutionary origins of the heart lie in much simpler biological pumps that pushed fluid around the body of a primitive creature by peristaltic motion – the same muscular ripple that squeezes food through our guts today. Two such tubes entwined through evolution and developed grossly expanded muscles to produce our four-chambered heart. During the time that Mankind has inhabited the Earth, much has been written about heartbeats, from tribal chants to poetry and pop songs. Now it is the turn of the mathematicians and scientists.

During the 1920s, seminal work in this field was done by two Dutchmen, B. van der Pol and I. van der Mark, which continues to provide valuable insights.[35] They were able to show that several different breakdowns in regular heart rhythms – arrhythmias – could be generated from their model as various parameters were adjusted; each was brought into play by a familiar feature in non-linear dynamics – bifurcations.

A major advance in the effort to produce mathematical models for physiological systems came in 1952, with Hodgkin and Huxley's pioneering work on the squid's giant axon, the long thread-like extension of a nerve cell that conducts impulses. Their effort won them a Nobel Prize. The methods they employed to tackle the resulting differential equations led to a quantitative description now routinely used to study the properties of equations which model the electrical activity of heart tissue.[36] In 1959 Huxley went on to show that inherent in the model was repetitive activity, regular nerve firing that occurred if the cell was deprived of calcium.[37]

Much more is now known about the way the heart works and the mathematical properties of the non-linear dynamical models used to describe it, which has led to impressive – and continuing – improvements in our ability to treat a variety of heart diseases. For instance, it is now possible to implant a device to watch over the heart so that if a potentially fatal rhythm is detected it can give a jolt to restore the normal rhythm. There is also the distinct possibility that the theoretical tools of irreversible non-linear dynamics may eventually prove fruitful for diagnosing the mechanism and guiding the therapy of complex cardiac arrhythmias. In the United States alone, some 400,000 people die each year when the heart develops a fatal arrhythmia without warning.[38] In Sudden Cardiac Death, the arrhythmia can be a slow heart beat (called a 'bradycardia') but most often it is fast, when it is dubbed a 'tachycardia'. Although this latter may be a fast and rhythmic beat, an electrocardiogram shows that it can degenerate into an irregular pattern when the heart fibrillates, or flutters. This represents an abnormal spatio-temporal organisation of heart cells.

Many physiological rhythms in the body are generated by a single cell or by a group of coupled cells. In the case of the heart, there are now known to be at least six different conductors of the heartbeat. A particular tissue, known as the Purkinje fibre, contains cells which are much larger than those from other tissues within the heart. Although the Purkinje cells do not provide the natural rhythm, they convey the firing of the sinus node 'pacemaker' to heart muscle. The rhythms of these pacemaker cells have to be excited by a so-called 'cardiac pacemaker current', analogous to the pacemaker centre in the Belousov–

Zhabotinsky reaction which can be established by touching the solution with a hot platinum wire or triggered by a piece of dust or a scratch in the dish the reaction is contained within.[39] In turn, similar spiral waves to those formed in the BZ reaction have been observed during heart attacks, in which vortices circulate round a region of healthy tissue.

There is a living heart cell – or at least a digital one based on a mathematical model – beating in a computer at Oxford University's physiology department, developed by Denis Noble and his group. Denis Noble, Dario Di Francesco in Milan and their colleagues are attempting to stitch together a numerical model incorporating the myriad chemical processes taking place in the cell. Refinements are being added as scientists worldwide uncover further details about what makes the heart tick. But to simulate even the motions of a single cell draws on a huge amount of computing time – it takes around 100 seconds on Noble's computer to work out one second of heartbeat. As a result, attempts to understand the cooperative action of many heart cells are severely hampered by the available computer time. Noble hopes to model heart tissue – up to half a million heart cells – on the University of Minnesota's Connection Machine, a high-powered computer consisting of 10,000 processors capable of carrying out calculations in parallel. Heavy-duty number-crunching exercises like this may help to link different types of heartbeat analysis and resolve divergent claims over such matters as fibrillation, which some scientists maintain is due to deterministic chaos and others explain in terms of phase singularities of rotating waves.

At the time of writing, the physico-chemical processes within the cell are described with the help of some thirty simultaneous coupled non-linear differential equations. Of all the processes, the most important are the channels that allow electrical signals to flash in and out of the cells. The signals are mediated by special proteins that shuttle electrically charged atoms called ions. Some ten channels and a handful of other processes transporting chemicals within the cell and on its surface are described in Noble's model. The most important of these is the calcium channel, which triggers the translation of ATP through a complex chain of processes into a heartbeat. In common with the Belousov–Zhabotinsky reaction, the heart is a dissipative system because the gradients of chemicals across the cell membrane run down as substances are transported in and out of the cell.

In chemical clocks, the dissipative structures were regular colour changes or beautiful spirals of colour. In the heart cell, one can view the dissipative structure as the beat formed by the coordinated movement of filaments of protein. Individual processes, such as the sodium pump, are crucial to maintaining this beat. Just as a pendulum in a grandfather clock is kept moving with a slowly falling weight, so these pumps ferry sodium ions across the cell membranes to ensure that the heart is held away from equilibrium. During each beat, calcium ions surge into the heart. Interacting with the channels ferrying the calcium ions into and out of heart cells, our old friend cAMP makes another appearance. As in the slime mould, it operates alongside the enzyme adenylate cyclase. In the heart the two act in concert in feedback processes which control how the calcium channels flip open and slam shut.[40] On the level of an individual cell, the calcium ions which surge into the heart cell trigger the contraction of special proteins by a kind of molecular 'ratchet' mechanism. Clearly, passage of calcium into the heart is only half a heartbeat. Before the beat is over the calcium must be extracted again by another transport mechanism so that the cell relaxes, a phenomenon that has been witnessed with the aid of special calcium-sensitive dyes.

The resulting picture depicts a heart that relies on a complex web of self-organising and irreversible interactions – a finely choreographed routine of messenger chemicals, proteins and enzymes conspire to produce each beat. As Denis Noble has shown with his models and the realistic-looking beats they produce, this cardiac dance can be portrayed mathematically by non-linear differential equations: a highly sophisticated physico-chemical 'clock reaction' makes our hearts beat.

Dynamical diseases

Besides telling us something about biological time, Noble's research offers practical benefits for understanding heart disease. Even among normal individuals there are marked fluctuations in heart rate; when there are several competing pacemaker centres in the heart, abnormal beats often arise, and these are sometimes referred to loosely as being 'chaotic'. Armed with a non-linear dynamical model of the beating heart, one can investigate whether such irregularities are due to deterministic chaos or the more traditional explanation – random noise. It is by no means trivial to separate the two in experimental data recorded on electrocardiograms. As a result, there is no general agreement on how these data are to be interpreted. Some writers have suggested that chaos will tell us the inside secrets of heart attacks, but this excitement may be misplaced. For example, ventricular fibrillation is an arrhythmia which leads to rapid death, but although it is frequently said by clinicians to be irregular (in the sense of being 'difficult to describe') there is still no indication of chaos in the technical sense (*see* Figure 25(a)).

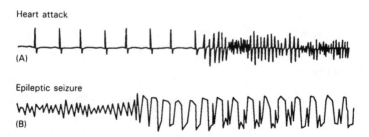

Figure 25. A change of rhythm appears in (A) a heart attack and (B) a *petit mal* epileptic seizure.
[Adapted from *Science* <u>243</u>, 604 (1989).]

Studies of embryonic chick heart cells performed by Leon Glass, Michael Guevara and Alvin Shrier of McGill University, Montreal, show probably the best-documented evidence of physiological chaos at the level of body organs.[41] Groups of ventricular cells beat spontaneously with a regular rhythm, but when given a strong electrical jolt, the next beat is pushed earlier or later than normal. If the jolt is periodic, the chick heart cells are driven by two different frequencies – their natural beat and that of the external pulse. This is a classic scenario which can lead to deterministic chaos. We came across it in the last chapter where it was noted that the equivalent of electrical stimulation in a BZ reaction – varying the way chemicals are added – can also lead to apparently random behaviour. It has also been observed experimentally in glycolysis by varying in a periodic manner the flux of sugar through the medium.[42] In a similar fashion, depending on the frequency of the electrical stimulus, the chick heart cells can either fire a regular number of times between pulses or in a chaotic fashion.

A critic would point out that laboratory studies of chick cells are somewhat artificial. Even if this is deterministic chaos, perhaps it does not occur naturally. Ary Goldberger of Beth Israel Hospital, Boston, believes that it does.[43] He even maintains, on the basis of an analysis of the electrical activity of the heart, that healthy human hearts are more chaotic than sick ones. This seemingly paradoxical conclusion is at odds with the common-sense view that the collapse of the rhythms of a healthy organism is at the root of many physiological disorders. By contrast, Goldberger thinks that the variability of chaotic dynamics is

Many scientists have made important contributions to our understanding of the mathematical properties of such period-doubling phenomena, notably P. Myrberg, A. N. Sharkovskii, Robert May, George Oster and Mitchell Feigenbaum. For us, the remarkable thing about the cascade is its universality. This means that, for any system in which chaos emerges from such a period-doubling sequence (and there are many throughout animate and inanimate nature), similar numerical scaling features are present. In experiments, this universality is of immense importance – it enables one to tease out from what might otherwise appear to be genuinely random noisy data the presence of deterministic chaos, a kind of latent orderly feature.

A proper understanding of chaos is thought by many scientists to be vital for the diagnosis of a whole variety of abnormal physiological conditions. In 1980 David Ruelle speculated about chaos in the beating heart, 'a dynamical system of vital interest to every one of us. The normal cardiac regime is periodic but there are many non-periodic pathologies which lead to the steady-state of death. It seems that great medical benefit might be derived from computer studies of a realistic mathematical model which would reproduce the various cardiac dynamical regimes.'[67] We will see to what extent his hunch has paid off in the next chapter.

There is a highly significant feature of the bifurcations, or crisis points, of non-linear systems which highlights very well the aspect of time as the 'medium of innovation'[68] which the philosopher Henri Bergson so strenuously sought to draw attention to in his writings.[69] The place on a bifurcation tree at which a system finds itself reflects also its particular history: just as any child knows that to reach an apple in a tree one has to climb the appropriate bough and set of branches, so, had the system not taken a quite specific route through the bifurcation diagram, it would not be in its present state. It is the crucial role of indeterminism, of random fluctuations, controlling the denouement at the crisis points, that makes time an innovative entity: between one stable state and the next, the system's entire future lies in the precarious hands of chance, unlike its past. One can see the asymmetry of time revealed in a bifurcation diagram in the same way as we experience it: a one-week-old baby may become a prince or a pauper, whereas the history of a 50-year-old man is fixed. Likewise, imagine that there is a beetle crawling up and down the bifurcation tree. It could have crawled from anywhere in the foliage to end up on the trunk. But to get to a particular twig from the trunk, it had to take a particular path through the branches. Thus even a beetle sitting on a twig in the bifurcation tree has a specific history.

212

more appropriate to a healthy body, while ailments are associated with a loss of this flexibility. His view is controversial.

Winfree adopts a different standpoint. He conjectured in 1987 that the heart muscle should be vulnerable to an arrhythmia consisting of rotating waves – similar to those excitations seen in the BZ reaction – if it is subjected to a stimulus of the right size and at the right time.[44] Since then, experiments to study the induction of such rotors in heart muscle, leading promptly to fibrillation, have backed his suspicion.[45] Both clockwise and counterclockwise rotating waves were observed when shocks were delivered at a critical phase of the cardiac cycle. According to Winfree: 'It is a fantastically satisfying instance (of which there are so pathetically few!) of physical theory leading the way in bio-medicine.'[46] But he remains sanguine about our current understanding. In the conclusion to his book, *When Time Breaks Down*, Winfree wrote: 'I first noticed the word "fibrillation" seven years ago and I still don't know what it means . . . It is a mystery. Somebody should solve it.'[47]

The breakdown of a regular heartbeat rhythm during a heart attack is one example of what Glass and Mackey call a 'dynamical disease' – a pathology which arises from a shift in the body's normal rhythms. This definition cunningly avoids the row over whether chaos is good or bad news – instead it is a change of rhythm that is linked with disease. Such dynamical disorders are well known to doctors; examples include epileptic seizures and respiratory arrhythmias such as 'Cheyne–Stokes breathing', a periodic quickening and slowing of respiration (often accompanying congestive heart failure). Viewed as dynamical disorders, comparisons can now be drawn between problems as diverse as heart attacks and epileptic seizures. This is evident from the electroencephalogram (abbreviated to EEG) in Figure 25(b), that shows the electrical buzz of a living brain, caused by electrical disturbances of the order of one-hundred-thousandth of a volt.

An EEG is recorded by placing electrodes on the scalp of the subject; it is used routinely in every hospital. The electrodes tune into the superficial area of the brain known as the encephalon, or cerebral cortex, which sits like the canopy of a mushroom above its stalk, hence the term electroencephalography.[48] Unlike a heartbeat, the EEG of a normal, healthy brain is irregular and rather quiescent. This activity undergoes a drastic change with the onset of an epileptic seizure, when the subject may enter a trance and lose consciousness. It is a condition which was thought in the Middle Ages to be due to possession by evil spirits or the Devil himself. But this demon does not seem to manifest itself as

241

disorder in the brain. For when we study an EEG recorded during a seizure it is remarkable to find that the electrical activity, although more violent in terms of amplitude, is actually rhythmically more regular.

A detailed study of the EEG recorded from brains undergoing epileptic seizure has been made by Agnessa Babloyantz and her colleagues at the Free University of Brussels. It is not clear how much can be read into EEGs, given that the electrodes used are measuring and averaging electrical activity over large parts of the brain. Nonetheless, using standard techniques for the analysis of dissipative non-linear dynamics, these workers claim to have found evidence for the presence of strange attractors – the tell-tale motifs of chaos – both in normal brains and during epileptic seizure. The fractal dimension of this mathematical abstraction gives an idea of how much chaos is present – the higher the dimension the greater the amount of randomness. From the EEG data they calculated that the fractal dimension of the attractor varied: for an alert normal brain they were unable to determine the dimension, although it was much higher than that measured during deep sleep, which was slightly greater than four. But during an epileptic attack this fell to nearly two dimensions as more ordered activity took place in the brain.[49]

The new insights into the body's rhythms provided by dissipative dynamical systems theory may well help to improve the effectiveness of current medicine. There is the intriguing question of the survival rate for patients with chronic myelogenous leukaemia, which is no better today than was the case 50 years ago. It has been suggested that this may be due to the fact that doctors do not take into account the oscillations known to occur in the number of white blood cells.[50] Glass and Mackey maintain that a deeper understanding of the control system governing these rhythms might lead to a more effective therapy.

Dynamical insights may help the many medical emergencies ensuing when a stimulus is delivered to the patient. That stimulus could be an electric shock to a heart which is out of control. It could arise from the use of a mechanical ventilator. Most common of all, it is a pulse of an active substance sent into the body, when an aspirin is taken or a drug injected on a regular basis. In these cases, it can be tricky to establish a stable relationship between the natural rhythm of the body and the imposed rhythm of the treatment.[51] This suggests a new way to improve the use of existing drugs, one currently under investigation by drug companies. In the not-too-distant future doctors will be able to make use of a substantially improved understanding of physiological rhythms. Indeed, it has even been suggested that properly timed doses of drugs may be able to suppress seizures in regularly cycling epileptics and that properly timed stimulation could suppress tremors.

Scientists are now in the process of acquiring many such insights from the use of non-linear dynamics. Who would have guessed that through abstract theoretical work major medical advances could have emerged? Certainly not the majority of the civil servants, bureaucrats and politicians who control research spending today.

Cycles of sex, or hard excitation

Even sex is succumbing to the arrow of time expressed by non-linear dynamics. The bifurcation diagram we encountered in Chapter Five showed in an idealised way the possibilities available to a system as it was driven further from equilibrium. In a chemical clock reaction, several states emerged which in practice could mean the transition between an oscillating state with regular colour changes and a non-oscillating state. One way in which this transition occurs has been dubbed 'hard excitation', also known as a 'subcritical Hopf bifurcation'. This means that as a parameter increases – for example the concentration of a chemical in the reaction – an oscillation or rhythm appears instantly where before there was none. In biology, an example is believed to be orgasm.[52]

A recording of electrical activity of muscle in the pelvic floor in a healthy male during ejaculation shows burst-like activity – a rapid firing of nerve impulses. Readers will be interested but not surprised to learn that this sudden onset of bursts, where there had been no obvious periodicity before orgasm, is consistent with hard excitation. Other examples of the abrupt onset of rhythms have been found, including the hot flushes experienced by post-menopausal women.[53] As well as being able to create a rhythm through hard excitation, the converse can occur when the oscillation is annihilated. This has been suggested as a mechanism for the cessation of a baby's breathing during cot death syndrome.[54]

Population dynamics

We shall now move from events within organisms to events between them. These, too, are governed by mathematics in which time's arrow is explicit. A good example is population growth, which was first modelled mathematically by Leonardo of Pisa (known as Fibonacci) in

1220. He made the terrifying prediction that a breeding pair of rabbits, if left to their own devices, could generate a volume of rabbits greater than the known universe in 114 generations. It has since been pointed out that 'long before that, the Earth would be submerged beneath a sphere of rabbits, expanding faster than light'.[55]

Because of predators, diseases, competition and social cooperation, nature is more subtle and less fertile than this. Populations of rabbits and the foxes who eat them evolve in a complex way in the same ecosystem. Life on Earth competes and evolves in a scheme as intertwined as the orchestrated interactions of molecules within cells. Microbes, mammals, plants and fish are cogs in what some people conceive of as a global living machine which cycles sunlight and nutrients.[56] In the nineteenth century, this interdependence began to be seen when entomologists recorded regular rhythmic changes in parasites and the hosts they inhabited. The balance of nature has to be well regulated, for otherwise one species will consume all the prey in its environment and then itself die out.

At the beginning of this century, people began to use mathematical models to describe how these populations vary with time. Such models inevitably contained time's arrow. They turned out to be coupled non-linear differential equations showing, for instance, the way in which the number of foxes and rabbits are dependent on one another and how each population changes with time.[57] The most important element in population dynamics (and indeed in evolution, which we shall come to shortly) is competition. With a limited number of rabbits, the growth of a fox population takes place at the expense of the rabbits, whose population diminishes. But with only a few rabbits, the fox numbers dwindle. This leads to a rabbit resurgence and the whole cycle can start again. Such competition for limited resources provides a regulatory feedback mechanism which, mathematically speaking, is once again due to non-linearities. It is the equivalent of autocatalysis in a chemical clock reaction. Alfred Lotka in the United States proposed the first model of an oscillating predator–prey system which was elaborated independently by Vito Volterra, an influential figure in Italian science in the years before the Second World War. It has come to be known as the Lotka–Volterra model.[58]

Yet this clock-like behaviour, and the regular waxing and waning it entails as time marches on, is by no means commonplace in animal populations. Probably the most detailed long-term records are of the Canadian lynx population, due to the unfortunate fact that the lynx has been hunted for its fur for more than two hundred years. These records

244

show that the lynx population has fluctuated wildly (*see* Figure 26). The complexity of this time-dependence led many to speculate that it mirrored irregularities in the population of the lynx's prey, the snowshoe hare, caused in turn by fluctuations in the hare's food supply. But the Lotka–Volterra model cannot describe this data – if one applies it injudiciously, one ends up concluding that the hares are eating the lynx.[59]

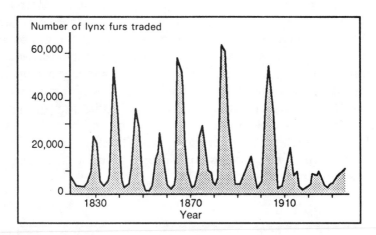

Figure 26. Canadian lynx populations fluctuated dramatically between 1820 and 1930, hitting peaks every nine to ten years and then dropping sharply. [Adapted from *Science* **243**, 310 (1989).]

Thanks to non-linear dynamics, an alternative explanation has been proposed that can be couched in terms of properties of the lynx–hare populations alone, without the need for mishaps in the snowshoe hare's food supply, weather fluctuations, disease or other external factors. In a non-linear dynamical system, the irregularities might owe their existence to chaos. Indeed, theoretical population biologists brought the notion of deterministic chaos on to centre stage for the very reason that erratic fluctuations in populations are rather commonplace and crave explanation. One theorist, Jim Yorke of the University of Maryland, coined the word 'chaos' in 1974 (in work that was published in 1975).

The mathematical properties of these chaotic phenomena were set out as long ago as 1962 by Myrberg and rediscovered independently by several people.[60] Having stumbled upon these bewildering possibilities of non-linear systems, Robert May was one of the first to bring them to the attention of population biologists in a paper published in the leading American journal *Science* in 1974.[61] 'In retrospect, it seems odd that such chaotic dynamics were not noted earlier,' says May, who is now at Oxford University.

May crawled along the non-linear dynamical path before he attempted to walk. He first studied probably the simplest of all models in population dynamics, the non-linear 'logistic equation' which gives an over-simplified description of the way a species changes its numbers when consecutive generations do not overlap.[62] A good example would be an insect population which hatches in the spring and dies after laying eggs in the autumn. Depending on the birth rate and the competition for food, the population can evolve in many different ways. The birth and death rates could be equal, so that the population could settle down in a steady-state, that is, remain constant. In the language of dynamical systems theory, this would be described by a fixed-point attractor. On the other hand, the population might show regular jumps, hopping among different fixed-point attractors, booming and busting between two, four or any other number of fixed values. Then again, it might hop about in a seemingly random fashion, corresponding to chaos described by a strange attractor (*see* Figure 27).[63]

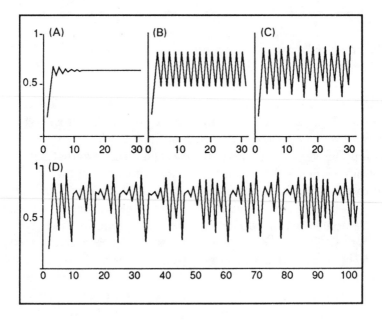

Figure 27. A model (the logistic map) shows a range of behaviour: (A) Steady-state. (B) A two-cycle. (C) A four-cycle. (D) Chaos.
[Adapted from *Science* <u>243</u>, 311 (1989).]

246

This spectrum of behaviour is contained within one single equation. Each type of behaviour can be summoned simply by varying a parameter fed into the equation. Overall, the options available to the hypothetical insect population as this parameter is varied can be illustrated with that familiar device, a bifurcation diagram.[64] Using computer graphics, one can draw a vast panorama of possibilities, an eerie landscape wherein each point is the answer to a single calculation (*see* colour plate section). Each shows the results of tens of thousands of calculations carried out by Mario Markus and Benno Hess at the Max Planck Institute for Nutritional Physiology in Dortmund, revealing what happens when non-linear equations are put through their paces. A tiny change in the parameters defining the equation produces an utterly different landscape.

May's work came as a surprise: complexity was an innate feature of a ridiculously simple model. Together with his colleague George Oster, he went on to study the bifurcation properties of members of the family of equations to which the logistic equation belongs, the so-called quadratic maps.[65] Some of these have a more plausible basis in biology;[66] in the 'trees' of cascading bifurcations which they drew, they found self-similarity on ever smaller scales, revealing the fractal geometry of a strange attractor.[67]

May recounts how the importance of this work – showing that simple models have unexpected consequences – suddenly dawned on him when writing reviews on the field: 'Once you have the quadratic map you have something that any schoolchild can see,' he remarked. 'You can iterate on a hand calculator. A twelve-year-old can do it. You can see a simple deterministic rule do something very weird.'[68]

In an influential review article published in the journal *Nature* in 1976, May noted that the mathematical models used by population dynamicists also occur within a vast range of other disciplines.[69] Aside from disparate fields within biology, including genetics, the same equations also arise in economics, where they are employed to describe such things as business cycles and the relationship between commodity availability and price. In the social sciences, they are used for the description of the propagation of rumours. We can expect these systems to show a similar range of behaviours that spread from regular oscillations to temporal chaos. May concluded: 'Not only in research, but also in the everyday world of politics and economics, we would all be better off if more people realised that simple non-linear systems do not necessarily possess simple dynamical properties.'[70] If deterministic chaos underlies the economy (rather than random or stochastic chaos) it would render Ministers of Finance and Chancellors of the Exchequer somewhat impotent in the long term. The Chancellor may claim that the adjustment of one parameter, such as the interest rate, can have a long-term effect on the economy. But remember the 'butterfly effect': according to dynamical systems theory, in the long term it could be just as likely that a pensioner removing her life savings from a bank in East Cheam could lead to economic collapse.

May's articles set in train a new way of thinking about the manner in which populations of creatures boom and bust. Yet when we look beyond the surgical predictive power of abstract mathematical models to 'real life' situations, such as the lynx–snowshoe hare data discussed above, it is difficult to prove whether genuine – deterministic – chaos is actually the explanation of the erratic population fluctuations that are observed. Indeed, despite fifteen years' effort, no one has provided a universally accepted example of chaos in population biology, largely because of the inadequacy of the data compared with what is necessary to achieve unambiguous answers.

In an effort to overcome this, fascinating experiments are being dreamed up to test the predictive power of non-linear dynamics. One such living experiment being directed by Mark Kot, an applied mathematician at the University of Washington, USA, bears a striking resemblance to the 'inorganic' Belousov–Zhabotinsky reaction.[71] Kot has proposed a predator–prey system where the predator is a simple single-celled animal called a protozoan and the prey a bacterium. Millions of these organisms are kept in an open reactor with a continuous influx of food for the bacteria to feed upon, and a constant outflow of waste products. In Kot's mathematical model of the experiment, for a steady influx of food the populations of the micro-organisms show quite straightforward dynamics. On the other hand, if the food supply is made to vary over a cycle so as to mimic the passage of the seasons, chaos is predicted to arise, just as we saw could occur in the BZ reaction (in Chapter Six). The experiment currently under investigation is designed to search for such chaos.

In other recent theoretical studies, Roy Anderson at Imperial College, London, and May believe that they have found evidence for chaos in AIDS-infected immune systems, where there is a kind of predator–prey relationship between white blood cells, the T4 lymphocytes, stimulated yet at the same time 'preyed upon' by the human immunodeficiency virus (HIV) responsible for the disease, via production of an immune response (B cells and antibodies) specific to HIV. More generally, epidemiologists are using chaos to analyse the spread of diseases such as measles and mumps.[72] William Schaffer of the University of Arizona is one of the most ardent advocates of chaos in this context. In an analysis of epidemiological data available for Copenhagen in Denmark, Schaffer and his collaborators claim to have found evidence of chaos for measles, mumps and rubella, while chicken pox follows a regular annual cycle. Again, whether such epidemics are intrinsically chaotic – or just the result of random external influence – is not yet clear.

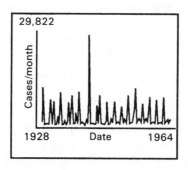

Figure 28. A mix of order and disorder: measles in New York City, 1928–64.

Evolution and deep time

It is now time for us to return to the question of evolution. The publication of Charles Darwin's *The Origin of Species by Means of Natural Selection* on 24 November 1859 marked an intellectual revolution that transcended biology and shook to the core the way thinking people looked at the world.[73] Darwin's evolutionary arrow, which points from simple single-celled organisms to complex creatures like Man, appears at first sight to contradict the thermodynamic arrow of time which Boltzmann hoped to explain. But more than anything else, evolution needed time to work. Had the Earth been around for long enough?

The Judæo-Christian religions portrayed a young Earth upon which Man appeared in the days following its creation. In the seventeenth century, the Archbishop of Armagh in Northern Ireland, James Ussher, maintained on the basis of the scriptures that the date of the Creation was 4004 BC. However, geologists glimpsed 'deep time' when fossils were recognised as the petrified remains of creatures that had once inhabited the Earth and when it was first suggested that geological layers were laid down in chronological order.

Deep geological time gained gradually in status from the seventeenth century and by the early nineteenth century estimates of the age of the Earth ranged from a million years to thousands of millions of years. In 1863, Lord Kelvin[74] made a famous estimate based on the rate at which the Earth should dissipate heat. Assuming among many other things that the Earth was at one time in a molten state from which it had cooled, he concluded that the Earth could only have supported life for 100 million years.[75] Refinements in the calculations followed, reducing the available time for which the Sun could have supported life on Earth down to as little as ten million years.[76]

The discovery of radioactivity in the last years of the nineteenth century resolved the question of the span of geological time. Radio-active decay in the Earth's crust could have been responsible for the Earth's temperature gradient from core to mantle, one that emphatically put paid to Kelvin's timescale.[77] Then followed the idea of using the remains of radioactive decay to date the Earth itself. The rate of decay of an atomic nucleus could be used to measure the passage of time, a notable example of a linear or non-cyclic clock. Given the rate of decay, the ratio of the amount of remaining radioactive material to the decay products provides an estimate of the Earth's age. In 1905 the British scientist John William Strutt produced on this basis an estimate of 2000 million years (from the decay of radium) and the American Bertram Boltwood suggested 2200 million years (from the decay of uranium).

The discovery of deep time exerted a profound effect, but it is a hard concept to grasp. In the words of Stephen Gould of Harvard University: 'An abstract, intellectual understanding of deep time comes easily enough – I know how many zeroes to place after the 10 when I mean billions. Getting it into the gut is quite another matter. Deep time is so alien that we can really only comprehend it as metaphor. And so we do in all our pedagogy. We tout the geological mile (with human history occupying the last few inches); or the cosmic calendar (with *Homo sapiens* appearing but a few moments before "Auld Lang Syne".'[78] Gould points out that John McPhee has provided the most striking metaphor of all in his book *Basin and Range*. He asks us to imagine the Earth's history as the old measure of the English yard, the distance from the king's nose to the tip of his outstretched hand. One stroke of a nail file on the king's middle finger would then erase human history. Such is the wonder of deep time, the foundation stone of evolution.

Darwin's theory

Darwin was born at Shrewsbury on 12 February 1809. He studied medicine at Edinburgh University, where he found lectures in geology 'incredibly dull . . . the sole effect they produced on me was the determination never as long as I lived to read a book on geology'.[79] He went on to Christ's College, Cambridge, in preparation for the Church. There he 'got into a sporting set, including some dissipated low-minded young men'.[80] Indeed, his father once told him: 'You care for nothing but shooting, dogs, and rat-catching, and you will be a disgrace to yourself and your family.'

When Darwin set sail on HMS *Beagle* on 27 December 1831, a five-year circumnavigation of the world lay ahead which was to sow the seeds of a new view of the living world.[81] Although he had been invited to join the cruise by Captain Robert FitzRoy, Darwin and the captain had an uneasy relationship that compounded Darwin's problems with seasickness and illness. FitzRoy was convinced that he could judge a man's character by the outline of his features: 'He doubted whether any one with my nose could possess sufficient energy and determination for the voyage,' said Darwin.[82]

On his return to Falmouth on 2 October 1836, Darwin had accumulated overwhelming evidence from a wide range of terrains, environments, plants and animals for an idea that had been suspected for some time – that over long periods, animals and plants can evolve into new species. But Darwin took another twenty years to publish his ideas. His hand was forced by the appearance of a paper by a young British naturalist, Alfred Wallace. Wallace had independently arrived at the theory of natural selection in a flash of inspiration during a fit of malarial fever on a tropical island. Just like Darwin, he had been pondering the population theories of Thomas Malthus.[83] Wallace wrote: 'While vaguely thinking how this would affect any species, there suddenly flashed upon me the idea of the survival of the fittest.'[84] Darwin was appalled and persuaded two influential friends to organise a joint presentation of Wallace's paper with extracts from an essay Darwin had written in 1844. Darwin's contribution was the first to be given and although some have maintained that Wallace was mistreated, he himself acknowledged the older man's priority.

Darwin became the butt of cruel caricatures by cartoonists for his 'monkey theory', enduring severe criticism from scientists and savage attacks from the religious establishment. His thesis was that throughout the time that life has existed on Earth, species have evolved from simpler to more complex forms, and that natural selection – survival of those species best adapted to their environment together with the elimination of those less so – provides an explanation for this evolution. The theory was more a hypothesis than a hard and fast description of the way the world is: there were no conclusive proofs which could be adduced in its favour. Its strength lay in the claim that it fitted the observed facts better than any other theories. It would be wrong to think of Darwinian evolution as rigid; no respected contemporary evolutionist accepts it precisely as it was first formulated. The theory evolves as our knowledge grows, but in the words of Robert May: 'All this work takes place within the sturdy framework erected by Darwin.'[85]

Darwin's recipe for the diversity of life on Earth receives a fascinating modern interpretation in the language of atoms and molecules. To a great extent, one can make sense of life – at least conceptually, if not in all its glorious detail – in terms of the underlying properties of those threads of life, the DNA molecules we encountered when discussing the origins of life towards the end of the last chapter. Richard Dawkins eloquently describes DNA's ability to self-replicate in *The Selfish Gene*. We are 'survival machines', robots whose job it is to protect and reproduce more of the same strands of DNA or, equivalently, the genes carried on these giant molecules which contain the blueprint for our traits and characteristics. The huge strides in molecular biology bear powerful witness to the veracity of this molecular picture of life. Breathtaking progress has been made in understanding what goes on in a living creature – down to the last molecule – and then in using this knowledge to make new generations of drugs, to redesign creatures by altering their genes and to formulate exquisitely sensitive tests for diseases and genetic disorders.

But as we have repeatedly observed, simple-minded reductionism has its limitations; if there is a criticism that can be addressed to present-day molecular biology, it is that through its complexity, the details may obscure the bigger picture. The concern with individual molecules, important though it is, may be taking precedence over the essential nature of the way these molecules interact and cooperate through time. These cooperative non-linear effects, so dramatically brought into play under far-from-equilibrium conditions, represent an essential element in the make-up of life.

There have been attempts to interpret evolution at a macroscopic level by calling on thermodynamics. Brooks and Wiley, in their otherwise controversial book *Evolution as Entropy*, make the telling remark: 'Everyone agrees, we presume, that organisms are far-from-equilibrium

252

dissipative structures.'[86] Unfortunately, their approach suffers from the muddling of distinct concepts – information-theoretic ideas and thermodynamic entropy, which as we observed in Chapter Five have been repeatedly confused since the time of von Neumann. What certainly does carry over from thermodynamics is the importance of irreversibility, the arrow of time.

One of the earliest suggestions that evolutionary processes are irreversible came in 1893 from French-born Louis Dollo (1857–1931), and is enshrined in Dollo's law. Yet according to one of the leading evolutionary theorists, John Maynard Smith, natural selection does not imply directionality in time.[87] Richard Dawkins has written of Dollo's law that it is 'often confused with a lot of idealistic nonsense about the inevitability of progress, often coupled with ignorant nonsense about evolution "violating the Second Law of Thermodynamics".'[88] There is no reason why general trends in evolution should not be reversed, he argues: 'If there is a trend towards large antlers for a while in evolution, there can easily be a subsequent trend towards smaller antlers again. Dollo's Law is really just a statement about the statistical improbability of following exactly the same evolutionary trajectory twice . . . There is nothing mysterious or mystical about Dollo's Law, nor is it something that we go out and "test" in nature. It follows simply from the elementary laws of probability.' The problem with Dollo's law is that it associates irreversibility with 'the evolutionary system' alone, while the Second Law is a global statement concerned with both the system and its surroundings (recall our discussion in Chapter Five).

Christopher Zeeman, for many years Director of the Mathematical Institute at the University of Warwick in England, and now Principal of Hertford College, Oxford, believes that the two ideas will never mix: 'It is the wrong argument to go from the Second Law to evolution. When you are modelling you work within a particular level like physics. The next level may be chemistry, then macromolecular structure, cells, creatures, ecology and then evolution. You have to start with a different modelling in each level. When you talk about life, evolution and the Second Law of Thermodynamics there is absolutely no connection between them whatsoever.' According to Zeeman, there are no grand theories of biology as there are in physics: 'The grand general theories of physics are useless in biology, save that gravity might affect embryo development or something like that. You want to bring in a large piece of machinery which does not really lead to predictions of interest to biologists. When faced with a large chunk of mathematics the biologists ask: "Well, is this really going to be of any use?" And by and large they do not waste the time to learn it.'[89] George Oster put it even more trenchantly: 'To think that thermodynamics will tell you how life evolved is like thinking that you can understand how a TV set works by burning it in a bomb calorimeter [a device for measuring the energy content of a substance].'[90]

On the other hand, many authors, including Maynard Smith himself, have asserted that increasing complexity is a hallmark of biological evolution, which is certainly a statement of a time-directed process. Nobel laureate Salvador Luria described the nature of evolution aptly when he wrote: 'Evolution, like history, is not like coin-tossing or a game of cards. It has another essential characteristic: irreversibility. All that will be is the descendant of what is, just as what is comes from what has been, not from what might have been. Men are the children of reality, not of hypothetical situations, and evolutionary reality – the range of organisms that actually exist – is but a small sample of all past opportunities.'[91]

Stephen Gould, an American palaeontologist, believes that he has found evidence for the arrow by studying the distribution through time of a group of organisms, descended from a common ancestor, which has since become extinct.[92] 'It appears that history is asymmetrical,' according to his colleague, Norman Gilinsky of Virginia Polytechnic Institute. Gould looked for evidence of the arrow in 'clades', segments of the branches of the evolutionary tree whose thickness denotes diversity of species. He found a decrease in the kinds of organic designs in the face of an increase in the number of species, an asymmetry revealed by a 'bottom heavy' clade shape. He explains it as early experimentation, when diversity quickly increases, and later standardisation, when it tapers off and the organism wanes to extinction. Gould believes it is the most outstanding trend of the fossil record.[93] As he puts it: 'This theme imparts a direction to time that is more clear and reliable than any statement we can make about change within lineages. It also probably reflects a more general and basic law about the history of change in natural systems.'[94]

At the very least we can say with conviction that the thermodynamic and evolutionary arrows of time are not in conflict. As it becomes progressively clearer that chemistry, physics and mathematics can inform biology, it is only natural that ideas from dissipative non-linear dynamics should begin to infiltrate the subject of evolution. Whether this will be beneficial to biology, only time can tell. Yet the seductive possibilities of self-organisation and chaos have already proved irresistible to many people, who have seen in evolutionary time the self-same arrow as is indicated by the Second Law of Thermodynamics.

The biochemist Arthur Peacocke wrote: 'Thus does the apparently decaying, randomising tendency of the universe provide the necessary and essential matrix (*mot juste!*) for the birth of new forms – new life through death and decay of the old.'[95] Further: 'The work of Prigogine and Eigen and their collaborators now shows how subtle can be the interplay of chance and law (or necessity), of randomness and determinism, in the processes that lead to the emergence of living structures. These studies demonstrate that the mutual interplay of chance and law is in fact creative within time, for it is the combination of the two which allows new forms to emerge and evolve – indeed, natural selection appears to be opportunistic. This interplay of chance and law appears now to be of a kind that makes it "inevitable" both that living structures should emerge and that they should evolve.'[96]

Those who advocate the use of thermodynamics in biology maintain that it provides the bigger picture rather than the details, a framework on which other ideas can be hung with the help of dynamical systems theory. This is entirely analogous to the way in which traditional thermodynamics gives way to irreversible non-linear dynamics when it comes to studying self-organisation in non-living things like the Belousov–Zhabotinsky reaction, as well as many biochemical processes. For example, Peacocke, among others, maintains that irreversible thermodynamics 'can serve to eliminate some putative models of biological situations as being incompatible with macroscopic physical laws, while permitting others, if not actually determining the choice between them'.[97]

The pace of evolution

The evolutionary arrow of time is a broken one – if we arrange all the available fossils in chronological order, they do not form a sequence of scarcely perceptible changes, like consecutive frames of a ciné film, but instead contain seemingly discontinuous leaps. Creationists and other extreme fundamentalists attempt to use this to discredit evolutionary theory, which they caricature by a continuous evolutionary arrow.[98]

The fossil record is, of course, incomplete. But these gaps in the record may also provide evidence of so-called punctuated equilibria, a term introduced by Niles Eldredge and Stephen Gould in 1972.[99] According to them, there is no such thing as a constant rate of evolution. Species tend to remain stable for a long time; if and when a species evolves, it does so in brief bursts, in evolutionary terms.[100] In fact, the fossil record consists of thick layers of land mass within which given species are uniformly distributed throughout, separated by thin 'surfaces' across which the species change abruptly. In general, at such a punctuation point a single species evolves into several new species, a process of multiple speciation as indicated by the multifurcation diagram shown below.

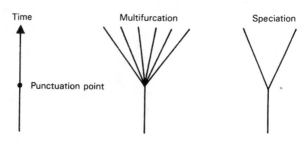

Figure 29. Multifurcation diagram.
[Adapted from E. Zeeman, *Dynamics of Darwinian Evolution*.]

Is there a conflict between Darwin's thesis and that of Eldredge and Gould? No. Zeeman has shown that Darwinian evolution can indeed explain the phenomenon of punctuated equilibria. His analysis of these evolutionary jumps is based on René Thom's 'catastrophe theory', designed to handle situations in which continuous causes – in this case, changes in the evolutionary landscape – produce discontinuous effects. The translation of Darwinian evolution – random small variations and natural selection – into the simplest mathematical form possible results in the punctuated equilibria shown in Figure 29 rather than anything else. As Zeeman put it: 'I have translated Darwin's words into the simplest possible mathematical function and then variation of the environment predicts these punctuated equilibria.'[101] It is interesting to note that, in the only diagram which occurs in *The Origin of Species*, Darwin chose Figure 29 to illustrate his ideas.

Despite the success of his model, Zeeman is quite pragmatic in assessing the role mathematics is likely to play in evolutionary theory. He believes that because much of evolutionary biology is and will remain inherently descriptive, mathematics is much more likely to play a qualitative than a quantitative role. However: 'We may expect it to be useful in precisely those counter-intuitive areas where verbal arguments lack conviction, such as when continuous causes produce discontinuous effects.'[102]

Chaos: a reason for having sex

Sex poses another challenge to evolutionary theory, but it is one that can be tackled with the idea of deterministic chaos emerging from the arrow of time. Much of the life which surrounds us reproduces sexually. Some creatures, like bacteria, reproduce asexually, that is they make copies of themselves without the attention of another bacterium. Still others, such as the thrips, tiny winged insects, can manage both. But why has sex evolved? In the most naïve view of evolution, genes are selected on the basis of how successful their survival machines are in the battle for limited resources. Those less able to adapt perish. However, Maynard Smith, in his book dedicated to the evolution and maintenance of sexual reproduction, opined that some element of our basic understanding of sex might still be missing. [103] What bothered him was that in the short term, asexual reproduction seemed to offer advantages over sex.

Sexual reproduction has some obvious disadvantages: finding a partner could prove a nuisance; genes from different individuals come together in a single descendant, thereby halving the effective gene content passed on by an individual of the species. This dilution of the individual's complement of genetic material appears to be at odds with the general notion of the 'selfish gene'. And sexually reproducing female animals on average produce one female offspring for every two offspring reproduced by an asexual animal; in a mixture of the two the asexual variety would quickly dominate.

But there are advantages to sex, aside from being 'the most fun I ever had without laughing', as Woody Allen put it. There have to be, or else we would not be here to consider the question. Sexual reproduction shuffles genes more effectively than asexual reproduction and introduces more genetic variation into a population. This flexibility makes the population better able to adapt to changes in the environment. And in the short term sexual reproduction may afford greater scope in dealing with new predators, such as viruses, since it makes evolving species less predictable. [104] Many species which reproduce sexually are known to be more resistant to invasion by parasites than asexual ones; for example, a rust fungus introduced to Chile severely reduced the growth of an asexually reproducing blackberry (*Rubus constrictus*) while having very little effect on the sexual species (*Rubus ulmifolius*).

Dynamical chaos may help to explain why sex should have evolved through natural selection to help us cope with parasites. At least, this is the view of Bill Hamilton of the Department of Zoology at Oxford University and his colleagues. [105] He is studying the role of sex in an evolutionary game where competing hosts try to develop strategies to outwit their parasites.[106] In his computerised simulation of the survival battle, Hamilton models the genetic make-up of the hosts and parasites. In its simplest terms, the aim of the parasites is to adapt so as to match the make-up of the hosts, which they can then home in on and devour. In the game, the closer the correspondence, the more 'fitness points' the parasite accrues and the more successful it is. Total correspondence is a disaster for the unfortunate host. Conversely, the host seeks to 'unmatch' its genetic complement with respect to that of the parasite so that it may better evade it. In Hamilton's model, the mathematical metaphor for this behaviour employs thirteen coupled non-linear dissipative equations which show how the colonies evolve in time. He thinks that, for a realistic choice of the parameters which enter the model, the result is deterministic chaos (in the strict sense in which we have been using the term).

In Hamilton's view, the presence of a strange attractor here, although not proven, would be 'good for sex'. For a strange attractor carries with it chaotic and unpredictable evolution, a crucial source of variability which serves to keep a sexually reproducing species one step ahead of its predatory parasites. In a creature whose behaviour is very heavily influenced by its genes (the environment, of course, also plays a role), chaos offers a means for prey to evolve much more unpredictably than by natural mutation alone.

Chaos is not a necessary reason for having sex, according to Hamilton; evolution through natural selection just works better in the long run if sex is present. And compared with parthenogenesis, sexual reproduction accelerates the move to a chaotic haven. Hamilton is no 'evangelist' of chaos; he talks disparagingly of his use of irreversible non-linear dynamics: 'My approach is extremely unintelligent really. I just put it into the computer, let it run and see what happens,' he says. It should also be added that, in a subject as intrinsically controversial as sex, there are many other current suggestions about the reasons for its having evolved.[107]

The beat of biology

In many ways, life on Earth resembles an orchestra playing to the beat of the heavens. The movements of the Sun and Moon are reflected in the ebb and flow of insect populations and the activity of living things around the globe. All are patterns in time. This rhythmic biology is explored further in the Appendix.

Within the theoretical framework used to describe the genesis of such patterns, time runs in only one direction. On this arrow of time is engraved endless variation. Some patterns are visible to the human eye, like the spots of the leopard or the stripes of the Kinabalu gold orchid. Others are invisible, like the beat of the heart, the firing of nerve cells or the onward march of cell division. These temporal patterns are not only part of life: they are fundamental to it. Even apparently random processes, like the spread of disease and the rise and fall of fish stocks, may conceal a deeper order. Contained within irreversible non-linear dynamics is the recipe for both self-organisation and dynamical chaos.

The intricately ordered yet fleeting patterns of life bubble up out of the general and irreversible passage of the universe through time towards a possible state of maximum entropy. To deny this arrow of time might seem to be convenient for the fields of relativity, quantum mechanics and classical mechanics — after all, in other contexts their principles work very well when wisely applied. But perhaps these huge theoretical edifices simply fail to tell the whole story.

To quote from the French theoretical physicist Bernard d'Espagnat: 'The step from the idea of "catastrophes" or "dissipative structures" to a quantitative and truly general understanding of how living things develop is still quite considerable . . . But there has undeniably been some progress. It is to be hoped that there will be more, and that in times to come we shall see the gradual emergence of a science which, far from reducing the *élan vital* to some humdrum mechanism, will reveal to us more clearly its profound beauty, much as classical astronomy revealed the beauty of the Universe to our forefathers.'[108]

A *unified vision of time*

A clash of doctrines is not a disaster, it is an opportunity.
Alfred North Whitehead
Science and the Modern World

OUR voyage through science must surely leave even the most hardened sceptic convinced of the power of irreversible dynamics. We have used analyses based on the arrow of time to offer compelling explanations of a vast panoply of phenomena, from the emergence of life to the appearance of a leopard's coat. If we dismiss the arrow of time as an illusion, we must forfeit all the insights we have gained. This would surely be an enormous sacrifice – and all we would gain are the absurdities of a world-view in which bowls of soup could heat up of their own accord, and snooker balls mysteriously pop out of their pockets. The objective existence of the arrow of time is an idea that cannot be denied. If it requires an overhaul of some conventional scientific notions, then so be it. In this, our last major chapter, we shall draw together the various hints of the reality of time's arrow so far recounted. We shall combine them with recent thinking on chaos, and in so doing uncover a direction to time that is capable of uniting our personal and scientific experiences.

A new kind of description is called for, in which the future is not uniquely fixed by the present or the past. The notion of strict determinism must be overturned in favour of a world-view more in keeping with our own experience of it, incorporating an open future wherein genuine evolution and innovation can produce the beautiful patterns that we see around us in nature, from slithering slime moulds to the intricacies of the world's weather systems and beyond, to the very processes by which the universe itself was born. This new view represents a genuine synthesis of the essential yet opposed concepts of chance (probability) and necessity (determinism). It is a view in which the arrow of time is accorded full objectivity, one that enables us to make sense of the myriad processes in the real world which have occurred, are occurring and will continue to occur whether we are here to see them or not. By rejecting the picture of strict determinism, we can also discard the age-old conflict between the mechanical universe, as portrayed by Laplace, and free will. This approach requires the full acceptance of the

260

Second Law of Thermodynamics as a statement of time's directed nature, and its incorporation from the outset into the description we seek. But first of all, let us briefly review our starting position.

The problem of time

In our quest to find the arrow of time we have explored all the major ideas of contemporary science. We saw that the so-called 'fundamental' theories of Newtonian, Einsteinian or quantum mechanics all deny a direction to time. We saw how determinism and causality are closely connected to the notion of reversibility in these theories. Indeed, Einstein's motivation in constructing his supremely successful geometrical description of gravity lay in his deep-seated belief in the primacy of causality. But in such deterministic theories, time is relegated to a subordinate role: whichever way time is taken to unfold, the entire future, as well as the past, is contained within the present – all three are in a sense aspects of one and the same thing.

Because such equations have no intrinsic arrow of time, there is no reason to choose one direction in time in preference to the other. But things are worse still: not only is time undirected, it should be cyclic and 'history' must repeat in keeping with Poincaré's return.[1] Just as a circle has no end, so this eternal return appears to rule out the existence of a beginning and an ending of time. The concept of equilibrium – the end point of temporal evolution, used in thermodynamics to describe the state for which processes have ground to a halt – would be completely undermined, for no instant in time would be different from any other. Indeed, if these theories were correct we should have to wave goodbye to all the concepts of thermodynamics, such as entropy, the sole quantity to have emerged in science which gives an authentic description of time as we perceive it.

If we comply with the view of a number of theoretical physicists and assert that the world is at root subject to deterministic and reversible laws, then we could conclude that our existence and all our actions can be traced to the initial or boundary conditions which gave birth to the universe. At the dawn of time, these conditions would have been set and the instructions for future behaviour executed according to Newton's lore. Perhaps it was God who selected those conditions – personally arranging the Big Bang so that life and Mankind would emerge.[2] In some respects, the former Cambridge mathematical physicist and Anglican priest, John Polkinghorne, would go further. In his opinion, God would not simply have lit the blue touch paper and stood back; His

261

role is immanent and active at every instant, for it is He who guarantees the validity of the laws of nature, keeping the show on the road by ensuring that the laws do not alter or become corrupted throughout time. Be that as it may, the actual direction of time is not explained by time-symmetric mechanics: it is an extra ingredient chosen by God, or whoever else selected the initial conditions. These conditions presuppose a direction to time. They are additional ingredients which say nothing about the nature of the theory in which they are used. By the end of this chapter, we will have shown that the crystal ball of Newton's determinism is badly cracked: these conditions can never be known exactly, even in principle.[3] Thus we must think again about what we mean by time evolution in mechanics.

Of all these 'timeless' theories, quantum mechanics is the most mysterious: its interpretation runs counter to almost all our received wisdom about the way the world works. Yet it is a superb tool for the working theoretical physicist, since it gives a mathematical recipe for predicting the outcome of experiments at the level of atoms and below. Provided that one does not try to understand what it means, one is fully entitled to sit back and enjoy the delights of quantum theory, our best existing theory of matter. But the problems surrounding the interpretation of the theory contain an important hint about the missing arrow of time. The central difficulty in quantum theory surrounds the act of measurement, when the microscopic atomic and molecular world confronts the macroscopic world, at the moment when a pointer on a measuring instrument trembles or a flash occurs on a phosphorescent screen. This fundamental act, necessary for studying the world, is irreversible. As we described in Chapter Four, in the orthodox Copenhagen interpretation the very process of recording the behaviour of a microscopic system like an atom causes its wavefunction – the quantum mechanical object containing all there is to be known about the atom – to collapse and thus produce a definite result. This collapse is irreversible: it lies outside the framework of the reversible Schrödinger equation.

We have mentioned that Roger Penrose is pursuing one of the 'Holy Grails' of cosmologists: a unification of Einstein's relativity and quantum mechanics in the search for a full quantum theory of gravity. By most theoretical physicists' reckoning, Penrose's approach is highly unconventional, for he recognises the importance of including time-asymmetry as a fundamental feature of the universe. Unification *à la* Penrose would offer several bonuses: it would contain an explicit cosmological arrow of time and so account objectively for irreversible wavefunction collapse;[4] it would eliminate the embarrassing singularities of general relativity; and it would also explain the highly improbable initial conditions of the Big Bang by itself (thus, rather than the initial conditions explaining the arrow of time, the arrow would account for these conditions). This approach, like other tentative descriptions of the birth of the universe from nothing, would perhaps have no need for a deity to start things off. However, a consistent fusion of quantum and relativity theories remains an elusive goal.

In the short run, it is simpler for physicists to dismiss such ideas — there are much easier pickings elsewhere. Indeed, the vast majority of physicists, pragmatists at heart, simply ignore the problems posed by time. They adopt the view expressed by Charles Lamb: 'Nothing troubles me more than time and space; and yet nothing troubles me less, as I never think about them.'[5] In the conclusion to his book, *The World Within the World*, the astronomer John Barrow registers his dissatisfaction with this kind of attitude: 'A common reaction to the problem of the interpretation of quantum mechanics is that of the physicist who says that quantum mechanics works, and that is all that matters. The question of the *meaning* of quantum mechanics is not one that physicists should worry about. However, this is not an attitude that we are happy to adopt elsewhere. If a student comes and asks how to solve a quadratic equation, and says he just wants to know the formula that extracts the solution but he does not want to know why it works or where it comes from, we would take a very dim view of that student. The whole scientific enterprise is based upon rejection of the view that if it "works" then that is good enough.'

Of the few physicists who are interested in the direction of time, we saw that many have tried to overcome the difficulties of the measurement process in quantum theory by asserting that wavefunction collapse does not really occur. They contend that when a measurement is made, it is only our knowledge that changes. The metamorphosis of the wavefunction is not something which describes the real world but rather one within our minds: irreversibility is brought about by our own intervention in proceedings. Thus 'delusion' and 'subjectivity' are watchwords frequently encountered in the limited response to the problem posed by the direction of time. Instead of recognising it as an opportunity to develop new and more profitable ideas, many scientists fall back on the subjectivists' response: irreversibility is an illusion. In a similar fashion, the Second Law of Thermodynamics is regarded as more of an irritant than an inviolable fact of nature. It is introduced and explained away on the basis of the coarse-graining arguments we encountered in Chapter Five, even though such arguments are generally quite incorrect. Another typically circular claim is that the tendency for change expressed by the Second Law is simply a consequence of the fact that the universe started off in a state with a capacity for change.

The 'arrow of time is illusory' argument sets in train a devastating chain of logic. Thermodynamics – and the Second Law in particular – is merely an approximation, a consequence of our own limitations or 'mistakes'. This in turn means that the apparent irreversibility of such manifestly one-way processes as life and death is simply the consequence of our own ignorance and failure to perceive their true time-symmetry. The logical extreme of the subjectivists' theory of knowledge is solipsism – that self-existence is the only certainty. It is a logically unassailable position, and there are a few scientists who adhere to it, presumably as a means of side-stepping the problems of irreversibility and measurement. Solipsism seems unlikely to be true in the face of all the evidence amassed by science in favour of an objective reality 'out there'. Moreover, solipsists usually lack the courage of their philosophical convictions: it is amusing to see that when they have children, solipsists invariably take out life insurance.

In fact, as we have stressed in the last two chapters, the arrow of time is a tool for expanding our knowledge, not a device for concealing ignorance. Patterns we see around us, from the simple geometries we find in pine cones to the complex markings on the coat of a cat, can all be explained in principle by means of equations which have a built-in arrow of time. Far enough away from thermodynamic equilibrium, the guiding principle of irreversibility, as enshrined within the Second Law, leads to the self-organising processes by means of which we can understand these ordered structures of nature. Indeed, without the irreversibility demanded by the Second Law of Thermodynamics we would not expect life to have emerged on Earth, nor the diverse forms of behaviour in space and time which characterise living things. It is only by virtue of irreversible processes that we can understand our own existence.

Perhaps it is not surprising how few scientists have felt the tremors of something revolutionary latent within this irreversibility paradox. Among the academic community, there is ever-increasing pressure to specialise in order to publish, to seek out the trees from the wood, which has led to the exponential growth of the scientific literature and the concomitant shift towards the sacrifice of understanding on the altar of calculation.[6] There is a wider panorama lying undiscovered before us, a luxuriant growth of possibilities to explore. The French mathematician

264

and Field's Medallist René Thom's apologia in his book *Structural Stability and Morphogenesis* made the plea that at a time when so many scholars in the world are calculating, is it not right that some, who can, dream?[7]

A new start on exploring the novel possibilities of time has been made, notably by Prigogine's school in Brussels. His group has asked how we can relate the various meanings of time – time as motion, as in dynamics; time related to irreversibility, as in thermodynamics; time as history, as in biology and sociology. He has written: 'It is evident that this is not an easy matter. Yet, we are living in a single universe. To reach a coherent view of the world of which we are part, we must find some way to pass from one description to another.'[8] We do not have to settle for a stand-off between the irresistible force of the Second Law and the immovability of reversible mechanics. An alternative point of view is possible, one that does not stretch credulity to the limit but instead takes the Second Law to be a foundation, not an approximation. This approach, pioneered by Prigogine and his group, is based on a radical reassessment of the microscopic world engendered by the recent recognition of the ubiquity of dynamical chaos in all but the most idealised situations. Although dynamics and the Second Law can never be reduced one to the other, both seem to be intrinsic elements of nature, in a manner reminiscent of quantum-mechanical wave–particle duality.

Determinism loses its grip

Let us first call on Newton in our search for the arrow of time and adopt his description of the microscopic world, as described in Chapter Two.[9] This, of course, marks an unpromising start for our quest. Newton's laws suggest that with sufficient information it is possible to determine the past and future of any system. For example, with complete knowledge of the positions and velocities of a huge number of molecules jostling around within a container at a particular instant, we could predict and retrodict the whole gamut of their behaviour through time. Moreover, it was shown by Poincaré that the molecules will repeat their motions again and again over huge periods of time. The arrow is lost in the maelstrom of activity within the flask. No one would be able to tell from a ciné film of these molecular motions – if they were discernible – whether the gas had reached equilibrium. But the world is not this simple: the limitations of Newton's dynamics can be turned to advantage in our quest to uncover the direction of time at the microscopic level.

In the early days, once Newton had delivered his equations of motion, people just assumed that any prediction of the movements of an apple or a planet was simply an exercise in applied mathematics. To find out the behaviour of the Moon orbiting the Earth, one would merely insert appropriate numbers into Newton's differential equations – the positions of the Moon and its speed at a given instant – and the answer would fall out after some calculation. The next 200 years were consumed by clever mathematicians and theoretical physicists seeking to find ways of solving these equations exactly when they were applied to describe increasingly complicated situations.

To use Newton's equations to find out the past or future of bouncing billiard balls or mice sliding down frictionless pendula, one had to discover so-called integrals or invariants of the motion. The success or otherwise of this mathematical procedure was thought to rest solely on one's ability to write down the mathematical expressions for colliding molecules or descending rodents. All mechanical descriptions were assumed to be integrable, that is, capable of yielding exact solutions (in terms of integrals of motion). Nobody stopped to think that it might be impossible to solve the equations exactly. Until Henri Poincaré came along.[10]

Poincaré was near-sighted, absent-minded and physically awkward. His contemporaries joked that he was ambidextrous: he performed equally badly with either hand. Indeed, at school he distinguished himself by scoring zero at drawing. However, he more than made up for these shortcomings with his mathematical dexterity. In 1872 his teacher, Elliot à Liard, wrote: 'J'ai dans ma classe à Nancy un monstre de mathématiques, c'est Henri Poincaré.'[11]

This monster published his first paper in *Nouvelles Annales des Mathématiques* the following year, at the age of nineteen. By the time of his death in 1913 he had written more than 30 books and 500 technical papers. He was probably the last universalist in mathematics, jumping from field to field in the lectures he gave at the Sorbonne. Poincaré contributed to optics, electricity, elasticity, thermodynamics, quantum theory, relativity and cosmology. A colleague remarked that he was a 'conqueror not a colonist' because he always brought new ideas to whatever area he investigated.[12] Poincaré was also a great populariser of science. In one of his books, *The Value of Science*, he remarked: 'If nature were not beautiful, it would not be worth knowing, and if nature were not worth knowing, life would not be worth living.'[13] He was honoured by being elected a member of the literary section of the Institut Français.[14]

Scientists were stunned in 1889 when Poincaré showed that, as soon as one tries to analyse the motion of as few as three bodies, such as the Sun, the Earth and the Moon, one is dealing with an intrinsically non-integrable system, a technical way of expressing the depressing fact that an exact solution cannot be found by mathematical analysis. Any motion involving more than three bodies, let alone the millions upon millions of molecules buzzing around in a gas, would be even more difficult to describe. This was the first of several cracks in Newton's crystal ball. These limitations provide a very powerful reminder of the danger of reductionists' attempts to render everything as simple as possible. By concentrating on oversimplified models which yield to the seductive power of mathematics, the whole richness of the real world is in danger of being overlooked; in particular, in peeling off layers of supposedly opaque behaviour to expose the 'fundamental' features beneath, one may be losing the very quintessence of time.

The search for simplicity is a trap which has snared scientists time and again. Newton wrote: 'Nature is pleased with simplicity, and affects not the pomp of superfluous causes.'[15] But the power of his mathematics was rendered impotent by Poincaré for some of the simplest situations imaginable. Poincaré remarked: 'A century ago it was frankly confessed and proclaimed abroad that nature loves simplicity; but nature has proved contrary since then on more than one occasion.'[16] In another context, the great pioneer of thermodynamics, Willard Gibbs, maintained: 'One of the principal objects of theoretical research in any department of knowledge is to find the point of view from which the subject appears in its greatest simplicity.'[17] The shortcomings of his idea of 'simplicity' can be appreciated by contrasting the equilibrium thermodynamics he developed with the more intrinsically complex non-equilibrium variety he neglected, which has so much more to do with reality. In all matters one does well to recall the injunction of the Cambridge philosopher and mathematician Alfred North Whitehead, 'to seek simplicity and to distrust it'.[18]

Ergodicity

Regardless of the impasse to Newton's mechanics erected by Poincaré, the founding fathers of statistical mechanics persevered with their task. Their avowed aim was to express the behaviour of the world in terms of atoms and molecules. Quite separately from Poincaré, Boltzmann and Gibbs devised an elegant way to emphasise the difference between a simple system with predictable behaviour and a complex one more typical of the huge numbers of molecules that make up everyday objects.

To visualise the behaviour of the countless millions of molecules, they depicted it in *phase space*, a pictorial device which plays a central role in all technical discussions on the subject. Portraits of behaviour in phase space can be extremely informative, revealing the underlying motions as surely as brush strokes identify an artist. They can also help to reveal the arrow of time.

To make a phase space 'portrait' of a single billiard ball moving around in a box, we must state its position, which requires three quantities or coordinates (conventionally labelled as x, y and z, representing 'left and right', 'up and down', 'backwards and forwards'), and its velocity, which is also expressed in terms of three components along the same three mutually perpendicular directions. That makes a total of six coordinates to map out on our portrait describing the ball's behaviour; for two balls, twelve such coordinates are required, for three eighteen, and so on. For N balls, there will be 6N position-plus-velocity coordinates: these 6N dimensions make up, in effect, the canvas of our phase space portrait. At any given instant in time, the state of the N balls can be described by a single point on the 6N-dimensional phase space canvas. Although it is impossible to visualise any dimensions above three, intuitively one can at least see that a representation of a million molecules by a single point in 6-million-dimensional phase space may offer advantages over a description of one million points milling around ordinary three-dimensional space. As the balls bounce around according to Newton's equations of motion, the point traces out a path or trajectory in phase space.

Consider the phase space portrait of a perfect pendulum, endlessly swinging to and fro. During each cycle it exchanges its energy between the potential form, at the end of each swing, and the kinetic form, when it reaches maximum velocity. For small swings, the pendulum constitutes an integrable system for which Newton's equations may be solved exactly. The phase space portrait of the pendulum shows a single point tracing out a closed, endlessly repeating loop (*see* Figure 30(a)). Each circuit of the loop corresponds to the pendulum completing a period of oscillation. If we made a ciné film of this idealised example and ran it in both directions, it would be impossible to distinguish the true direction in which the phase space portraits were running — both motions are allowed. Clearly time's arrow has been left out. And the very fact that the trajectory traces out a closed loop is a graphic illustration of Poincaré's cyclic time and the eternal return. Using our analogy, the artist responsible for this portrait can be said to come from the school of predictable integrable systems.

Now let us compare this work with something more complex that arises when depicting the behaviour of molecules in a gas. The millions of molecules are not tethered like the pendulum bob. Instead they are constantly colliding with one another and with the walls of their container. With every collision, the phase space coordinates of the gas undergo rapid and substantial changes. Boltzmann and Gibbs reasoned that given a large enough number of molecules and sufficient time, the entire phase space canvas would be explored. Intuitively it seemed reasonable to expect a single molecule, with the passage of time, to explore the entire container. This behaviour they called a property of *ergodic* systems: an immortal 'ergodic monkey' that freely tries every key of a typewriter would, after an unimaginably long time, type the entire works of William Shakespeare (as well as those of Charles Dickens, Jeffrey Archer and everybody else). If a system such as an isolated container of gas is ergodic, the molecules will explore every point in phase space available to them (*see* Figure 30(b)). Compared with the pendulum, the erratic nature of the trajectory indicates how the gas molecules exhibit random motion.

Boltzmann, Gibbs and Maxwell were of the opinion that the only constraint on the behaviour of the gas was that of its total energy (which remained constant, since they considered isolated systems): 'The only assumption which is necessary for a direct proof of the problem of thermodynamic equilibrium is that the system, if left to itself in the actual state of motion, will sooner or later pass through every phase which is consistent with the equation of energy.'[19] Could it be that these ergodic systems contain the key to understanding the arrow of time?

The strikingly different phase space portraits of a pendulum, which is an integrable system, and a collection of gas molecules, which are supposedly ergodic, furnish an invaluable visual aid for distinguishing between these two behaviours.

During the 1930s, mathematicians such as John von Neumann, George Birkhoff, Eberhard Hopf and P. R. Halmos began to lay down a mathematically rigorous framework for the theoretical treatment of ergodic systems, known as ergodic theory, which has since developed into an entire field of pure mathematics.[20] A group of Soviet mathematicians, originally inspired by Aleksandr Khinchine and more recently led by Andrei Kolmogorov, D. V. Anosov, Vladimir Arnold and Yasha Sinai have come to dominate this field.

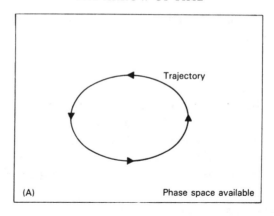

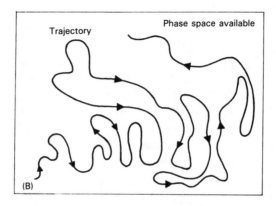

Figure 30. (A) Phase space portrait of a pendulum for small swings, which constitutes an integrable dynamical system. The trajectory is confined to a very small region of phase space. (B) Phase space portrait for a collection of molecules in a gas. Here the trajectory probes every part of phase space – the motion is ergodic.

Their work has revealed that there is a whole hierarchy of behaviours within ergodic systems – some simple, some complex, some paradoxically both complex and simple at the same time – which can likewise be expressed by means of a gallery of phase space portraits. Just as an art historian classifies trends and movements into 'Classical', 'Impressionist', 'Cubist', 'Modernist' and so on, so the classification of these phase space portraits uncovers the true character of dynamical instability – deeply related to chaos – which is central to an understanding of the arrow of time at the atomic and molecular level.

One problem has to be overcome if we are to find that arrow – for even in ergodic systems, there remains the spectre of Poincaré's return to exorcise. There can be no arrow of time for a system locked in a cycle of

270

eternal return. As Poincaré pointed out, there cannot be remorseless entropy increase associated with irreversibility in any dynamical system, no matter how complicated, if it is destined to repeat its behaviour.[21] One word stands between Poincaré's return and a timeless universe: chaos.

The electron at the end of the universe

A sports journalist at a cricket match telephoned a report back to the office which read: 'Smith hit the ball over the slips and into the deep for three.' When his article was published the following day, the account of the incident read: 'Smith hit the ball over the cliffs and into the deep blue sea.'[22] A better known example of the same *genre* is the commanding officer's order, 'Send reinforcements, we're going to advance,' which when whispered along a human chain through the trenches emerged as 'Send three-and-fourpence, we're going to a dance.'

These two anecdotes show how relatively small changes to a sentence can completely alter its meaning. This is analogous to dynamical chaos, where small changes in the initial conditions lead to drastically different time evolution. The term dynamical chaos is used to distinguish it from stochastic chaos, which is truly random behaviour arising from external influences and is not considered in this chapter (the distinction was explored in Chapters Six and Seven). In the context of dissipative systems from which energy may be lost, we have already come across the butterfly effect used by Edward Lorenz to account for the vagaries of the weather. It underlines how far we have come from the dream world of Laplacian determinism.

'The electron at the end of the universe' is a cosmic version of Lorenz's butterfly effect popularised by Mike Berry of the University of Bristol (but now applied to a conservative classical dynamical system for which there is no loss of energy).[23] Imagine trying to follow the behaviour of a single oxygen molecule as it moves frantically around in a small region of a room. This is extraordinarily tricky because of its collisions with the millions upon millions of other molecules it encounters. Each time it hits another molecule, it is deflected through an angle which is calculable in principle. For simplicity, we may pretend that the molecules behave like billiard balls. But the slightest uncertainty in the knowledge of the specified molecule's initial position and velocity causes a rapidly growing uncertainty in its angle of deflection as it collides with the other molecules. For example, instead of hitting another molecule, it may miss altogether, drastically

271

altering its subsequent trajectory. Suppose that, like God, we did know the initial conditions precisely. Even then, if an effect were neglected, as infinitesimal as the gravitational interaction of our oxygen molecule with a single electron sitting at the remote 'edge' of the universe, all hope of predicting where the oxygen molecule is going will have vanished.

We do not need to concern ourselves with electrons at the edge of the universe, nor even with those a few metres away from the laboratory, except in ultra-sensitive work. There is enough to worry about already in anticipating the antics of the oxygen molecule. Mike Berry's story relies for impact on his pretending that at a given moment we already knew exactly the state of the selected oxygen molecule. But as we shall emphasise, even in Newtonian mechanics its state can never be known; within this intrinsic uncertainty, the effect of the electron at the end of the universe is insignificant.

Chaos cracks determinism

When chaos is present there is exquisite sensitivity to the initial conditions. Newton's equations can predict what will happen over short times but not in the long run, unless the initial conditions are known precisely. This aspect of the real world can be used as a weapon to put paid both to determinism and the eternal return implied by Poincaré. A small change in the initial conditions of the gas in our example would mean that molecules which would have collided no longer do so, and *vice versa*. The trajectory of the single point in a phase space portrait representing all those molecules would rapidly change. Thus, although Newton's equations should be able to describe the behaviour of the gas, one would have to know the initial conditions (all those millions of numbers) with infinite precision to make any certain predictions about the future from the deterministic equations of motion. Even in principle this task simply could not be performed by any brain or calculating process of less than infinite capacity. Saying that the search for precision would be infinite means exactly that – it could go on for ever, and still not be over. Determinism can only exist if one enters the realm of religion. For verily, only a being as omniscient as God Himself could hope to handle such a literally limitless amount of information.

The gas is an example of an unstable dynamical system: a small change in the initial conditions leads to a large change in its long-term behaviour. In a similar manner, a pinball machine always gives a quite different game no matter how carefully a pinball wizard attempts to reproduce the push given to the steel ball. A dynamical system is said to be chaotic if it is highly unstable: in such cases, trajectories emanating from neighbouring initial conditions separate extremely quickly (exponentially fast) in time. Yet the problem of describing time evolution goes much deeper than the mere practical difficulty of garnering the necessary numbers to insert as initial conditions into Newton's equations of motion. Suppose that the speed – one of the initial conditions – of a steel ball as it shoots into a pinball machine is given by a single number between zero and one (it could be a number between 1 and 100, or indeed between any pair of numbers, without affecting the argument). This sounds perfectly innocuous but actually hides a fundamental problem. For any number we take to describe that initial condition is very special and atypical. This is because we can only work with what are called rational numbers, precisely defined by the ratio of two integers, while mathematics reveals an abundance of irrational numbers, much more unpleasant quantities described by a strictly infinite sequence of randomly occurring digits. The blow against determinism arises because, although there is an infinite number of rational numbers between zero and one, there are infinitely many more irrational numbers. Thus the rationals themselves, which are all we are able to handle (the irrationals must always be approximated by rational numbers), form a highly abnormal selection.[24] It is infinitely more probable that when the steel ball is set in motion its speed will be given by an irrational number. We can never hope to be able to describe exactly how it rattles through the pinball machine. This feature is a matter of principle not just of practice. Even rational numbers can be very long and may require an infinite number of digits. For example, $\frac{1}{3}$ is 0.333333 recurring *ad infinitum* and to be handled numerically must be rounded off, say to 0.333. But any such rounding-off will soon lead to a completely different behaviour of the steel ball and a different game of pinball from that generated by the 'true' value of the initial condition. We must face up to the fact that there will always be some uncertainty about the starting conditions.

The instability of classical systems governed by the three-centuries-old Newtonian equations of motion has only very recently begun to be appreciated, and the shortcomings of Newtonian determinism taken on board. Sir James Lighthill, a leading authority on the once-traditional subject of fluid dynamics, recently made a moving statement of contrition on behalf of the many workers who have over the centuries vainly attempted to realise the deterministic dream: 'We are all deeply conscious today that the enthusiasm of our forebears for the marvellous achievements of Newtonian mechanics led them to make generalisations in this area of predictability which, indeed, we may have generally tended to believe before 1960, but which we now recognise were false. We collectively wish to apologise for having misled the general educated public by spreading ideas about the determinism of systems satisfying Newton's laws of motion that, after 1960, were to be proved incorrect.'[25] Sir James's own field of expertise, formerly the exclusive preserve of engineers and applied mathematicians, has now been radically transformed into a novel and fertile branch of mathematical physics by the new approach to dynamical systems.

Problems set for students in countless examination questions concerning the application of Newton's equations to the motion of colliding balls and neatly orbiting planets are the exceptions, not the rule. They would be a source of powerful ammunition for those who contend that formal education is divorced from the 'real world'. For although the equations used to tackle these problems are simple, the grand irony of the concept of chaos is that complex behaviour can also sprout from the same source.

With the development of ergodic theory, it became easy to visualise complexity and the arrow of time in phase space (*see* Figure 30). The artistry of chaos can be displayed by phase space portraits. We have seen how the initial conditions, even in a game of pinball, cannot in general be described exactly. This is taken into account on the phase space portraits by using a blob rather than a single point. The blob contains a bundle of possible trajectories (of the gas molecules, or of the steel ball in a pinball machine), all of which are compatible with the uncertainty in the initial conditions: the blob contains the spread of possibilities.[26] What happens when we study the blob's subsequent motion in time? The rules of the game are simple: the blob evolves according to the Liouville equation, which we discussed in Chapter Five in relation to non-equilibrium statistical mechanics; and the volume of the blob (but not necessarily its shape) must be preserved. The reason for the latter rule can be illustrated by thinking of gas in a container. No matter how many ways there are in which the energy can be distributed among all the gas molecules, the probability of finding the gas in the container has to stay the same, that is, be unity. It cannot simply disappear in a puff of smoke.[27] Thus the blob, containing all the possible trajectories of its molecules, can be thought of as a drop of incompressible fluid which always maintains its volume – and thus overall energy – but can vary its shape (in phase space the overall shape reveals the way in which the motion may be carved up among the molecules).

Now let us stroll down the gallery of phase space portraits where a blob is used rather than a point (*see* Figure 31) and see if we can spot

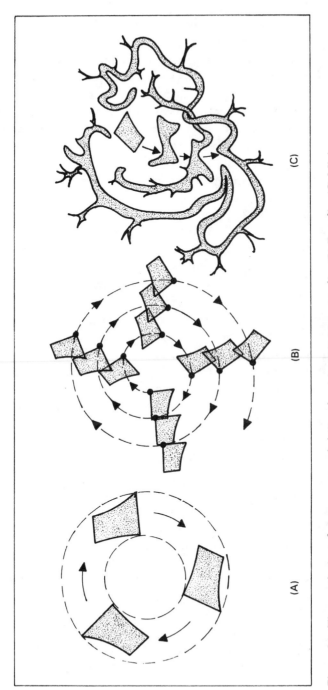

Figure 31. Time evolution of phase space probability densities: (A) Non-ergodic. (B) Ergodic. (C) Mixing. [Adapted from R. Balescu. *Equilibrium and Non-Equilibrium Statistical Mechanics*, p. 718.]

different hands at work. In (a), we see the examiner's delight, the simple and aesthetic lines of an integrable (non-ergodic) system which is predictable and well behaved: the blob moves around in a periodic fashion through only a small part of the available phase space (the 'canvas') and retains its shape. This school of art clearly influenced (b), an ergodic system where again all trajectories within the blob remain close to one another as time passes (if this did not happen the blob would become smeared out), but the blob probes the entire phase space. Portraits (a) and (b) represent stable (or regular, or non-chaotic) dynamics, because the small uncertainty in the initial conditions leads to an equally small uncertainty in the state of the system at a much later time.

Portrait (c) reveals a potentially more interesting situation, as frenzied-seeming as a Jackson Pollock painting: the motion explores the entire phase space, and is thus ergodic. But the blob, whose volume must remain constant, spreads out into ever finer strands or fibres, like a drop of ink spreading in water or a dollop of cream diffusing into coffee. Eventually it invades every region of the canvas, after which no further change can occur. This is very significant, for it shows that, at the level of the probability distribution function (the blob), equilibrium and thus an end state of time evolution, can be reached. Because this portrait has much in common with the way liquids mix by diffusion, it is called a mixing ergodic flow, a kind of time evolution first investigated in mathematical detail by Hopf,[28] although such behaviour was already conceived of by Gibbs.[29]

The effects of chaos can be seen at work in Figure 32 by looking in detail at the 'brush strokes' of the phase portrait, investigating what happens to neighbouring points within the blob of initial conditions.

No matter how close together the points are initially, they always move apart exponentially fast as time goes on. In the very short term, there will be little difference between the behaviour of any such pair of points. But as time passes, the long-term behaviour of one such trajectory in the initial blob becomes utterly different from the other and they sample completely different regions of phase space. This is precisely what we mean by chaotic time evolution, for only if we knew with infinite precision the initial conditions would we be able to make use of Newton's deterministic equations to calculate the future behaviour. In the language of phase space artistry, if we start with a single point we can invoke Newton. But for chaotic systems, the omnipresent uncertainty at time zero means that predictability – the cornerstone of Newtonian physics – crumbles.

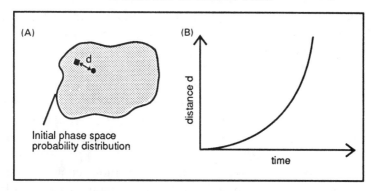

Figure 32. The dynamical instability of mixing flows. (A) The initial phase space probability distribution showing points a distance d apart. No matter how close together these two points are, they diverge exponentially fast as time goes on, as shown in (B). *See* Figure 31 (C) for comparison.
[Adapted from P. V. Coveney. *La Recherche* **20**, 190 (1989).]

Cast-iron determinism must give way to probabilistic statements for these mixing flows. It means that from the outset we must give up deterministic trajectories and work instead with probabilities – precisely the kind of description employed in statistical mechanics. This conclusion has considerable significance for time's arrow, since statistical mechanics was one of the principal routes by which scientists, most notably Boltzmann, have attempted to express the irreversible Second Law of Thermodynamics in terms of molecular motions. And now this conclusion has been reached without any reference to subjective arguments.[30]

This portrait of mixing is only one member of a gallery of chaotic phase space portraits. Mixing flows are intermediate in the hierarchy of increasingly unstable and thus chaotic ergodic dynamical systems found there. Even more chaotic behaviour occurs in the so-called K-flows. The K stands for Kolmogorov, who studied their properties with Sinai. The behaviour of K-flows is at the limit of total unpredictability, even though the 'underlying' equations of motion are again deterministic. K-flows have the remarkable property that even an infinite number of prior measurements cannot serve to predict the outcome of the next one – unless the prior measurements were of infinite accuracy, which is physically impossible. These flows are intrinsically random.[31]

All these kinds of behaviour may sound rather abstract. What do ergodic systems have in common with the real world?[32] As a matter of fact, many people had begun to wonder whether there was any connection at all between reality and the abstract mathematics embedded within ergodic theory, until Sinai's ground-breaking contribution. In 1962, Sinai announced that he had proved that a box filled with two or more billiard balls described by Newton's equations has the property of mixing shown in Figure 31(c). Sinai's result came as another blow against determinism. Although somewhat idealised, the motion of just two balls typifies the kind of behaviour studied in statistical mechanics. Yet until then, people had widely thought that ergodicity (let alone mixing, which is a stronger property) could only be a property of large numbers of atoms or molecules, such as the millions upon millions of molecules in a gas. Sinai showed that if there are more than two balls in the box, the dynamics degenerates still further into that of a K-flow. Thus, even a game of snooker or billiards is actually chaotic and unpredictable. A small error in describing the way in which the cue hits the ball leads to total uncertainty in its position in the long run. Fortunately for snooker *aficionados*, the unpredictability only becomes evident over an interval of time which is greater than the duration of their shots, and could only be realised if the effects of friction were absent.

Since Sinai's pioneering work, a number of other idealised situations, some involving billiard balls bouncing off one another or convex obstacles, have been shown to possess the mixing and indeed the K-flow properties. The drawback from the theoretical point of view is that it is notoriously difficult to establish these ergodic properties, rigorous proofs typically running to scores of pages. The cases for which ergodic properties have been proved are all highly 'singular' in the sense that billiard-ball-type interactions are rather atypical between molecules, because the balls are blithely ignorant of each other's existence until impact. Interactions in the real world are invariably smoother. It is believed by many people, however, that this is only a technical difficulty, and that most real systems are indeed best modelled by ergodic K-flows.[33]

Chaos and the arrow of time

Let us try to develop a firmer understanding of how the phase space portrait gallery fits in with time. We saw how the probability distribution of a chaotic mixing system, for instance a gas in a box, starts out as a blob and sprouts finer and finer tendrils as time goes by. This growth and exploration represents the investigation of the dizzying number of possibilities open to the gas molecules. It is meaningless to envisage a definite trajectory for a single point in phase space which represents the motion of all the molecules in the gas. All we can say is that the gas molecules, having been somewhere in a small volume in phase space (represented by a single point in the phase space) could now be anywhere within one of those strands in the phase space portrait. Enormous choice is available. Herein may lie the microscopic arrow of time, for the inexorable trend is for the blob to smear out to the dead state of equilibrium, when the phase space is replete with fine strands, a cotton wool of uniform probability.

Once we relinquish the deterministic description based on trajectories, we have made a radical theoretical and philosophical reassessment based on the kind of knowledge available to us even for the simplest of situations. In place of a rigid deterministic framework and inflexible predictive power, we are 'reduced' to a statistical level, wherein determinism can find no place, and the future behaviour is unpredictable in anything other than a probabilistic sense.

There is room for irreversibility where before there was none, and in a fully objective way which is intimately connected with the instability of the dynamics, an intrinsic or objective property of a given system. Now we can see that there is a deeper reason for resorting to the use of probabilities and regarding this description as fundamental, even for a handful of billiard balls or a small number of molecules. It rests on intrinsic properties of the system concerned, not on our intervention in the proceedings, the technique used in coarse-graining and other 'subjective' approaches.

The use of probability theory to describe unstable ergodic systems makes many things possible that were thought before to be prohibited — including irreversibility and thus an arrow of time. What then of the apparent spanner thrown into the works by Poincaré's return, which seems to say that there is no escape from the eternal return? Baidyanath Misra, who formulated the quantum Zeno's paradox of Chapter Four, has shown that Poincaré's return is not relevant when one gives up portraits based on points for those arising from blobs, and trades a trajectory-based picture for a probabilistic one.

Crucially, it is only in probabilistic approach, rather than from Newton's deterministic equations, that one can legitimately seek an entropy-like quantity which furnishes the Second Law's arrow of time.[34] In an important paper published in 1978, Misra showed that such an entropy can be constructed for the K-flows.[35] As we have seen, the K-flows include billiard balls and ideal gases, and are believed (but not proved) to be very widespread in nature, particularly where macroscopic systems are concerned. Just like the thermodynamic entropy, Misra's entropy assumes its maximum value at equilibrium, when all evolution of the probability distribution stops.

With this dynamical non-equilibrium entropy we have lifted up Newton's ideas and found that underneath they are compatible with the thermodynamic concept of equilibrium. It is a vital discovery in our quest for the direction of time at the microscopic level. For, as the Second Law indicates, the state of equilibrium amounts to the target for the arrow of time. There still remains the problem of whether this arrow points 'forwards' or 'backwards'. Remember that the classical mechanics upon which this analysis is based is time-symmetric: there are two distinct time evolutions theoretically possible, one evolving to thermodynamic equilibrium in the future, the other towards equilibrium in the past.

The question as to which of these evolutions should be chosen is too deep to be solved by means of our existing first principles alone. Its answer may quite possibly be at root cosmological, perhaps along the lines proposed by Roger Penrose and discussed in Chapter Five when we looked at the cosmological arrow of time. But this is speculation. Instead, we appeal to the Second Law as a phenomenological fact of nature to select evolution towards future equilibrium in accordance with what we observe. By this simple yet enormously significant step, the Second Law's arrow of time is incorporated into the structure of dynamics, which thereby undergoes a quite radical alteration. Such a resolution of the irreversibility paradox, which puts the Second Law and mechanics on an equal footing, is quite different from the unsuccessful route taken by those who wish to see mechanics explain thermodynamics in the time-honoured reductionist spirit.

The approach has received further support from Misra in later work done in collaboration with Ilya Prigogine and Maurice Courbage. They showed that if an entropy-like quantity exists which increases with time, then the reversible trajectories cannot be used.[36] Moreover, they discovered a new formulation of time consistent with irreversibility. This quantity, called the internal time, represents the age of a dynamical system. One can think of the age as reflecting a system's thermodynamic aspect, while the description of the system held in Newton's equations portrays purely dynamical and reversible features.

Thermodynamics and mechanics have been pitted against one another for well over a century, since the advent of the irreversibility paradox. But the Brussels group has established a remarkable relationship between them. In close analogy with the incompatibility of observable quantities in quantum mechanics, as expressed by the Heisenberg uncertainty principle (*see* Chapter Four) and highlighted by Bohr in his interpretation of quantum theory, we find that a complete knowledge of the thermodynamic properties of a system – that is, knowledge of its irreversible age – renders the reversible dynamical description meaningless, whilst complete certainty in the dynamical description similarly disables the thermodynamic view.[37] Prigogine has interpreted this state of affairs according to his view that 'The world is richer than it is possible to express in any single language. Music is not exhausted by its successive stylisations from Bach to Schoenberg. Similarly, we cannot condense into a single description the various aspects of our experience.'[38]

It seems that reversibility and irreversibility in the guise of dynamics and thermodynamics are opposite sides of the same coin. As we found with quantum mechanics, the full structure of the world is richer than our language can express and our brains comprehend. There are two essential yet complementary aspects of this new vision of time which are as striking in contrast as heaven and hell. Heaven is ruled by dynamical equations that are reversible and 'timeless'; their simplicity ensures stability for eternity. Hell is more akin to the real world, where fluctuations, uncertainty and chaos reign. It is a world of instability and decay towards death and equilibrium.

KAM *cometh*

A new version of time expressing the concept of age has thus been discovered within the heart of classical physics; this time is associated with an entropy which increases with time. Thus irreversibility and entropy appear to be basic properties of sufficiently unstable dynamical systems, even comprising as few as two colliding bodies. Have we therefore made a universal link between the thermodynamic arrow of time and the reversible equations which govern the microscopic world? Not quite: after all, we did not consider all possible kinds of motion in our gallery of phase space portraits, but only examples of integrable and ergodic flows. There is also the case of non-integrable systems involving the motion of three or more bodies, discovered by Poincaré and mentioned earlier in this chapter.

The phase space portraits we have so far encountered depict only the extremes of behaviour, contrasting 'simple' and 'complex' dynamical systems. Ergodic systems, particularly mixing or K-flows, are to be regarded as 'complex', for even two or three billiard balls in a box explore all the phase space available, as expressed by the 'hairy' type of phase space portrait. As we have seen, an ideal pendulum or a single planet circling the Sun are examples of simple (integrable) systems, in the sense that their motions are highly restricted, free from collisions, predictable, and hence not ergodic: the monotonous regularity of their behaviour demonstrates this. Phase space portraits of these integrable systems show confined (non-ergodic) trajectories which periodically return to their starting point, biting their own tail as the behaviour is repeated. And, in a qualitative way, this dovetails with our own experience of simple dynamical systems: that bite on the tail corresponds to the Earth starting a new period of motion around the Sun, or the pendulum a new swing.

This may seem puzzling, even though a stable orbit of the Earth is profoundly reassuring. Given what we have learned so far, the motion of the Earth around the Sun would appear to have all the ingredients of a delinquent and chaotic system: it is not really an integrable 'two-body problem' since there is also the gravitational tug of the Moon and other planets in the solar system. In reality, celestial motions are a consequence of many-body effects (more than two bodies are present) and are therefore, according to Poincaré, non-integrable. And it may seem that we have shown that in all but the simplest of situations, chaos should ensue.

Fortunately, we know we can still count on the rising of the Sun each day thanks to something known as the Kolmogorov–Arnold–Moser (or KAM) theorem. This theorem arose from work on Poincaré's non-integrable systems begun by Kolmogorov in 1954 and developed by his colleague Arnold during the early 1960s. Work along similar lines was simultaneously being performed by Moser in Germany and the USA. Their research established that such Poincaré systems represent a kind of intermediate situation between complete regularity and utter chaos.

This twilight zone of behaviour is reached by adding a new ingredient to a simple situation. For instance, in our model of the Earth orbiting the Sun we may introduce a small additional force or forces to take into account the effects of the gravitational pulls of the Moon and the other planets. Originally, people thought that such perturbations would automatically convert the non-ergodic to ergodic behaviour, turning a simple phase space portrait, like the loops described by a pendulum, into a very complex one, for instance the 'hairy' mixing flows we have encountered.

But the KAM theorem shows that this does not always happen: the small extra perturbation has different effects in different parts of the phase space canvas. There are regions of the canvas which display regular features, just as existed prior to the introduction of additional inter-actions, while in other parts of the portrait quite irregular, chaotic behaviour now appears. In the former, the blob of initial conditions remains confined to relatively simple closed loops; in the latter, it sprouts creepers and tendrils. Figure 33 shows a simplified example. The regular regions correspond to stable motion without collisions between bodies; the chaotic ones are where the bodies experience the randomising effects of collisions. In effect, it is as if there were two artists at work on the phase space portrait, each working on different parts of the canvas. However, these regions generally interlock in a very complicated manner.

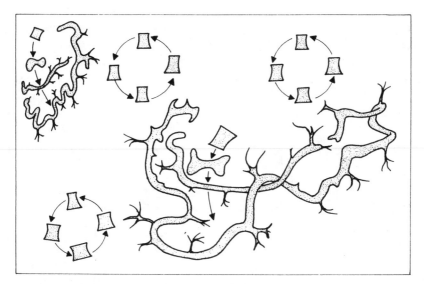

Figure 33. Simplified representation of the phase space behaviour of a non-integrable system. Regions of regular and irregular behaviour co-exist side by side. Compare with Figure 31.

Various factors determine how the canvas is divided into regular and irregular regimes. Suppose we start with a very small extra pertur-bation. As more energy is put in to the system, beyond a certain threshold one observes a so-called stochastic transition from predomin-antly highly regular motion to predominantly random motion. This is because the greater the energy, the more scope there is for collisions to disrupt the peaceful regular motions. Moreover, as the strength of the additional perturbations is increased, beyond a certain point all the dynamical motion becomes chaotic, again for the same reasons.

Within the world of celestial mechanics, astrophysicists received the tidings of the KAM theorem with great joy, for it goes a considerable way to proving the stability of planetary motion, a still unsolved problem. If planetary motion were unstable, we could expect a planet, such as the Earth, to go flying off into the abyss without any notice. Life might never have emerged. Even if it had, the ancients would have had few clues from the chaotic behaviour of the heavens about the 'underlying regularities' of the universe. Without this inspiration, which as we saw in Chapter Two played a profound role in the development of intellectual thought (and reversible time), Newtonian science itself might never have achieved ascendency.

But there *are* irregularities in the solar system just as the KAM theorem predicts would occur in some regions of phase space. Celestial chaos can be found in the motion of Hyperion, a satellite of Saturn which is shaped rather like a potato. Overall, its orbit is regular, but as it moves it tumbles in the same irregular and unpredictable way that a potato does if rolled along the ground. There is similar behaviour in the orbit of Pluto;[39] and gaps in the asteroid belt between Mars and Jupiter have also been explained on the basis of chaotic dynamics.[40]

Figure 34. The challenge posed to the Second Law of Thermodynamics by the KAM theorem.
[Adapted from Figure 12.14 of J. Ford's article in P. C. W. Davies (ed.). *The New Physics*, p. 364.]

For those restlessly searching for an arrow of time within mechanics, the KAM theorem appears unwelcome at first sight. Just as we thought we had managed to ground our description of time on dynamical instability and randomness, the KAM theorem has spoilt it all by showing that complex systems can display simple timeless behaviour in parts of the phase canvas just as surely as the arrow of time will appear in others. However, important work (particularly by

Joseph Ford and his co-workers at the Georgia Institute of Technology in Atlanta) has shown in computer simulations how it may come about that for a 'sufficiently large number of bodies' – and almost certainly for the macroscopic level consisting of millions upon millions of molecules – all hint of regular and periodic behaviour gets drowned out. The KAM theorem is then deprived of relevance – ergodicity, instability and thus irreversibility reign supreme once more. Although it should be stressed that no one has proved these claims in a mathematically rigorous way,[41] time's arrow is thus resurrected through computational number-crunching. This fits in with our own experience, since it is precisely at the level of everyday objects and events, such as melting snowmen, that we are so aware of the arrow of time.

However, there seems to be a cut-off here somewhat like that in coarse-graining, to which we gave short shrift in Chapter Five because it said that the arrow of time would mysteriously appear as we moved up the size-scale from atoms to apples. Does the same arbitrary appearance of the arrow occur when we move from a few atoms to a 'sufficiently large number of bodies'? Statistical mechanics suggests a means of handling this problem using a device called the thermodynamic limit. Rather than imagining an arbitrarily large assembly of molecules, we can consider the properties of a hypothetical system for which the number of molecules (N) and the volume containing them (V) both become infinitely large, but in such a manner that the density or concentration of the molecules (N divided by V) remains finite. The thermodynamic limit avoids the appearance of the arrow at an arbitrary length-scale. It is an essential element in statistical mechanics and is widely used, one other example being to obtain melting points in theoretical calculations of the temperatures at which solids turn into liquids.[42] In the thermodynamic limit, the KAM theorem ceases to be relevant; any 'timeless' regular behaviour is washed out and only the irregular, chaotic motion persists. In this way, the arrow of time can be uncovered.

Relativistic chaos

A little is known about chaos in relativistic mechanics. In Chapter Three we mentioned that in general relativity, motion takes place along geodesics, paths of shortest distance between neighbouring points. The French mathematician Jacques Hadamard showed at the end of the nineteenth century that geodesic motion on a surface of constant negative curvature[43] was highly unstable, and it was later

285

proved to be an ergodic K-flow. It can be shown that in certain cosmological contexts, such geodesic motion does occur and hence enables an internal time and the notion of age to be established. Moreover, it even appears that relativistically invariant fields – of the kind used to describe electromagnetic phenomena – can have K-flow properties and are therefore intrinsically irreversible.[44] But the general problem of establishing the existence of such quantities as a dynamical entropy and age in the context of gravitational interactions is a formidable one, for which no answers are known at present.

Quantum theory and irreversibility

Our attention so far has been confined to describing how irreversibility may step upon the stage if the classical, Newtonian view is taken of events at the microscopic level. Yet contemporary evidence (discussed in Chapter Four) indicates that the correct microscopic description of matter, in spite of its many difficulties, must be couched in the language of quantum theory. Thus we should really seek to base our theory of irreversibility and time on quantum rather than classical theory. However, similarities remain with what we have already discovered in classical mechanics.

In quantum theory, the uncertainty principle guarantees that there is always an intrinsic imprecision in the values of observable quantities (such as position and momentum) embedded within the wavefunction describing the state of a system. Yet the evolution of the wavefunction given by the Schrödinger equation is reversible, just like the trajectories in Newtonian mechanics. The natural question to ask therefore is whether there is such a thing as dynamical instability of the wave-function, analogous to chaos in classical mechanics. This would necessitate the introduction of the quantum mechanical density matrix mentioned in Chapter Five as the primary object in place of the wavefunction, in the same way that the probability distribution function becomes the basic object in the classical mechanics of dynamically unstable systems.

Despite intensive efforts made over several years by many people, there is (as yet) no analogue of chaos known to occur in small quantum systems.[45] This feature is related to the existence of a strong form of Poincaré's return in quantum theory which means that all finite isolated quantum systems, for instance the apparatus including Schrödinger's cat, are periodic – they suffer from the eternal return.[46] The cat would be stuck in limbo as a combination of live-and-dead states for ever. The notion of an irreversible approach to equilibrium appears to be undermined in quantum mechanics. However, we can introduce an entropy into quantum mechanics simply by acknowledging the presence in macroscopic objects of vast numbers of molecules. This may be done by taking the thermodynamic limit in the same way as for large non-integrable Poincaré systems in classical mechanics. It is a powerful approach that works for both disciplines. Of course, as before, irreversibility can only be made to emerge in this way when one considers macroscopic systems rather than microscopic ones.

This conclusion is of significance for one of the major difficulties in quantum theory – the measurement problem discussed in Chapter Four. While the pragmatic application of quantum mechanics is a routine matter, its interpretation remains open to dispute. The principal difficulty concerns the means by which information is extracted from the theory about events occurring in the real world, which necessitate wavefunction collapse during the act of measurement. Quantum theory has been highly successful at the microscopic level, so what is all the fuss about?

John Bell, a theoretical physicist at the European centre for particle physics CERN (Conseil Européen pour la Recherche Nucléaire) and one of the deepest thinkers on the intrinsic problems of quantum mechanics, adopts a 'softly, softly' approach. In his view, 'progress is made in spite of the fundamental obscurity in quantum mechanics. Our theorists stride through that obscurity unimpeded . . . the progress so made is immensely impressive. If it is made by sleepwalkers, is it wise to shout "Wake up"? I am not sure that it is. So I speak now in a very low voice.'[47]

If one likes to live in the microscopic world, then sleepwalking is probably a fine pursuit. However, the conclusion to be drawn from Chapter Four is that quantum theory is flawed at the macroscopic level when it comes to describing our own world. The difficulties are illustrated by quantum paradoxes such as Schrödinger's cat, in which 'quantum reality' has to be described by in-limbo states such as 'live-and-dead cat'. Bell states the situation succinctly: 'How exactly is the world to be divided into speakable apparatus that we can talk about and unspeakable quantum system that we can not talk about? The mathematics of the ordinary theory requires such a division, but says nothing about how it is to be made.'[48]

He continues: 'Was the world wavefunction waiting to jump [collapse] for thousands of millions of years until a single-celled living creature appeared? Or did it have to wait a little longer for some more

287

highly qualified measurer – with a Ph.D.? If the theory is to apply to anything but idealised laboratory operations, are we not obliged to admit that more or less "measurement-like" processes are going on more or less all the time more or less everywhere? Is there ever then a moment when there is no jumping and the Schrödinger equation applies?'[49]

Conventional wisdom requires that when a measurement is made the wavefunction will collapse irreversibly in a completely unspecified and random way – after all, it lies outside the scope of Schrödinger's equation. But as Niels Bohr and his student Leon Rosenfeld repeatedly emphasised, the measurement process occurs at the 'interface' between the macroscopic and microscopic worlds, by means of a macroscopic measuring apparatus. Some insight into this difficulty can be gained from the introduction of entropy and irreversibility into a quantum theoretical framework along the lines mentioned above. Thus, if an entropy-like quantity can be found, the reversible wavefunction can be discarded in favour of a probabilistic approach, just as Newton's equations needed to be replaced by the Liouville equation.[50] As soon as irreversibility is admitted, wavefunction collapse ceases to be the great mystery it was before. From this viewpoint, the measurement process has nothing particularly remarkable about it at all – it is just a typical example of an irreversible process in keeping with the Second Law of Thermodynamics.

The introduction into quantum mechanics of intrinsic irreversibility has considerable appeal, for it incorporates the Second Law of Thermodynamics *ab initio* and explains the measurement process. However, in our view it remains only a partial theory, because reversible quantum laws and irreversible thermodynamics are still conjoined in an *ad hoc* way. We would concur with Bell in his judgement that further fundamental steps have still to be made: 'The new way of seeing things will involve an imaginative leap that will astonish us. In any case, it seems that the quantum mechanical description will be superseded. In this it is like all theories made by man. But to an unusual extent its ultimate fate is apparent in its internal structure. It carries in itself the seeds of its own destruction.'[51] Perhaps Roger Penrose's speculation about the time-asymmetric nature of a satisfactory future theory of quantum gravity may provide the 'radical conceptual renewal' required.[52] If successful, we could expect such a theory to spirit away the spacetime singularities of general relativity, account for the Second Law of Thermodynamics and explain wavefunction collapse. But this is work for the future.

Ludwig Boltzmann.

James Clerk Maxwell.

Albert Einstein.

Max Planck.

Nicolas Leonard Sadi Carnot.

William Thomson, Lord Kelvin.

J. Willard Gibbs.

Lars Onsager.

Ilya Prigogine.

Rudolf Clausius.

James Prescott Joule.

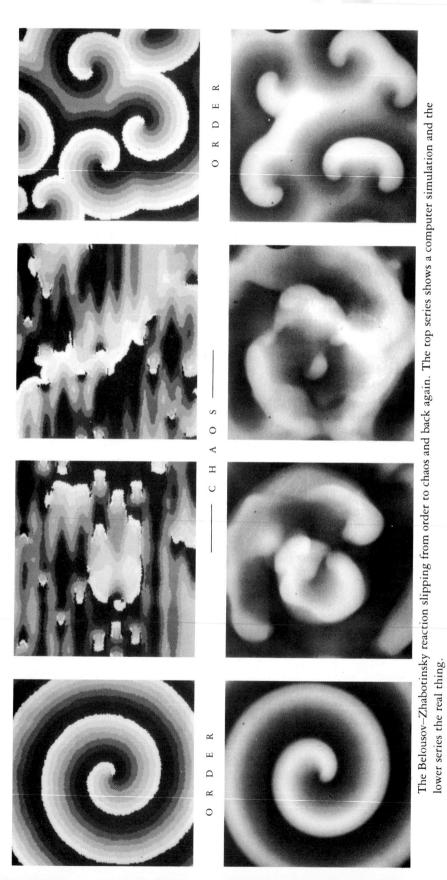

The Belousov–Zhabotinsky reaction slipping from order to chaos and back again. The top series shows a computer simulation and the lower series the real thing.

(Below) A fractal object. Close-ups reveal the same shape, no matter the magnification.

O R D E R

━━━ C H A O S ━━━

O R D E R

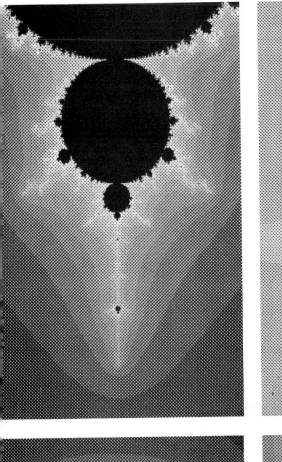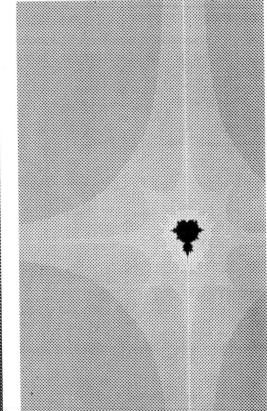

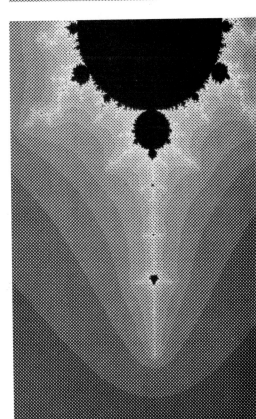

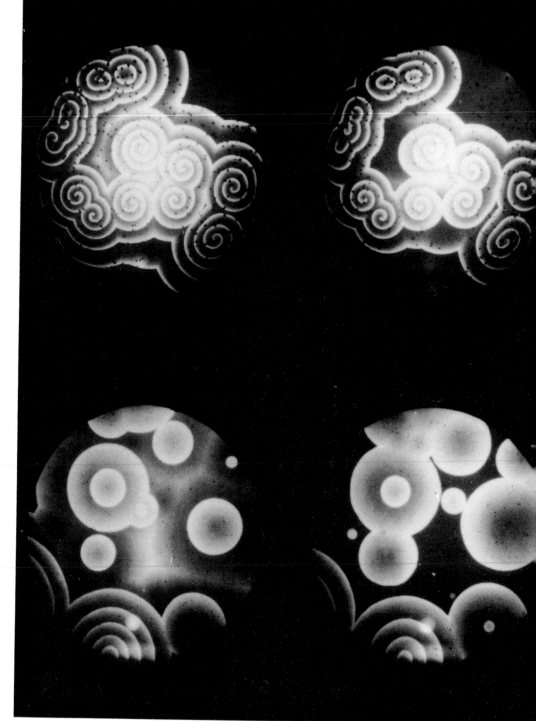

Spiral patterns formed by the Belousov–Zhabotinsky reaction.

Aggregation of the slime mould, *Dictyostelium discoideum*.

Time and leopard spots

What does all this mean for those equations we used in previous chapters to describe how leopards get their spots and slime moulds aggregate? What is the dynamical origin of these non-linear 'kinetic' equations which contain the arrow of time?

In physics, such equations – which include the famous Boltzmann equation mentioned in Chapter Five – have been successfully employed to describe transport processes such as viscosity, diffusion and thermal conductivity. Underlying the familiar irreversible process of diffusion is the flow of matter from regions of higher concentration to those of lower concentration. Similarly, viscosity results from a sort of fluid friction, the process by means of which ordered mechanical energy within the flowing fluid is dissipated as heat, corresponding to randomised molecular motion.

Measurable quantities, such as the viscosity or thermal conductivity of a fluid, are regarded by most people as being 'objective' properties of matter. But according to those who adhere above all else to atomistic reductionism, these everyday properties are under threat of being cast off as 'illusions'. Can properties such as viscosity really be mere figments of our over-weening imaginations in a fundamentally timeless world? Are we still obliged to invoke subjective elements in order to explain the world we see around us? Fortunately, one has reason to believe that the answer to both these questions is no. For there is a rather general way of deriving these kinetic equations from reversible mechanics, in a manner which seems to be deeply connected with the origins of irreversibility. The general approach is beset by mathematical difficulties but nevertheless suggests a remarkable clarification of the problem of the arrow of time. In particular, Claude George and Françoise Henin, working in close collaboration with Prigogine, showed during the late 1960s and early 1970s that the time evolution of a large dissipative system, formulated in terms of probabilities, divides uniquely into two totally independent components, which they called sub-dynamics.[53]

There is a 'kinetic' part which describes how the system evolves over long intervals of time, and contains the approach to thermodynamic equilibrium, while the remaining 'non-kinetic' part describes the transient behaviour starting from the initial conditions, but which vanishes during the evolution. The precise conditions that govern this division between long- and short-term behaviour remain unclear, that is at what point the molecules 'forget' the initial conditions and head for equilibrium under the influence of the arrow of time. The problem is still under active investigation.[54] It certainly requires a high degree of dynamical instability (chaos). One thing is sure: if and when such behaviour can be derived from microscopic dynamics, the kinetic component turns out to obey a very general time-asymmetric kinetic equation. In this way, we have managed to uncover the arrow of time in statistical mechanics.

There is fertility in the sub-dynamics approach, since it also enables old and new kinetic equations to be derived in a systematic manner to describe a wide range of phenomena. Radu Balescu, J. H. Misguich of the CEA-Euratom Association at Fontenay-aux-Roses, south of Paris, Vladimir Škarka of the Institute of Physics in Belgrade and one of us (PVC), have extended the analysis to study the evolution of systems subjected to external fields that vary in time, such as the electromagnetic fields generated by lasers. The latter is a problem of interest in the quest for controlled fusion power, in which plasmas consisting of ionised atoms are trapped in magnetic fields in the hope that positively charged atomic nuclei can be squashed together to release vast amounts of nuclear energy. The time evolution of such plasmas can be accurately described by non-linear kinetic equations which may be derived by the sub-dynamics method. Moreover, it is possible to use the same technique to obtain solutions of these equations describing how specific kinds of plasmas evolve in spatially and temporally varying electromagnetic fields.[55]

It would be wrong to give the impression that the Brussels group are alone in their efforts to avoid subjective ways of deriving irreversible kinetic equations from mechanics. Other approaches have been attempted along different, but still microscopic and objective lines. Oscar E. Lanford III, a mathematician at the University of California, Berkeley, made the most mathematically rigorous derivation of the Boltzmann equation to date in 1975.[56] However, he was only able to demonstrate its validity for very short times, whereas we would expect it to be most appropriate for describing the long-term behaviour of dilute gases. Methods based on 'scaling' techniques have been developed by several others, among whom we might mention Herbert Spohn of the University of Munich.[57] But these methods suffer from similar limitations to those of Lanford. While such approaches have the advantage of greater mathematical rigour, they lack the scope and philosophical content of those pioneered by the Brussels group.

Entropy and the creation

Irreversibility and entropy also seem to be unavoidably linked with the greatest happening of all, the birth of time and of the universe itself.

290

This event to begin all events – the process of creating a universe from nothing – is irreversible, being inextricably bound up with the production of entropy. We find this entropy everywhere: as we mentioned in Chapter Three, the ubiquitous microwave black-body radiation background, which was discovered by Penzias and Wilson in 1965, is believed to be a relic from the Big Bang. This black-body radiation has associated with it an abundant supply of entropy, comprised of a thin gruel of low-energy photons which pervades the universe.

Unfortunately, it is impossible to link this entropy with Einstein's relativistic equations describing the large-scale structure of the universe. As these equations can only describe reversible processes,[58] there is no way to provide an explanation whence this entropy came, save by the ill-defined coarse-graining procedure we have already encountered. If we reject this approach on the grounds of its *ad hoc*, not to say subjective nature, the question remains: how did the entropy arise?

An intriguing, albeit speculative resolution to this problem was proposed in 1987 by Edgar Günzig, Jean Géhéniau and Ilya Prigogine.[59] They built on the ideas of several other cosmologists including, in addition to Edgar Günzig, Robert Brout and François Englert, whose ideas concerning the creation of the universe *ex nihilo* we discussed in Chapter Four.

Once upon a time in their view, it will be recalled, spacetime was empty and flat – there was no mass in the universe to warp and bend it, in line with Einstein's interpretation of gravity. Heisenberg's uncertainty principle allowed energy to be lent for nothing over a short period of time to create the universe. According to Einstein's mass–energy relationship, this energy produces matter (in the form of black holes) which causes curvature of spacetime, better known to us as gravity. From nothing we get a very substantial something; the overall energy cost of creating the universe is nevertheless zero, since the energy of all the gravitational forces in the universe is negative and exactly cancels the (positive) energy of the mass produced.

We saw earlier in this chapter that the idea of chaos and dynamical instability is consistent with thermodynamic irreversibility when we were attempting to express the Second Law in the language of atoms and molecules. Here it is proposed that the instability and unpredictability of the flat ('Minkowski') spacetime quantum vacuum leads to irreversibility by seeding the very birth of a matter-filled universe.[60] The process of matter formation is thus taken to be irreversible on a cosmological scale, and it is this primeval process which produces the entropy in

291

black-body radiation. The birth of time then becomes an inevitably one-way process. It is the ultimate manifestation of time's arrow.

After its creation, the universe is considered to undergo a period of inflationary expansion (during which the black holes evaporate) until it switches over to a universe composed of a mixture of matter and radiation of the kind familiar today.[61] An interesting feature of this model is that once again we can find the two facets of time: irreversibility and repetition. For if the universe were open, in the sense that there were insufficient matter to drag it towards a Big Crunch, then, as it continued to expand, it would lead in the long run to a universe in which matter becomes extremely dilute. This state of affairs would correspond to a flat spacetime, so that the whole show would be repeated but on an enormously greater scale.[62] It seems that the dual features of time which we have emphasised throughout this book may exist at the most breathtaking level of all.

The unended quest

> Have we reached some unity of knowledge or is science broken into various parts based on contradictory premises? Such questions will lead us to a deeper understanding of the role of time.
>
> Ilya Prigogine
> *From Being to Becoming*

A NUMBER of problems remain in our attempt to establish the arrow of time. Many depend for their solution on highly involved mathematics but there is also room for bold insight and intuition, which may lead to new concepts barely conceived of today.

We have been cursed for more than a hundred years by the irreversibility paradox: the failure of the so-called 'fundamental' theories to distinguish between past and future. To be sure, these reversible theories have given us powerful insights into the world but in that very same world, death and decay remind us of the importance of the irreversible and of time's arrow.

The birth and expansion of the universe also bear testimony to time's arrow. So do the curious decays of the long-lived kaon; the predilection for light waves to spread into the future, not the past; the tendency of things to mix, cool, and decay; and the asymmetry of diversity in the evolutionary tree. Only in an irreversible world are cause and effect distinct, so that a logical narrative of events can be laid down.

Irreversibility permits a spectrum of exciting new possibilities, for it is also crucial to creation and life. Instead of trying to minimise its role, as many have done, we now know how to exploit it. Irreversibility can help explain much of the varied patterning we see in time and space. This creativity is what lies behind such contemporary watchwords as 'dissipative structures', 'self-organisation', 'chaos', 'complexity', and so on. From today's standpoint, it is impossible to relegate irreversibility to the realms of the illusory, since its genesis appears to lie within the dynamical instabilities present around us in nature.

As we have seen, the key to understanding all this complex behaviour is the branch of mathematics that deals with non-linear dynamics. Its applications today extend well beyond the province of the 'hard' sciences. Without doubt, economies must also have chaotic aspects

293

(although they are driven by random external elements as well).[1] It is no wonder that economic forecasting appears to be more a black art than a science.[2]

Human and animal societies also form patterns on the arrow of time, from mass migrations to crowd behaviour at football matches. Societies can be regarded as open and highly non-linear dynamical systems, in which feedback loops and competition abound. Scientists have begun to draw parallels between the self-organisation and chaos which can be seen in, for example, chemical reactions, and the phenomena which develop in human and animal societies, typified by words like 'revolution', 'riot' and economic 'crash'.[3]

It is impossible to forecast where this novel approach might lead, but such interdisciplinary cross-fertilisation between physical and social scientists is likely to benefit all sides of the exchange.[4] On the one hand, physical scientists will learn more respect for the intrinsic complexity of the natural world; while on the other hand, social scientists (and even politicians?) will come to appreciate the benefits of a mathematical approach over vague, hand-waving explanations.

Some scientists believe that the simple admission of time's arrow 'opens up new perspectives on many economic processes'.[5] Others maintain that the evolutionary laws found in non-equilibrium thermo-dynamics help 'grasp large-scale patterns in the functioning and devel-opment of human societies'.[6] Bruce West of the La Jolla Institute and Jonas Salk of the Salk Institute in California are persuaded that by fusing the reductionist approach with the new integrative approach, 'we will come to recognise the existence of the laws of nature that determine human conduct and human evolution. When we do so we will then develop the capacity to better understand human nature.'[7]

What has emerged from non-linear dynamics and non-equilibrium thermodynamics has already shaken twentieth-century science and offers a sophisticated reassessment of time. Non-linear equations show us that thermodynamics can account for both linear and cyclical time. While the arrow of time points unequivocally towards thermodynamic equilibrium, the very process of sweeping towards that goal can spawn repetitive behaviour, be it in the colour changes of a chemical clock, the ripples of chemical messages sent out by a slime mould, or the beat of a human heart. We have also discovered that these non-linear equations contain the recipe for both order and chaos. In the chemical clock, both regular and irregular colour changes could arise from the same set of mathematical expressions: chaos is nothing but a delinquent form of self-organisation.

Chemical clocks are not just idealised and colourful laboratory experiments. They throw light on the processes by which natural systems grow and develop. Our cells contain chemical clocks which gently turn to regulate a range of vital cycles on the linear track that runs from conception and birth to decay and death. The branch of mathematics that describes these clocks contains more than the arrow of time – it contains the recipe for life.

We have reached a turning point. Instead of starting from the abstract world of atoms and molecules and trying to derive laws which rule the macroscopic world of snowmen and bulls in china shops, why not do the opposite? Why not believe what our senses tell us and what poets have described for millennia? The fundamental reality in our world is not to be found in its smallest parts but in the capacity of vast assemblies of atoms and molecules for change, as expressed in the Second Law of Thermodynamics.

This is a daring imaginative leap, but we saw good grounds for taking it when we looked at phase space portraits. These showed us how the working of simple 'reversible' mechanics produced chaotic and irreversible behaviour. While the phase space portrait of a swinging pendulum is a clearly defined dot trapped in an orbit, which gives no clue to the passage of time, we get quite different results as soon as we consider marginally more complicated systems. With just three bouncing billiard balls, the dynamical behaviour is so unstable that it is impossible to predict their behaviour. The phase space portrait becomes a blob of possibilities, sprouting tendrils which creep and spread, exploring myriad opportunities. The passage of time corresponds to this increasing hairiness.

The onset of chaos, and thus the arrow of time, is overwhelmingly evident when we deal with the world around us, where objects contain countless atoms. We are unable to predict the behaviour of a gas molecule with any certainty after only a few collisions. For more massive objects, like the trajectory of a planet orbiting the Sun, in some aspects we can predict behaviour for much longer, even for millions of years. But even that will eventually become unpredictable, and we will only be able to predict gross features of the universe, such as temperature and density.

These new ideas about complexity would seem to turn conventional science upside down. The Second Law of Thermodynamics, which describes the behaviour of vast numbers of molecules, is one element of the fundamental truth. Reductionism, with its attempts to explain the world in terms of the behaviour of its microscopic components alone, is invalid. Rather than maintaining that the arrow of time is an illusion,

295

we must ask whether it is the time-symmetric 'fundamental' laws that are approximations or illusions. After all, it seems that they only apply to extraordinarily simple systems.

We are no longer faced with the possibility that time's arrow appears arbitrarily at some given scale of length. Instead, we can see how it reveals itself as complexity develops. Only its role at the idealised level of absolute simplicity remains a mystery.

What will happen next? It seems that we are groping towards a new vision. For, as Prigogine has put it: 'Irreversibility corresponds not to some supplementary approximation introduced into the laws of dynamics but to an embedding of dynamics within a vaster formalism.'[8] It should never be forgotten that the microscopic quantum world as we know it is riddled with ambiguities — typified by the measurement problem — requiring reference to the *a priori* existence of the macroscopic level to make a modicum of sense. And we have mentioned that a more general theory embracing both quantum mechanics and general relativity may well be intrinsically time-asymmetric. Far from being on the verge of a complete understanding of nature, as some theoretical physicists would have us believe, we may be on the threshold of a radically new framework in which time occupies a central rather than a marginal role.

At the present time, most scientists advocate the reductionist view. But our journey through science has shown that we need to consider the universe in its totality and so adopt a more holistic outlook. As the writer Alvin Toffler commented: 'One of the most highly developed skills in contemporary Western civilisation is dissection: the split-up of problems into their smallest possible components. We are good at it. So good, we often forget to put the pieces back together again.'[9] A new theory is needed, one that will provide a deeper understanding of time. From the arguments presented throughout this book, we can see clearly what that new theory should be able to do. It would banish the singularities that undermine Einstein's relativity. It would end the argument over what the act of measurement means in quantum theory. It would definitively outlaw time travel, auto-infanticide and white holes (the time reverse of black holes), while rendering nugatory recourse to arguments based on subjectivity and illusion.

The inescapable conclusion we have reached is that the traditional methods of the physicist, induced by an undue emphasis on very simple or idealised models, are too narrow to make sense even of everyday phenomena. We must recognise the intrinsic complexity of reality and accept a radical reconceptualisation.

Today's endeavours reach far beyond Boltzmann's original programme; yet his profound insight remains with us. We inhabit a world in which the future promises endless possibilities and the past lies irretrievably behind us. The arrow of time is essential for preserving the integrity of science. It is the medium of creativity in terms of which life can be understood. Only through the recognition of these facts can we begin to make an intellectual *rapprochement* between our human and scientific experiences.

Biological rhythms

Every bodily process is pulsing to its own beat within the overall beats of the solar system.

Michael Young
The Metronomic Society

T H E beat of biology has been clear to Man for hundreds of thousands of years. More than three hundred years before the birth of Christ, Aristotle noted the swelling of the ovaries of sea urchins at full Moon. Hippocrates observed daily fluctuations in the symptoms of some of his patients and thought that regularity was a sign of good health. Herophilus of Alexandria recorded daily changes in pulse rate.[1] Cicero mentioned that the flesh of oysters waxed and waned with the Moon, an observation confirmed later by Pliny.[2] However, the recognition that this behaviour was driven by an internal process rather than being a direct response to sunlight, darkness and other environmental factors, did not come until many centuries later. Today many biologists regard these beats and rhythms as being the most conspicuous features of natural ecosystems.[3]

In 1729 the first evidence of a natural 'clock' was reported in a paper that appeared in the *Comptes Rendus de l'Académie Royale des Sciences de Paris*. It described a landmark experiment conducted by the French astronomer Jean de Mairan on *Mimosa pudica*. During the plant's regular daily cycles, its leaves and stems close and contract when darkness approaches. But to his surprise, de Mairan found that the cycles of movement continued when he placed the plants in darkness. Other scientists described related phenomena, including Darwin, who discussed them in his book *The Power of Movement in Plants* in 1880. But it took until the middle of the twentieth century for the subject of chronobiology – the study of the biology of time – to flourish.

Many examples of 24-hour-long rhythms have been discovered, although they are very rarely of exactly 24-hour duration. These *circadian rhythms* – a term first coined by Franz Halberg of the University of Minnesota in 1959 (from the Latin *circa* meaning 'approximately' and *dies* meaning 'day') – are said to be endogenous; that is, they are generated internally by organisms rather than being driven externally by the cycle of day and night, as might at first be thought. However, as

298

we shall see, these clocks are synchronised with the environment and had plenty of time to cope with the changes wrought by the gradual slowing of the Earth's rotation.[4]

Circadian rhythms abound in nature. In protozoa and algae there are 24-hour periods regulating photosynthesis, cell division, movement and even luminescence, where, for example, a sea teeming with the single-celled green organism *Gonyaulax polyedra* produces a shimmering bluish light. In some fungi, there is a fixed daily hour for the discharge of spores; other plants possess diurnal rhythms governing the movement of leaves and the opening and closing of flowers, indeed to such an extent that in 1751 the great Swedish naturalist Carl Linnaeus (who pioneered taxonomy, the classification of plants and animals) proposed a floral clock. The various blooms revealed the time of day in colour – from the spotted cat's ear at 6 a.m., through the passion flower at midday, to the evening primrose at 6 p.m.

Firm proof of a biological clock came in the 1950s as a result of studies carried out by Gustav Kramer at the Max Planck Institute for Marine Biology in Wilhelmshaven. He was interested in how birds navigate and found that they rely on the position of the Sun. But its position alone is not enough. Kramer showed that birds compensate for the Sun's movement by using an internal body clock rather as a ship's navigator uses a chronometer with a sextant: although the Sun arcs across the sky, the bird's clock compensates for its changing position so as to maintain its intended flight path.

Navigation by the Sun is also used by the wolf spider and some varieties of frogs. Here an internal clock is essential to compensate for the movement of the Sun during the day. Bees also use their clock in order to help follow a 'bee line' back to the hive and to anticipate when plants will release pollen, which may only occur at certain times of the day. This time sense was first recorded by the Swiss naturalist August Forel in 1910 who noticed while breakfasting on the veranda of his chalet in the Alps that bees came to feast on his marmalade at the same time each day, even when the sticky jam had been put away.[5] The existence of the bee's internal clock was confirmed in 1955 when 40 bees trained to forage in Paris at a given time of day were flown to New York, where they continued their French rhythm on arrival. It must have been the first case of bee jet-lag.[6]

This 24-hour cycle is only one 'note' of many that can be played by such biological orchestras. One influential rhythm that strikes out a slower beat arises from that other prominent object in the sky, the Moon. The combined movements of the Moon and the Earth and of both around the Sun produce the complex rhythms of moonlight and tide. Body clocks can strike out a lunar rhythm: the human female menstrual cycle has roughly the same period as the lunar month, 29.6 days; the emergence of the mayfly on Lake Victoria is synchronised by the full Moon and occurs at about the same time as the efforts of the ant-lion reach a peak: these insect larvae dig pits and lie in wait at the bottom, jaws uppermost, for a passing ant to drop in.

Twice every lunar day (24.8 hours) tidal cycles tend to occur, bringing with them variations of temperature, pressure, wave disturbance and food supply. Some creatures are able to predict the time of an outgoing tide in order to avoid the danger of drying out: the green flatworm wriggles to the surface of the sand at high tide and then buries itself again as it dries. The rhythm continues even in an aquarium, confirming that it is driven by an internal (endogenous) clock.

Lower 'notes' – longer cycles – are mostly designed to anticipate the fluctuations in weather that accompany the change in seasons caused because of the tilt in the Earth's axis of rotation. The North Pole is tilted towards the Sun from March to September, so the northern hemisphere bathes in more sunlight per day than the southern, while from September to March the southern hemisphere receives more. It is generally believed that long-lived species do possess an endogenous annual clock, usually 'set' by changes in day length. Only in a few cases, such as the hibernation of ground squirrels and the reproduction of European starlings and cave crayfish, has evidence of a true endogenous circannual clock been found which is independent of light and temperature.[7] Some animals use the effects of the seasonal changes in day length to register the march of the seasons rather than employing another 'annual clock' to keep track of these longer period cycles – the annual cycle can be compressed in experiments by changing the effective day length in an artificial environment. The colour and thickness of pelts, seed germination, testis growth in warblers, body weight changes, oestrus cycles and antler growth all display this annual seasonal cycle.[8]

There are even lengthier rhythms. One of the strangest of all is struck out by periodic cicadas, little insects that chirp incessantly. Some remain underground as nymphs for as long as seventeen years.[9] In Ohio in the United States, thousands emerge at exactly the same moment at a given time seventeen years after going to ground; they crawl out of their pupal skins, climb the trees, live for a further few hours to mate, then all drop dead and the cycle begins again.

In the struggle for survival, these in-built clocks confer advantages on living organisms, enabling them to avoid predators or competition and to predict major changes in the environment. The ebbs and flows of tides, the rising of the Sun, and the passing of the seasons may all demand a response in the behaviour of organisms if they are to survive the rigours of natural selection. The 'timetables' that govern feeding, mating, and the birth of young support this view. Moreover, it seems that major advantages accrue to any creature which can anticipate changes in its environment so that it can be prepared, for instance, for arousal or sleep. This is particularly useful if the creature is cut off from the environment during its daily routine, for instance if it enjoys scurrying around deep in a hole. But no really clear reasons for the evolution of circadian clocks have been discovered; even one of the leading experts on biological time, Winfree, believes that it is 'a substantial mystery'.[10]

In search of the biological clocks

The biochemical mechanisms of circadian clocks remain unknown, although important clues are emerging. Experiments show that they are controlled by different organs and diverse biochemical pathways, depending on the organism, from fruitfly to Man. On the other hand, there are common features regardless of organism: the period of around twenty-four hours persists down to the smallest biological unit – the cell; the artificial injection of most chemicals has no effect on the rhythm, apart from some notable exceptions which include inhibitors of protein synthesis, hormones like melatonin, and substances that affect the structure of cell membranes. Finally, many circadian rhythms are sensitive to light of an appropriate colour; Winfree has found that in two species, continuous light suppresses rhythms even at intensities well below that of a full Moon.[11]

There is evidence that circadian clocks in more complex organisms consist not of a single clock but many, spread through organs and tissues. Jon Jacklet, of the University at Albany, New York, and co-workers showed, for example, that each eye of a mollusc – *Aplysia* – has its own circadian rhythm. Components of dissected fungi show similar rhythms, as do individual leaves and flower petals in plants. Circadian rhythms of photosynthetic activity have even been reported in the cytoplasm – the cell contents – of a single-celled alga called *Acetabularia mediterranea*.[12]

Different circadian rhythms can beat within a single organism. Individual leaves and petals follow different rhythmic timetables on the same plant, while the mollusc's oscillators appear to be quite independent. Nevertheless, there are other cases where the individual oscillators *do* interact in order to stay in synchrony. This is true, for example, within *Acetabularia*, even though the isolated fragments beat separately. Earlier we encountered the tick of the sugar clock. Interactions between yeast cells ensure that their individual sugar clocks tick in unison. Chemicals passed through seawater synchronise a soup of *Gonyaulax* mentioned previously in this appendix.[13] Individual cells of the heart's pacemaker achieve the same end by means of electrical interactions, while Southeast Asian fire-flies take account of the flashes of their neighbours to wink in unison.

Setting the body clock

Our dominant rhythm does not tap out an exact 24-hour cycle but actually a 25-hour one. This innate tendency to let schedules drift later and later is restrained each day as our circadian body clock is reset to the right time by the cycle of day and night through a process known as entrainment. This type of timekeeping – the equivalent of continually readjusting a cheap watch – is more robust than if nature attempted to evolve the equivalent of an atomic clock: over a lifetime, the tiniest inaccuracies would wreak havoc; the cumulative slip in the biological clock would turn the biological cycle upside down from time to time, as occurs in some blind people.

The role of this natural cue to reset the human body clock was dramatically demonstrated in an experiment carried out by a Frenchman, Michel Siffre, in 1972. A fanatical cave explorer, he spent seven months living alone in the Midnight Cave near Del Rio in Texas. As expected, time passed for him in alternating periods of activity and sleep. However, without external cues to entrain his body clock it ran freely, gradually falling out of step with the passage of time on the Earth's surface, cycling instead every 25 hours.

The internal body clocks of plants and animals vary over periods of between 22 and 28 hours, these extreme values occurring in plants: spore discharge of one alga, *Oedogonium*, shows a 22-hour period while the daily leaf movements in the bean *Phaseolus coccineus* run up to 28 hours. Typical of circadian rhythms, the mechanism that determines these periods is cunningly designed by nature to be independent of temperature, unlike most other physiological processes which can double their rate with a 10 degree Celsius rise in temperature. This ensures that biological timepieces remain reliable under a wide range of climatic conditions.

302

Daylight and darkness are the most powerful cues used to reset body clocks. The German physiologist Jürgen Aschoff coined the word *Zeitgeber* as the term for a cue or time setter. 'The' human body clock is also ruled to a large extent by light. It is possible that Man may have more than one *Zeitgeber*, although none has been definitely established.

Michel Siffre is only one of many people who have experienced the effects of a free-running body clock by being placed in an environment stripped of every possible *Zeitgeber*. Bunkers in the Max Planck Institute for Behavioural Physiology near Munich, and at Manchester University; caves in the Swiss Alps; special flats at Stanford University and at New York's Laboratory of Human Chronophysiology: all have been used for human experiments. Variations in light, television, radios, watches and even social cues like 'Good morning' and 'Good night' are forbidden. Astonishingly, some people are happy to stay in these isolated or artificial environments, subjected to 'timeless' conditions, for days or months under constant medical supervision, performing frequent tests, and suffering such delights as rectal temperature probes. One research worker, Richard Coleman of Stanford University Medical School, wrote: 'As a psychologist, one of my tasks was to find someone crazy enough to live in this environment who could still be considered normal.'[14]

These and other studies have yielded a wealth of information on the rhythms that rock and rule the body.[15] Franz Halberg has used computer analysis to tease out some of the details of body rhythms and has found that they cover a number of tempi. He likes to compare the discovery of this spectrum of rhythms with the discovery by Newton in 1666 that white light can be carved up into a spectrum of colours by a prism. Just as we use the terms ultraviolet and infrared to describe radiation at either end of the visible part of the electromagnetic spectrum, so Halberg refers to 'ultradian' and 'infradian' to describe cycles shorter and longer than the circadian period. In nature, the ultradian notes are played by the nervous system, where oscillations may only last one-tenth of a second, moving down in frequency as we encounter heart beats at around one cycle per second and breathing at one cycle every four seconds, to longer hormonal and metabolic oscillations, and on to the 24-hour cycle, the middle C of the biological ensemble which even seems to rule our sense of timekeeping.[16] The menstrual cycle and the annual hibernation of some mammals are infradian. Computer analysis of body rhythms has even suggested evidence for weak circaseptan (approximately weekly) cycles in blood pressure, heart beat and response to infection.[17] However, this weekly rhythm is probably induced by the environment – for instance the Sunday morning lie-in – rather than being a natural cycle of the body clock.

The time machine experiment

Experiments testing the awareness of fruitflies to the passage of time provide another scenario in which patterns arise of a similar nature to those observed in the Belousov–Zhabotinsky reaction encountered in Chapter Six. In the BZ reaction, spatial patterns arise from the billions of individual molecular chemical reactions which can undergo the same clock-like behaviour but become out of phase with each other in different regions of the liquid. This lag can set up beautiful spiral patterns, the spiral arms of colour and thus different phases converging to a phase singularity at the centre.

A phase singularity also appears in Winfree's 'pinwheel experiment' on pupating fruitflies. In the experiment, Winfree used a single stimulus to probe the body clock of the fruitfly. This creature was in his view an obvious choice for the experiment, for he wrote: 'Martin Luther thought flies were created by the devil, since they had no possible practical use, but biologists have since found marvellous uses for them.' Since geneticists adopted the fruitfly in the early 1900s, it has become 'the most studied multicellular organism on the planet'.[18] Winfree fattened fruitfly maggots on enriched mashed potatoes. They were then ready for his experiment, in which he studied the effects of light on their body clocks by using the time of emergence from the pupae as the zero setting of the clocks.

Lightly glued on a dish, his pupae were placed in the 'time machine', when they were exposed to blue light to reset their internal clocks. Winfree wrote: 'The pupae lie motionless, one side of each sparkling in the blue light, the other side still red in shadow. A second passes, or even hours, depending on the programmed protocol, while the animals' brains absorb the light and they are projected ahead or thrown back in their circadian cycles.'[19]

Winfree plotted the way the phases of the flies' clocks were thrown around in time by various bursts of light. This elaborate three-dimensional plot, which he called a time crystal, exhibits a special geometrical feature – a phase singularity. At this particular point, a precisely timed burst of blue light causes a complete breakdown of the flies' clocks.

What does this mean in practical terms? In the laboratory, exposure to blue light just adequate to read by turns out to be roughly equivalent to one minute's exposure close to midnight in the usual circadian cycle. Not surprisingly, it corresponds to a time in nature when a wild fruitfly would expect to see little light. Exposure to this artificial stimulus puts the flies into a 'timeless' state, where their circadian clocks lie dormant. In this state the flies hatch at all hours, not in rhythmic bursts.

The overall result is a 'wave' of pupae that all hatch at the same time, spinning like a pinwheel around a fixed spot on the lawn of pupae. This spot, like the eye of a hurricane, is the phase singularity where they hatch continuously or not at all. The phase singularity (observed by Winfree in each of three successive cycles) is a focal point or 'organising centre' for a wave of hatching pupae; it bears a striking resemblance to the pacemaker centres in the Belousov–Zhabotinsky reaction and in the cardiac rotating waves we encountered in Chapters Six and Seven.

Winfree has constructed time crystals and located phase singularities in a variety of species – insects, plants, and single-celled organisms. Outside the domain of circadian rhythms, they have also been found in the biochemistry of sugar metabolism, as well as in contractions of the leech's and the cat's heart.

Circadian clocks: programming

Winfree's experiments show how the clocks behave when perturbed. One can ask a different question: what is the molecular mechanism which mediates these changes? At the most basic level lies the biological programming of the body clock in genetic material. Genetic programming can be investigated systematically using primitive creatures with short life spans where mutants displaying unusual timing can be made with the help of X-rays, ultraviolet radiation or mutagenic chemicals. Fungi and algae have been used but fruitflies are again by far the biologists' favourite.

Components of circadian clocks are encoded within specific genes. Wild fruitflies' clocks have a period between 23.8 and 24.2 hours but mutants with 19- and 28-hour periods as well as arrhythmic varieties have been found. The gene involved in the latter mutations is encoded on the X chromosome, suggesting that there is indeed a 'clock gene' and that these three varieties of mutants are alleles – defective versions of one and the same gene. The gene responsible, called 'periodic', abbreviated as *per*, was discovered in 1971 by Ronald Konopka working in Seymour Benzer's laboratory at the California Institute of Technology.[20] Its action is not limited to the circadian rhythm, indicating that the protein which the gene codes for is a useful component of several additional mechanisms, including the song of the fruitfly.

The love songs of fruitflies have proved a boon in the hunt for the clock genes. It is fortunate, as geneticist Charalambos Kyriacou of Leicester University points out, that 'if a male fruitfly wishes to mate, he does not simply jump on the first female he sees. The female has to be serenaded with a love song which is produced by the male extending and vibrating his wings.' The song itself consists of a hum together with species-specific bursts of pulses. It is used by the male to identify his species to the female. If he sings the wrong song, it simply appears to go in through one antenna and out the other. The normal love-song rhythm lasts a minute but the different *per* mutants sing with a 40-second cycle, an 80-second cycle or are arrhythmic.[21]

It seems that the more active the *per* protein which the fly makes, the faster its clocks run. The protein has also been likened to sand in an hourglass.[22] *Per* was one of the first behavioural genes to be identified by mutation, then cloned and read. The gene encodes a protein of 1200 amino acids. In each of the fast and slow mutants one amino acid is substituted for another. In the arrhythmic mutant only the first 400 amino acids are made, giving a useless *per* protein.

Once the *per* genetic code is read into a protein, it becomes possible to speculate about its role in the body. The *per* protein is similar to the proteoglycans, a class of proteins which are often found at cell surfaces in mammals. This observation has led Michael Young at the Rockefeller University, New York, to speculate that the protein may affect communication between nerve cells and therefore that the song rhythm may be produced by a network of interacting nerves. And indeed, tests with a *per* mutant which produces a faster tick showed that it did have faster intercellular communication properties.

The *per* gene is significant but unless it can be shown that a similar gene exists in other organisms than the fruitfly, producing a similar clock protein, interest in it will be limited. Hence it is remarkable that a clock gene called 'frequency' (*frq*) has been found in bread mould (*neurospora*), its structure being similar to the fruitfly's *per* gene. Also, antibodies directed against the fly's *per* protein recognise similar proteins in organisms ranging from moths to mammals. Even more remarkably, these *per*-like proteins are found in anatomical locations which are known to house the clock. There is evidence that the gene plays a role in cellular division and is similar to clock genes in other creatures, such as the *frq* gene in the bread mould.

But, in spite of all these investigations, the cellular basis of circadian rhythms even in the best-studied species is still not well understood. Speculative suggestions have been made which include the idea that these rhythms are associated with respiration and mitochondria, with ions and membranes, with glycolysis, or with DNA transcription and RNA translation. Since all of these are normally present in circadian organisms, it is at present difficult to choose one in preference to any other. Indeed, there seem to be almost as many models as investigators.[23] However, it is of interest to note that Goldbeter's 1975 model of cAMP rhythms in slime moulds, which we discussed in Chapter Seven, has been adapted to describe circadian rhythmicity.[24] In Winfree's view it is probable that the actual mechanisms, when eventually elucidated, will turn out to be diverse rather than of universal character.[25]

Circadian clocks: location

Where is our dominant clock? There is still a great deal of debate over how many central circadian clocks there are in Man, but a common view is that we are controlled by two main circadian clocks operating in parallel: one ruling the sleep–wake cycle, the second controlling body temperature and other aspects of our physiology. This explains why people attending all-night parties may experience a 'second wind' around dawn. Kept awake all night, their body temperature still drops as it does during normal sleep. There is a link between temperature cycle and a person's alertness, resulting in a drop in performance as they dance through the small hours. Once a partygoer has survived the nadir of body temperature, alertness rises once again.

But these rhythms are dragged out of synchrony by perturbations such as jet-lag or shift work. It is the dislocation between the sleep–wakefulness rhythm, the body temperature cycle, hormone cycles and local time which is thought to lead to the discomforts of jet-lag and shift work. Rhythmic dislocations are also caused in those exposed to a constant visual environment, such as the blind. The elderly may also be affected, since their body clock is more prone to be out of step with their environment, perhaps because they do not get a regular daily dose of sunlight to entrain it. Because the effective maintenance of body temperature requires accurate timing of the various physiological systems within the body, some believe that the breakdown of the body clock may make the elderly prone to hypothermia.

The dominant clock is thought to be that which controls sleep and wakefulness and governs a hierarchy of clocks at every level of the body, from cell to organ to organism. The first clue to the existence and location of this central circadian clock came in 1972 when Fred Stephan and Irving Zucker of the University of California at Berkeley found that the circadian rhythms of rats could be disrupted by damaging a small clump of nerve cells in their brains.[26] Indeed, when microscopic electrodes are placed in this part of the rat's brain, the nerve cells are found to fire in a regular pattern, like the tick of the second hand on a watch. We too have this area in our brains. It resides within the hypothalamus which is found in the middle of the brain, just above the roof of the mouth. These specialised groups of nerve cells, called collectively the suprachiasmatic nuclei (SCN), are linked to the retina of the eye and seem to organise cycles of rest and activity, skin temperature, and the secretion of hormones.

The position of the SCN underlines the importance of light as a cue for our master body clock. It is located above the junction of the two optic nerves in an area called the optic chiasma (hence the name suprachiasmatic nucleus). Fine nerve fibres reach from the optic nerves and plug into the SCN, providing a connecting route between signals triggered by sunlight and the clock that ticks in the middle of our brain.

The SCN has now been shown to be a principal circadian pacemaker in a highly significant experiment conducted by Martin Ralph of the University of Virginia.[27] He succeeded in changing the circadian-activity rhythms of a normal hamster by transplanting the SCN from a mutant hamster with an unusually short daily rhythm. Conversely, he found that the periodicity of mutant hamsters can be lengthened by replacing their own SCN with that of a wild-type.[28] In fact, the golden hamster (*Mesocricetus auratus*) is to chronobiologists what the fruitfly is to geneticists, according to Nicholas Mrosovsky of the University of Toronto: the humble hamster has circadian rhythms of wheel-running activity which are easy to measure, easy to quantify and highly predictable. The mutant (*tau*) hamster (analogous to the *per* mutant in fruitflies) should also provide insights at the molecular level when scientists discover the genetic basis for its rhythm disorder. This in turn can be compared with the genetic clock machinery of the fruitfly.

The pineal

A vital part of the machinery governing our master clock is found in another small outgrowth of the brain called the pineal gland. Colloquially it is referred to as the 'third eye' because people used to think that it was the vestige of an extra eye. Known since antiquity, the pineal was described by the seventeenth-century philosopher René Descartes as the place where the soul exercises influence over the body. The gland is light-sensitive and possesses many similarities to the retina; nevertheless, while in non-mammals it responds directly to light, in mammals it receives information on day and night indirectly through a special pathway leading from the retina via the SCN. The pineal differs so much in structure, in mode of operation, and in size between species that while studying it the biologist S. D. Wainwright remarked that it is 'like seeing evolution in progress'.[29] In Man it is about five millimetres long and is located near the centre of the brain.

It is the SCN that are thought to determine the various cycles in the body by striking out a master rhythm. This biological pendulum is translated by the pineal into surges of the hormone melatonin. The pineal can be thought of as the clock's mechanism, while the melatonin it exudes is the equivalent of the hour and minute hands which are 'read' by the body. There is a feedback loop here, with melatonin in turn affecting the activity of the SCN.

Regardless of the species of vertebrate, from lamprey to Man, melatonin is only produced during the hours of night; for this reason it has sometimes been called the darkness hormone. The light–darkness cycle causes[30] rhythmic increases in a nocturnally active enzyme[31] in the pineal which converts a neurotransmitter,[32] a chemical messenger between nerves, into melatonin. 'Free running' and entrainment experiments have shown that one or more oscillators maintain the periodicity of melatonin production. The number of such 'clocks' depends on the species: for example, in pike there is a single, intra-pineal clock; in chickens, there are both intra- and extra-pineal oscillators, the latter located in the SCN; and in rats, the manufacture of melatonin in the body is governed solely by the SCN beat, in turn synchronised by light via the retina. From what little is known, the same appears true for humans as for rats. These observations indicate how, through evolution, the mechanism for decoding information on day length has changed from direct phototransduction to an indirect method via nerve messages.

Through the action of secreted melatonin, the pineal provides vital information on time, not only regarding day–night cycles but also about the seasons. A well-known example is the change in coloration of cold-blooded animals: in the larvae of fish and amphibians, and in adult fish and some amphibians and reptiles, the dorsal skin becomes lighter during the night.[33] The pineal complex exerts control over body temperature (thermoregulation). In mammals, the pineal also seems to play a role in the daily processes of initiating, maintaining and leaving the state of lethargy as well as hibernation. For example, melatonin is known to increase the incidence and length of squirrel hibernation.[34]

In many species, reproduction is seasonal, subordinated to the lengthening or the shortening of the day in the summer and winter respectively. Natural selection has evolved such temporal behaviour for good reasons: birth is timed so as best to ensure the survival of the young. Decreasing day length in the autumn, together with increase in the nocturnal production of melatonin, triggers breeding in sheep. However, the annual reproductive cycle is completely disrupted in hamsters and sheep which have had their pineal removed (by pinealectomy).[35] The injection of melatonin into these animals restores the seasonal cycle. In this way, it is now possible artificially to stimulate sexual activity and the growth of fur and fleeces in a variety of mammals, a discovery with significant implications for farmyard economics. Feeding sheep on minute quantities of melatonin in mid-afternoon has been used to advance the breeding season, even completely reversing it from the natural pattern in some instances.[36] The sites of action of melatonin have now been located. By uncovering the structure of these receptor sites, scientists will be able to design potent drugs to eliminate seasonal breeding altogether.

Less is known about how the human pineal works and what it does: experiments are limited for obvious reasons. However, it is reasonable to suppose that it is related to what has been found in other animals. Changes in the melatonin profile have been measured. Levels are at their highest in the first years of infancy and fall during development until adult age, when they remain constant until declining again in old age. It has been suggested that the first fall-off is associated with the onset of puberty; in separate studies, melatonin levels have been claimed to vary in women during the menstrual cycle, although this is contentious.

Artificially administered melatonin has been found to induce sleep and short-term tiredness in Man, just as its natural nocturnal secretion may do. These effects have led to its use in test trials for the treatment of the deleterious effects of jet-lag, for instance by Josephine Arendt at the University of Surrey.[37] The results indicate substantial benefits for around two-thirds of passengers in terms of improved sleep, mood and vigiliance, with rapid resynchronisation of the normal circadian rhythm. For other cases where circadian rhythms are disrupted, such as blindness, shift work and insomnia, one may anticipate the therapeutic value of doses of melatonin. However, because there is vast individual variability between circadian clocks, it requires a great deal of further

310

work to see when and how much melatonin should be administered to give the best results.[38]

Exploiting the biological clock

Bio-time is evidently of more than academic interest. Hippocrates wrote: 'Whoever wishes to pursue the science of medicine in a direct manner must first investigate the seasons of the year and what occurs in them.'[39] Improved understanding of biological rhythms will inevitably improve the lot of Mankind and lead to a multi-billion-dollar industry. These rhythms give us insights into why we perform better at certain times than at others. There is increasing evidence that the bogeyman clocks on early in the morning, the time at which we are most likely to have accidents. Perhaps it is no surprise that the nuclear accidents at Chernobyl and Three Mile Island happened in the small hours and that the Bhopal disaster occurred just after midnight.

Cycles influence susceptibility to disease: people suffering from allergies are most sensitive to substances such as pollen at night and in the early morning. These are also the times when asthma is likely to strike.[40] Parasitic infections take advantage of good timing: schistosome eggs excreted in the urine of bilharzia sufferers peak in the morning, when they are more likely to encounter the snails that play host for the next part of their life cycle. The microfilariae that cause elephantiasis roam in the blood at night, when the mosquitoes which transmit the disease bite. And tape worms migrate up the gut at certain predetermined times to make the most of their host's food.

A new door may have been opened to the treatment of some psychiatric illnesses from the investigation of the human body clock. The notion that light affects our moods can be traced back at least two thousand years but has recently gained scientific credibility. At one extreme there is the phenomenon of 'arctic hysteria', caused by the Arctic midwinter when there is almost continual darkness. Close to the Arctic circle, the indigenous population is subjected either to continuous night or continuous day. During the periods of continuous darkness, people suffer from a form of insomnia known as 'big eye'. They can also suffer large mood swings, called the mörktid syndrome.

Even in sub-arctic latitudes seasonal depressions can occur, with depression descending with the onset of winter, then lifting with the arrival of spring. This is a seasonal affective disorder known as winter depression, marked by overeating, oversleeping and carbohydrate craving. The National Institutes of Health (NIH) in the United States are attempting to treat this depression by using a burst of artificial sunlight to reset the body clock. The mirror-image pattern called summer depression has also been recognised for hundreds of years. Hippocrates noted that 'some are well or ill adapted to summer, others are well or ill adapted to winter'.[41] According to Thomas Wehr and Norman Rosenthal of the NIH, winter depression appears to be linked to light deficiency, while anecdotal evidence links heat with summer depression.

Exposing these depressed people to bright artificial light seems to improve their condition, according to Wehr, who notes that the intensity, duration, colour and the way light is administered are important. One suggested regime of 'phototherapy' for winter depression is to expose patients' eyes to diffuse visible light for two hours in the morning every day throughout the winter. However, there is still a question mark hanging over the precise effect of the intense light used. One line of reasoning maintains that the treatment is successful because bright light mimics the natural length of daylight in spring, modifying the secretion of the hormone melatonin. Others, including Rutger Wever of the Max Planck Institute for Psychiatry in Munich, West Germany, claim that bright light has a synchronising effect, acting as a cue for rhythms that have slipped out of synchrony. Another suggestion is that the light somehow amplifies the circadian rhythms. Perhaps several effects are at work. Or perhaps sunshine makes people cheerful for reasons unrelated to clocks: in Antarctica, moods improve with the first sight of the Sun, which is not very bright.[42]

Many of the findings relating to non-human circadiana (including Winfree's work on fruitflies and other species) can be translated wholesale into the human realm,[43] according to research by Chuck Czeisler and colleagues at the Center for Circadian and Sleep Disorders Medicine in Boston. Their findings confirm work by Al Lewy of Oregon Health Sciences University School of Medicine that the sensitivity of the human circadian pacemaker to light is far greater than previously recognised: exposure at one time to light brighter than that used to illuminate a room can reset the clock in one direction while exposure at another moment can reset the clock in the opposite direction.[44] By exposure at the phase singularity, Czeisler's team even managed to bring the circadian clock to a halt. This work confirms that light could prove a useful tool to treat disorders of circadian regulation, such as jet-lag. 'With the right kind of exposure we could stop someone's clock, then with another kind of exposure, start it again, at any time you want, exactly as Winfree predicted,' said Gary Richardson, one of Czeisler's colleagues.[45] And studies on disruption of body rhythms caused by night-shift work, with the associated decline in alertness and performance, showed that this maladaption can be treated effectively with scheduled exposure to bright light at night and darkness during the day.[46]

A traditional theme found in medical textbooks is homeostasis – the notion that the body's environment is constant. Now that so many biological rhythms have been found, it is simpler to assume that everything is rhythmic unless proved otherwise. Not only can doctors shift biological time with drugs, human body rhythms can make a vast difference to the efficacy of those drugs. Wide variations in the lethal doses of some drugs in animal tests have been reinterpreted with the help of chronobiology. It has been found that such variations are actually linked to the time the drug was administered. Chronobiology has revealed many examples of the influence of the body clock. At the Masonic Cancer Center in Minneapolis, scientists have discovered that giving the cancer drug adriamycin at six in the morning dramatically reduces side effects. Cisplatin, another anti-cancer drug, is best administered in the late afternoon. These cancer treatments appear to be taking advantage of the rhythmic regulation of cell division, which contrasts with the unpredictable division of cancerous cells. Anti-cancer drugs often interfere with cell division so that to be of greatest effect they are applied when normal cells are not dividing; the cancerous cells, which are still dividing, are thereby selectively attacked.[47]

Biorhythms

There is one pseudo-scientific theory that deserves a mention, albeit a brief one. For some mysterious reason, many people have encountered biorhythm theory, which was originated around the turn of the century by Wilhelm Fliess, a nose and throat surgeon in Berlin. Winfree has mischievously pointed out that Fliess was also responsible for a monograph entitled *Relations Between the Nose and the Female Sex Organs* . . . During the 1970s, some 40,000 suicides and accidents were recorded and compared with the predictions of biorhythm theory, which marks out certain days as dangerous, depending on the state of an individual's physical, emotional and intellectual cycles. No correlation was found.[48]

Notes

PROLOGUE

1. J. T. Blackmore. *Ernst Mach. His Work, Life and Influence* (University of California Press, 1972), p. 212.
2. *Die Zeit* No. 1420, Abendblatt, 7 September 1906, Column 2, *Weitere Nachrichten die letzen Tage.*
3. D. Flamm. *Studies in History and Philosophy of Science* (hereinafter *Stud. Hist. Phil. Sci.*) 14, 274 (1983).
4. G. Jaffé. *Journal of Chemical Education* (hereinafter *J. Chem. Ed.*) 29, 230 (1952).
5. E. Broda. *Ludwig Boltzmann, Mensch, Physiker, Philosoph* (F. Deuticke, Vienna, 1955), p. 27. The translation was provided by Doris Highfield.
6. D. Flamm. *Stud. Hist. Phil. Sci.* 14, 267 (1983).
7. D. Goodstein, *States of Matter* (Prentice-Hall, Englewood Cliffs, N. J., 1975). We are indebted to Dilip Kondepudi, of Wake-Forest University, USA, for drawing this quotation to our attention.

CHAPTER ONE

The introductory quotation appears in *The Sonnets of William Shakespeare* (Shepheard-Walwyn, London, 1975).
1. I. Prigogine. *From Being to Becoming* (W. H. Freeman, San Francisco, 1980), Preface.
2. In *some* ways space and time are analogous. *See* R. Taylor. *Journal of Philosophy* (hereinafter *J. Phil.*) 52 (1955).
3. Gifford Lectures at the University of Edinburgh.
4. H. Meyerhoff. *Time in Literature* (University of California Press, 1955); C. A. Patrides. Part III, (Time in Modern Literature), *Aspects of Time* (Manchester University Press, 1976); S. L. Macey. Part III (The Influence of Literature) *Clocks and the Cosmos* (Archen Books, Hamden, Connecticut, 1980); P. Landsberg (ed.). *The Enigma of Time* (Adam Hilger, Bristol, 1984); R. Glasser. *Time in French Life and Thought* (Manchester University Press, 1972).
5. E. Fitzgerald. *The Rubáiyát of Omar Khayyám* (Fifth edition, 1889. *Rubáiyát of Omar Khayyám and Other Writings by Edward Fitzgerald*, Collins, London and Glasgow, 1953).
6. A. Eddington. *The Nature of the Physical World* (Cambridge University Press, 1928), p. 91.
7. R. Morris. *Time's Arrows* (Simon & Schuster, New York, 1985), p. 22.
8. *Ibid.*, p. 20.
9. G. J. Whitrow. *Time in History* (Oxford University Press, 1989), p. 43.
10. M. Eliade. *The Myth of the Eternal Return* (Routledge & Kegan Paul, London, 1955), p. 86.
11. *Ibid.*, p. 137.

12. J. T. Fraser. *Time, the Familiar Stranger* (Tempus, Washington, 1987), p. 20; G. J. Whitrow. *Time in History, op. cit.*, p. 51.

13. G. J. Whitrow. *The Natural Philosophy of Time* (Oxford University Press, 1980).

14. Plato. *Timaeus and Critias* (Penguin, Harmondsworth, 1971, trans. D. Lee), p. 40.

15. It is in fact only referred to as the 'Achilles argument' by Aristotle, the primary extant source on Zeno.

16. Simplicius' account of the Achilles paradox, in which mention *is* made of the tortoise, in H. P. D. Lee, *Zeno of Elea* (Cambridge University Press, 1936), quoted in G. J. Whitrow. *The Natural Philosophy of Time, op. cit.* p. 196.

17. The latter position fits in with the 'process philosophy' (the philosophy of 'becoming' rather than of 'being') of Alfred North Whitehead, the Cambridge mathematician and philosopher who maintained that 'There is no nature apart from transition, and there is no transition apart from temporal duration. This is why an instant of time, conceived as a primary simple fact, is nonsense.' (A. N. Whitehead. *Modes of Thought* (Cambridge University Press, 1938), p. 207, quoted in G. J. Whitrow. *The Natural Philosophy of Time, op. cit.*, p. 200).

18. J. T. Fraser. *Time, the Familiar Stranger, op. cit.*, p. 42.

19. Quoted in J. B. Priestley. *Man and Time* (Aldus Books, 1964), p. 70.

20. D. Flamm. *Stud. Hist. Phil. Sci.* <u>14</u>, 261 (1983).

21. Ludwig Boltzmann. Populäre Schriften Essay 19, *Ludwig Boltzmann, Theoretical Physics and Philosophical Problems*, B. McGuinness (ed.) (Reidel, Dordrecht, 1974), p. 164.

22. D. Flamm. *Stud. Hist. Phil. Sci.* <u>14</u>, 257 (1983).

23. G. J. Whitrow. *The Natural Philosophy of Time, op. cit.*, pp. 59, 220.

24. L. Mumford. *Technics and Civilization* (Routledge & Kegan Paul, London, 1934), p. 15.

25. In keeping with Occam's razor, he made a minimum of assumptions.

26. I. Barrow. *Lectiones Geometricae* 1735 (trans. E. Stone) Lect. I, London.

27. *Sir Isaac Newton's Mathematical Principles (Principia Mathematica 1687)* trans. A. Motte, revised by F. Cajori (University of California Press, 1947), p. 6.

28. *Albert Einstein and Michele Besso, Correspondence 1903–1955*, P. Speziali (ed.) (Hermann, Paris, 1972).

29. Recently, in conjunction with quantum theory, it has become fashionable to talk in terms of many more dimensions than just four, even to the extent of including more than a single time dimension. The extra spatial and temporal dimensions which we do not see are supposed to be 'compactified', rolled up so tightly that they cannot be observed in the present-day universe. These ideas are extensions of the Kaluza–Klein theory proposed in the years following Einstein's theory of relativity, but were originally resisted by him.

30. Dennis Sciama, quoted in J. D. Barrow. *The World Within the World* (Oxford University Press, 1988), p. 306.

31. Interview with the authors during a conference entitled 'Self-Organisation, a Scientific Paradigm: Western and Eastern Perspectives', Brussels, 27 September 1989.

32. A. Eddington. *The Nature of the Physical World, op. cit.*, p. 74.

33. The 'Heat Death' of the universe was first articulated by the German von Helmholtz and later popularised by Eddington and Jeans.

34. At present, astronomical data are not good enough to decide whether the universe is open or closed.
35. P. T. Landsberg. *The Study of Time III*, J. T. Fraser, N. Lawrence and D. Park (eds) (Springer-Verlag, Heidelberg, 1978), p. 118.
36. D. Flamm. *Stud. Hist. Phil. Sci.* <u>14</u>, 265 (1983).
37. It appears there in a form first written down by another scientist who will make an appearance in this book, Max Planck.
38. Private communication from Professor Dieter Flamm (14 December 1989).
39. M. Waldorp. 'Spontaneous Order, Evolution and Life', *Science* <u>247</u>, 1543 (1990): a report on the second 'Artificial Life' workshop, held at Santa Fé, describing how some scientists believe that the chemical building blocks that formed on infant Earth organised themselves through chemical catalysis.
40. *Science* <u>245</u>, 26 (1989).
41. It has been claimed that this may lead to 'time-symmetric indeterminism' (H. Price. *Nature* <u>348</u>, 356 (1990)). This erroneous view arises from a lack of understanding of what is meant by entropy: for highly chaotic systems for which an entropy can be defined, one might *a priori* attempt to describe time evolution either in terms of probabilities or with deterministic equations. But only the stochastic description can break the time-symmetry and reveal the arrow of time, as demonstrated by P. Coveney, *Nature* <u>333</u>, 409 (1988) and S. Goldstein et al. *J. Stat. Phys.* <u>25</u>, 111 (1981).

CHAPTER TWO

The introductory quotation appears in W. Gratzer (ed.). *The Longman Literary Companion to Science* (Longman, Harlow, 1989), p.203.
1. J. K. Fauvel, R. Flood, M. Shortland and R. Wilson (eds). *Let Newton Be!* (Oxford University Press, 1988), p. 12.
2. Newton died between 1 and 2 a.m. on 20 March 1727.
3. J. K. Fauvel, R. Flood, M. Shortland and R. Wilson (eds). *Let Newton Be!*, *op. cit.*, p. 1.
4. *Ibid.*, p. 14. R. Westfall. *Never at Rest: A Biography of Isaac Newton* (Cambridge University Press, 1980).
5. An epitaph written by Fatio was used on the monument: *Qui genus humanum ingenio superavit* – who surpassed all men in genius. However, Pope published a modified version of his proposal in 1735. The original was: All nature and its laws lay hid in night/God said: Let Newton be! and all was light. D. Gjertsen. *The Newton Handbook* (Routledge & Kegan Paul, London and New York, 1986), p. 439.
6. J. K. Fauvel, R. Flood, M. Shortland and R. Wilson (eds). *Let Newton Be!*, *op. cit.*, p. 43.
7. I. Newton. *Principia Mathematica*, 1687, trans. A. Motte, rev. F. Cajori (University of California Press, 1962), vol. II, p. 397.
8. *Philosophiae Naturalis Principia Mathematica*, published in three editions (London 1687, Cambridge 1713 and London 1726).
9. B. Russell. *History of Western Philosophy* (Allen & Unwin, London, 1979), p. 512.
10. D. Gjertsen. *The Newton Handbook*, *op. cit.*, p. 557.
11. J. T. Fraser. *Time, the Familiar Stranger* (Tempus, Washington, 1987), p. 77.
12. F. A. B. Ward. *Time Measurement* (Science Museum, London, 1970), p. 9.

13. *Ibid.*, pp. 16, 35.
14. G. J. Whitrow. *Time in History* (Oxford University Press, 1989), p. 66.
15. Private communication to the authors by Gerald Whitrow.
16. M. Young. *The Metronomic Society* (Harvard University Press, 1988), p. 63.
17. Some biologists believe the week is self-determined after the discovery of evidence for a seven-day biorhythm in the body. *See* Appendix and A. Aveni. *Empires of Time* (Basic Books, New York, 1989), p. 100.
18. M. Young. *The Metronomic Society, op. cit.*, p. 65.
19. G. J. Whitrow. *Time in History, op. cit.*, p. 32.
20. *Ibid.*, p. 120.
21. J. K. Fauvel, R. Flood, M. Shortland and R. Wilson (eds). *Let Newton Be!*, *op. cit.*, p. 103.
22. *Ibid.*, p. 124.
23. G. Childe. *A History of Technology* (Oxford University Press, 1954), p. 187.
24. C. Ronan. *The Astronomers* (Evans, London, 1964), p. 152.
25. Between 600 and 300 BC. M. Kline. *Mathematical Thought from Ancient to Modern Times* (Oxford University Press, 1972), p. 3.
26. However, Copernicus himself believed that there were difficulties with his model. A moving Earth implied that the stars should shift in an annual cycle, yet they were not observed so to do, owing to the vast distances involved which were not appreciated at the time. A moving Earth that did not shudder nor respond to winds and tidal waves also seemed hard to believe.
27. W. Gratzer (ed.). *The Longman Literary Companion to Science, op. cit.*, p. 163.
28. T. Kuhn. *The Copernican Revolution* (Harvard University Press, 1957), p. 191.
29. C. Ronan. *The Cambridge Illustrated History of the World's Science* (Cambridge University Press, 1983), p. 331.
30. J. K. Fauvel, R. Flood, M. Shortland and R. Wilson (eds). *Let Newton Be!*, *op. cit.*, p. 185.
31. T. Kuhn. *The Copernican Revolution, op. cit.*, p. 200.
32. *Astronomia Nova*, Heidelberg.
33. C. Ronan. *The Cambridge Illustrated History of the World's Science, op. cit.*, p. 339.
34. Thomas Harriot came to similar conclusions on uniform acceleration at roughly the same time; J. K. Fauvel, R. Flood, M. Shortland and R. Wilson (eds). *Let Newton Be!*, *op. cit.*, p. 50.
35. Hans Lippershey of the Netherlands, Giambattista della Porta of Naples, Italy and Leonard and Thomas Digges in England have all been credited with its development.
36. E. Rosen. *The Naming of the Telescope* (Henry Schuman, New York, 1947), p. 3.
37. H. C. King. *The History of the Telescope* (Charles Griffin, London, 1955), p. 34.
38. A. Einstein. Foreword to G. Galileo. *Dialogue Concerning the Two Chief World Systems* (University of California Press, 1953), p. xvii.
39. C. Ronan. *The Astronomers, op. cit.*, p. 154.
40. F. A. B. Ward. *Time Measurement, op. cit.*, p. 18.
41. R. Penrose. *The Emperor's New Mind* (Oxford University Press, 1989), p. 166.
42. J. K. Fauvel, R. Flood, M. Shortland and R. Wilson (eds). *Let Newton Be!*, *op. cit.*, p. 187.
43. *Ibid.*, p. 194.
44. I. Newton. *Principia Mathematica, op. cit.*, p. 13.

45. First formulated by Descartes in his *Principia philosophiae* (1644). D. Gjertsen. *The Newton Handbook*, op. cit., p. 297.

46. However, Newton had threatened to suppress the third book, when Robert Hooke claimed primacy for the notion of universal gravity; D. Gjertsen. *The Newton Handbook*, op. cit., p. 459.

47. From Voltaire's 'Essay on the Civil War in France', 1727, quoted in D. Gjertsen. *The Newton Handbook*, op. cit., p. 30.

48. *Ibid.*, p. 300.

49. R. Westfall. *Never at Rest*, op. cit., pp. 272, 696, 853.

50. S. Hawking. *A Brief History of Time* (Bantam Books, New York, 1988), p. 181.

51. We are here ignoring the possibility that time may be discrete rather than continuous, a point already mentioned in Chapter One when discussing Zeno's paradoxes. Moreover, it may indeed be that here a crucial oversight was made in the construction of classical mechanics. For, in Newtonian theory, we are supposed to be able to measure both the position and the velocity of a body with infinite precision. But a body in motion is not likely to have a well-defined location in space. This point is discussed by Peter Landsberg in *Foundations of Physics*, 18, 969 (1988); he draws parallels with the situation in quantum mechanics, where simultaneous measurement of both quantities is denied in principle. We are grateful to Peter Landsberg for drawing our attention to this paper.

52. S. Hawking and W. Israel (eds). *300 Years of Gravitation* (Cambridge University Press, 1989), p. 11.

53. B. Russell. *History of Western Philosophy*, op. cit., p. 522.

54. For further discussion of this, *see* e.g. D. Bohm. *Causality and Chance in Modern Physics* (Routledge & Kegan Paul, London and New York, 1984).

55. D. Gjertsen. *The Newton Handbook*, op. cit., pp. 231, 232.

56. F. Manuel. *A Portrait of Isaac Newton* (Harvard University Press, 1968), p. 124.

57. J. K. Fauvel, R. Flood, M. Shortland and R. Wilson (eds). *Let Newton Be!*, op. cit., p. 173.

58. D. Gjertsen. *The Newton Handbook*, op. cit., p. 209.

59. G. J. Whitrow. *Time in History*, op. cit., p. 165.

60. J. K. Fauvel, R. Flood, M. Shortland and R. Wilson (eds). *Let Newton Be!*, op. cit., p. 128.

61. *Ibid.*, p. 135.

62. In modern parlance, we would say that Maxwell *unified* electricity and magnetism.

63. I. Tolstoy. *James Clerk Maxwell* (Canongate, Edinburgh, 1981), p. 126.

64. F. A. B. Ward. *Time Measurement*, op. cit., p. 36.

65. G. J. Whitrow. *Time in History*, op. cit., p. 167.

66. I. Tolstoy. *James Clerk Maxwell*, op. cit., p. 125.

67. J. T. Fraser. *Time, the Familiar Stranger*, op. cit., p. 41.

68. H. Poincaré. *Comptes Rendus de l'Académie des Sciences (Paris)* 108, 550 (1889).

69. The figure is in the order of 10^{24}.

70. Cyclic time does not necessarily imply that there is no associated temporal arrow; nevertheless it is highly restrictive as to its nature.

71. It applies to isolated and finite non-gravitating systems.

72. I. Tolstoy. *James Clerk Maxwell*, op. cit., p. 141.

318

73. The Vienna Circle was originally known as the 'Mach Circle'.
74. S. Hawking and W. Israel (eds). *300 Years of Gravitation, op. cit.*, p. 13.
75. S. Hawking. *A Brief History of Time, op. cit.*, p. 156.
76. W. Gratzer (ed.). *The Longman Literary Companion to Science, op. cit.*, p. 128.
77. J. Keats. *Lamia* Part II (1820), *The Poetical Works of John Keats*, W. Garrod (ed.) (Oxford University Press, 1958), p. 212.
78. Quoted from I. Prigogine. *From Being to Becoming* (W. H. Freeman, San Francisco, 1980), p. 3; *see* e.g. H. Bergson. *L'évolution créatrice*, in *Oeuvres* (Editions du Centenaire, Presses Universitaires de France, Paris, 1963) and 'Durée et simultanéité', in *Mélanges* (Presses Universitaires de France, Paris, 1972).
79. A. Koyré. *Etudes Newtoniennes* (Gallimard, Paris, 1968), quoted in I. Prigogine. *From Being to Becoming, op. cit.*, p. 2.
80. J. Fauvel, R. Flood, M. Shortland and R. Wilson (eds). *Let Newton Be!, op. cit.*, p. 27.

CHAPTER THREE

The introductory quotation appears in W. Rindler. *Essential Relativity* (Springer-Verlag, Heidelberg, Second Edition, 1977), p. 203.
1. A. Eddington. *The Nature of the Physical World* (Cambridge University Press, 1928), p. 43.
2. R. W. Clark. *Einstein. The Life and Times* (Hodder and Stoughton, London, 1973), p. 45.
3. A. Pais. *'Subtle Is the Lord' . . . The Science and Life of Albert Einstein* (Oxford University Press, 1982), p. 20.
4. Fizeau had earlier (1851 and 1853) performed similar experiments on the speed of light in moving fluids.
5. Michelson was once described by Einstein as the artist in science: 'His greatest joy seemed to come from the beauty of the experiment itself and the elegance of the method employed.' (A. P. French (ed.). *Einstein, a Centenary Volume* (Heinemann, London, 1979), p. 38. A scientist who described one effort to measure the speed of light over a 22-mile path between Mount Wilson and Mount San Antonio as 'such good fun', Michelson devised experiments to an accuracy of one part in around four thousand million and 'taught the world to measure', according to Einstein. Michelson seems to have expired in the pursuit of the velocity of light. He died in 1931 after several strokes during preparations to measure its speed in a three-foot diameter pipe measuring a mile in length.
6. By the use of interferometry.
7. A. P. French. *Einstein, a Centenary Volume, op. cit.*, p. 77.
8. Newton's mechanics left no room for absolute space, because it was invariant under Galilean transformations between observers moving with uniform relative velocity; electromagnetism was not invariant under these changes of perspective.
9. Both these quotations were made by Henri Poincaré in his address to the International Congress of Arts and Science at St Louis in 1904 and were published in *The Value of Science*. Reprinted in *The Foundations of Science*, trans. G. R. Halsted (Science Press, New York, 1913).
10. A. Pais. *'Subtle Is the Lord' . . . The Science and Life of Albert Einstein, op. cit.*, p. 170.

11. *See* M. Wertheimer (ed.). *Productive Thinking* (Tavistock Publications, London, 1961), Chapter 10. We are indebted to Gerald Whitrow for pointing this out to us.
12. A. Einstein, letter to F. C. Davenport, 9 February 1954, quoted in A. Pais. *'Subtle Is the Lord'* . . . *The Science and Life of Albert Einstein*, op. cit., p. 172.
13. A. Beck and P. Havas (eds). *The Collected Papers of Albert Einstein* (Princeton University Press, 1987), Vol. 1, p. xviii.
14. R. W. Clark. *Einstein. The Life and Times*, op. cit., p. 35.
15. *Ibid.*
16. *Ibid.*, p. 127.
17. A. French. *Einstein, a Centenary Volume*, op. cit., p. 54.
18. In December 1901, several years before his first paper on special relativity, Einstein wrote to his sweetheart, Mileva Marić: 'I am now working very eagerly on an electrodynamics of moving bodies, which promises to become a capital paper. I wrote to you that I doubted the correctness of the ideas about relative motion. But my doubts were based solely on a simple mathematical error. Now I believe in it more than ever.' (A. Beck and P. Havas (eds). *The Collected Papers of Albert Einstein*, op. cit., p. 187.) Einstein went on to marry Mileva in January 1903, against his parents' wishes. They were divorced in 1919.
19. A. P. French. *Einstein, a Centenary Volume*, op. cit., p. 27.
20. A. Beck and P. Havas (eds). *The Collected Papers of Albert Einstein*, op. cit., p. 158.
21. A. Einstein (1921), quoted statement chiselled in the stone fireplace in the common room of Fine Hall, the Department of Mathematics at Princeton University.
22. 'What led me more or less directly to the special theory of relativity was the conviction that the electromagnetic force acting on a [charged] body in motion in a magnetic field was nothing else but an electric field [in the body's rest frame]'. (A. Einstein (1952), quoted in R. S. Shankland, *American Journal of Physics* 32, 16 (1964).)
23. G. Galileo. *Dialogue Concerning the Two Chief World Systems*, trans. S. Drake (University of California Press, 1953), p. 187.
24. Poincaré built on their work simultaneously to Einstein.
25. Einstein wrote in his paper that 'the introduction of a light-æther will prove to be superfluous'.
26. Through the study of orbiting muon decay in particle accelerators.
27. S. Weinberg, in *300 Years of Gravitation*, S. Hawking and W. Israel (eds), (Cambridge University Press, 1989), p. 7.
28. Relativistic optics is discussed, for example, by Wolfgang Rindler in his *Essential Relativity*, Chapter Three.
29. More correctly, mass here means *inertia*.
30. Another example is the neutrino, still generally thought to be massless, although some theories and experimental data suggest otherwise.
31. J. Taylor. *Special Relativity* (Oxford University Press, 1975), p. 16.
32. T. Wilkie. *Nature*, 268, 295 (1982).
33. An alternative discussion of the twin paradox is presented by G. J. Whitrow. *The Natural Philosophy of Time* (Oxford University Press, Second Edition, 1980), pp. 264–6.
34. S. Hawking. Halley Lecture, Oxford University, June 1989.
35. H. Minkowski. *Annalen der Physik* 47, 927 (1915) in an address to the 80th

Congress of German Scientists and Physicians, quoted in A. Pais. *'Subtle Is the Lord'* . . . *The Science and Life of Albert Einstein, op. cit.*, p. 152.

36. A. Eddington. *The Nature of the Physical World, op. cit.*, p. 68.

37. Thus, while Einstein's theory of special relativity made the Maxwell æther redundant, it did nothing to explain away absolute space as the standard of non-acceleration, based firmly on the privileged role of inertial frames.

38. Reported in the *New York Times*. R. W. Clark. *Einstein. The Life and Times, op. cit.*, p. 238.

39. *Albert Einstein and Michele Besso, Correspondence 1903–1955*, P. Speziali (ed.) (Hermann, Paris, 1972), 26 March 1912, quoted in A. Pais. *'Subtle Is the Lord'* . . . *The Science and Life of Albert Einstein, op. cit.*, p. 210.

40. The Tower of Pisa experiment was actually a *Gedankenexperiment* – a thought experiment dreamed up by Galileo as an argument for the linear superposition of gravitational forces.

41. There is of course a long story associated with the genesis of the equivalence principle. The Eötvös experiment played a role in the development of general relativity analogous to that of the Michelson–Morley experiment in special relativity. In the same way, Einstein's own statement of the equivalence principle owed very little to these results.

42. L. Kollros. *Helvetica Physica Acta Supplement* $\underline{4}$, 271 (1956), quoted in A. Pais. *'Subtle Is the Lord'* . . . *The Science and Life of Albert Einstein, op. cit.*, p. 212.

43. The word parsec derives from parallax and seconds, referring to the fundamental method of measuring astronomical distances. As the Earth moves in its orbit around the Sun, the apparent position in the sky of a star, as seen from the Earth, varies. Using trigonometry, the amount of this variation, called parallax (measured in seconds of arc), gives the distance to the star relative to the Earth's orbit. A parallax of one second corresponds to a distance of one parsec, by definition, which is 3.26 light years.

44. A. Einstein. *Relativity. The Special and General Theory* (Methuen, London, 1954), p. 1; his sentiments may no longer hold true in today's educational system.

45. E. A. Abbott. *Flatland: A Romance of Many Dimensions* (Dover, New York, 1952).

46. The metric gives a complete description of the local properties of the space. For a complete global description, the topology – how things are connected together – is also needed.

47. However, there is documentary evidence that Riemann was attempting to make a unified description of light, electromagnetism, heat and gravitation. 'Throughout his life, he regarded this as his main work (*meine Hauptarbeit*) and therefore, if the question is asked why Riemann's "purely mathematical" developments are so wonderfully appropriate to the most advanced developments of general relativity, the answer is straightforward: not only was Riemann deeply involved in coping with the same general type of problems as Lorentz, Poincaré, Einstein *et al.* half a century later, but the scope of his project of a "unified field theory", as it would now be called, undoubtedly exceeded that of Einstein, employing a methodological rigour which Einstein approximated only on a few rare occasions.' From 'The Concept of the Transfinite', in *The Campaigner* $\underline{9}$, 6 (1976).

48. The geodesic law of motion was originally introduced as a separate axiom in general relativity. A proof that the geodesic law follows from the Einstein equations was given by Einstein and colleagues in a series of papers starting in 1927; while it is

certainly correct for test particles of negligible mass, to this day some mathematicians still argue over whether the proof is rigorous enough in general. *See* W. Rindler. *Essential Relativity* (Springer-Verlag, Heidelberg, 1977), p. 182.

49. More precisely, at low spacetime curvature.

50. This is enshrined in Einstein's famous 'household' law $E = mc^2$.

51. A. V. Douglas. *The Life of Arthur Stanley Eddington* (Nelson, London, 1956), p. 40 (the poem quoted appears on p. 43). However, Hawking points out (*A Brief History of Time* (Bantam Books, New York, 1988), p.32) that later examination of the photographs taken on the expedition showed that the errors were as great as the effect sought. 'Their measurement had been sheer luck, or a case of knowing the result they wanted to get, not an uncommon occurrence in science. The light deflection has, however, been accurately confirmed by a number of later observations.' This makes Einstein's recalculation of the effect prior to the trip arguably irrelevant.

52. C. Will, in *300 Years of Gravitation*, S. Hawking and W. Israel (eds) (Cambridge University Press, 1989).

53. R. W. Clark. *Einstein. The Life and Times*, op. cit., p. 317.

54. W. Rindler. *Essential Relativity*, op. cit., p. 21.

55. *Proceedings of the 33rd Annual Symposium on Frequency Control 1979*, p. 4.

56. 16 picoseconds for a one-and-a-half-hour rotation period: *Journal de Physique, Colloque* (hereinafter *J. Phys. Colloq.*) 42, C-8, 395 (1981).

57. K. Popper. *The Logic of Scientific Discovery* (Hutchinson, London, 1959), p. 15.

58. The density is of the order of 10^{-29} gram/cm^3.

59. The spiral arms are roughly speaking only a 10 per cent modulation in the density of the disk, which is otherwise fairly smooth. They happen to look more prominent because most of the young stars, which are brighter, are located in these arms. Most of the mass of the Milky Way is thought to be in a 'dark halo' – the mass of this halo is inferred from its gravitational effects on objects in and around the galaxy.

60. However, there is the problem of dark matter, whose presence can be inferred in galaxies and clusters of galaxies from its gravitational effects (the velocities of orbiting particles – gas, stars and galaxies – are larger than can be accounted for if the only mass present is what we can see as stars and gas). If neutrinos do have a rest mass, they would contribute to this dark matter (or even provide all of it). Observations indicate that the dark matter in galaxies can account for only about one-tenth of the critical density required to make the universe just closed. However, there could be dark matter distributed more smoothly between the galaxies, which could change the balance sheet in favour of collapse.

61. C. Will, in *The New Physics*, P. C. W. Davies (ed.) (Cambridge University Press, 1989), p. 31.

62. *Ibid.*, p. 7.

63. H. Bondi, T. Gold and F. Hoyle. *Monthly Notices of the Royal Astronomical Society* 108, 252 (1948).

64. Penzias remarked in an interview with RRH in November 1989 that with the pigeons in place 'the temperature was about one degree too high'. In a BBC radio interview Penzias made the point that 'one half of one per cent' of the noise that a listener can pick up tuning between radio stations is due to this radiation.

65. G. Gamow. *The Creation of the Universe* (Viking Press, New York, 1952), p. 64.

66. S. Hawking, in *The New Physics*, P. C. W. Davies (ed.), *op. cit.*, p. 64.

67. It should be pointed out, however, that logically an extra term allowing for the cosmological constant can be included in the Einstein equations: it is a minimal modification, yet it still amounts to a complication. Through the cosmological constant, it is possible to make a fascinating connection between the very large and the very small scales. For it is also possible to 're-interpret' the cosmological constant as representing the energy–momentum tensor of the vacuum (for example, if the vacuum (*see* Chapter Four for further discussion) has a positive energy-density, then the pressure is negative, and conversely). Indeed, there is no reason to suppose that the vacuum has zero energy-density. Moreover, some types of matter (such as the hitherto unobserved scalar Higgs particles which cause symmetry breaking in the Glashow–Weinberg–Salam electroweak theory) can in some circumstances produce an *effective* cosmological constant. It is this possibility that enters the 'inflationary universe scenario' – requiring at very early times a universe whose expansion proceeds at an accelerating, exponential rate – for which the cosmological term is necessary. The universe later exits from this phase (there are various models of how this might happen) and enters the present phase, where the effective *net* cosmological constant is small. *Why* the present effective net cosmological constant should be so small (or zero) is currently a hot topic in particle physics.

68. The possibility of a bouncing universe, which goes through the singularities at either end and reappears, is not taken very seriously, owing to the problem of understanding what goes on at the singularities themselves.

69. S. Hawking, in *300 Years of Gravitation*, S. Hawking and W. Israel (eds.), (Cambridge University Press, 1989), p. 651.

70. C. Will, in *The New Physics* , P. C. W. Davies (ed.), *op. cit.*, p. 29.

71. Certain specific conditions have to be fulfilled to guarantee this result, including the requirement that time travel be impossible. This is discussed later in the text.

72. Interview of R. Penrose with PVC and RRH, June 1989.

73. The repulsive forces are ultimately provided by the quantum mechanical Pauli exclusion principle.

74. C. Will, in *The New Physics*, P. C. W. Davies (ed.), *op. cit.*, p.26.

75. *Ibid.*, p. 25.

76. This is because of the gravitational redshift.

77. S. Hawking. Halley Lecture, Oxford University, June 1989.

78. *Ibid.*

79. Things are not that simple, because again the gravitational redshift would prevent all the light signals from reaching us without degradation.

80. At least some heavy elements can be produced in ten million years by nuclear reactions in stars greater than ten times as massive as the Sun. In fact, most of the elements such as oxygen, silicon, sulphur, magnesium and calcium are thought to be produced in this way, together with some carbon and iron. Most carbon, iron and nitrogen are reckoned to be formed by stars with masses only a few times that of the Sun, having lifetimes of 100–1,000 million years. The actual timescale for producing heavy elements depends on the rate of processing gas into stars. Since star formation is still continuing in galaxies like our own, so too is the production of heavy elements. Various pieces of observational evidence suggest that the timescale necessary for producing half the heavy

elements now existing was a few billion years – but this is controlled by the rate of star formation rather than the lifetime of individual stars.

81. J. D. Barrow. *The World Within the World* (Oxford University Press, 1988), p. 355.

82. B. Carter, in *Confrontation of cosmological theories with observation*, M. S. Longair (ed.) (Reidel, Dordrecht, 1974), p. 291. The theologian F. R. Tennant also used the term in his *Philosophical Theology*, Vol. II (Cambridge University Press, 1930).

83. G. J. Whitrow. 'Why physical space has three dimensions', *British Journal for the Philosophy of Science* 6, 13 (1955); *see also* G. J. Whitrow. *The Structure and Evolution of the Universe* (Hutchinson, London, 1959), pp. 199–201.

84. Yet even the final anthropic principle has a very precise scientific formulation in terms of information processing. Its validity depends on the laws of physics and it might eventually be tested with the help of modern developments in the field of algorithmic complexity. The scientific, non-teleological, interpretation of the strong and final anthropic principles is promoted by Barrow in various writings listed in the bibliography; e.g. *The Anthropic Cosmological Principle, The World Within the World* and in *The Anthropic Principle*, F. Bertola and V. Curi (eds). For a technical discussion of algorithmic complexity, the reader should consult G. Chaitin. *Algorithmic Complexity Theory* (Cambridge University Press, 1987).

85. Kurt Gödel was born the year after special relativity emerged but died in strange circumstances. He was one of the world's greatest logicians and a close colleague of Einstein at the Institute for Advanced Study in Princeton. But as a child (called Herr Warum, or 'Mr Why' by his parents) he was a hypochondriac. Nervous depression and sojourns in sanatoria were to follow. In later life he showed a profound distrust of doctors and of food, which he believed was poisoned. He died on 29 December 1977 of malnutrition caused by this personality disorder.

86. K. Gödel. *Reviews of Modern Physics* 21, 447 (1949).

87. Gödel's model is of interest in cosmology as an example of a universe which is homogeneous but not isotropic. However, there are other such models – anisotropically expanding versions of the usual Friedman models – which make a lot more sense.

88. M. Morris, K. Thorne and U. Yurtsever. *Physical Review Letters* 61, 1446 (1988).

89. C. Sagan. *Contact* (Simon and Schuster, New York, 1986).

90. J. Friedman. *Nature* 336, 305 (1988).

91. Technically, the spacetime analogue of a Newtonian trajectory is known as a 'worldline'.

92. Interview of R. Penrose with PVC and RRH, Oxford, May 1989.

CHAPTER FOUR

The introductory Schrödinger quotation is from a letter to Willy Wien, dated 26 August 1926; quoted in W. Moore. *Schrödinger, Life and Thought* (Cambridge University Press, 1989), p. 225.

1. Nevertheless, Aristotle is also called 'the father of modern biology' since he was the first to undertake the systematic study of flora and fauna.

2. Quoted in N. Herbert. *Quantum Reality. Beyond the New Physics* (Anchor Press/ Doubleday, New York, 1985), p. 9.

3. Einstein's Ph.D. thesis, completed on 30 April 1905, had dealt with a new method for calculating the sizes of molecules and the number of molecules present in one mole of substance, called the Avogadro number, after Count Amedo Avogadro, a lawyer turned mathematical physicist who was born in Turin in 1776. Eleven days after Einstein's thesis was finished, the prestigious German physics journal *Annalen der Physik* received the manuscript of his first paper on Brownian motion.

4. *Dictionary of Scientific Biography*. (Scribner, New York, 1981), Vol. 11, p. 31.

5. D. Wilson. *Rutherford. Simple Genius* (Hodder & Stoughton, London, 1983), p. 295.

6. D. Flamm. 'Boltzmann's statistical approach to irreversibility', *University of Vienna Theoretical Physics Report: UWThPh 1989–4*, p. 8.

7. *Ibid.*, p. 8.

8. G. Gamow. *The Thirty Years that Shook Physics* (Dover, New York, 1966), p. 22.

9. *Albert Einstein and Michele Besso, Correspondence 1903–1955*, P. Speziali (ed.) (Hermann, Paris, 1972), p. 453, 12 December 1951, quoted in A. Pais, *'Subtle is the Lord . . .' The Science and Life of Albert Einstein* (Oxford University Press, 1982), p. 382.

10. G. Gamow. *The Thirty Years That Shook Physics*, *op. cit.*, p. 81.

11. W. Moore. *Schrödinger, Life and Thought*, *op. cit.*, p. 187.

12. G. Gamow. *Biography of Physics* (Hutchinson, London, 1962), p. 237.

13. Professor Peter Landsberg told us: 'I met Bohr and he did make a heavy, slow impression on me, interminably and largely unsuccessfully lighting his pipe. However, his ideas in physics had subtlety.' (Personal communication with the authors, December 1989.)

14. More specifically, the allowed orbits are constrained by the quantisation of the electron's angular momentum in a well defined but nevertheless somewhat arbitrary way.

15. E. Schrödinger. *Sitzber. Preuss. Akad. Wiss. Phys. – Math. Kl.*, C-CII (1929), quoted in W. Moore. *Schrödinger, Life and Thought*, *op. cit.*, p. 39.

16. *Ibid.*, p. 2.

17. He was also impressed by some applications of de Broglie's work made by Einstein in 1924 to describe quantum effects in gases.

18. W. Moore. *Schrödinger, Life and Thought*, *op. cit.*, p. 196.

19. *Ibid.*, p. 200.

20. G. Gamow. *The Thirty Years that Shook Physics*, *op. cit.*, p. 3.

21. He neglected the effects of relativity, though he had attempted to include them at the outset. This work had to wait for Dirac.

22. Corresponding to the limit of vanishing de Broglie wavelengths.

23. P. Dirac. *Proceedings of the Royal Society of London* (hereinafter *Proc. R. Soc. London*) A123, 713 (1929).

24. A. Pais. *'Subtle Is the Lord' . . . The Science and Life of Albert Einstein*, *op. cit.*, p. 443.

25. *Ibid.*, p. 15.

26. W. Heisenberg. *Der Teil und das Ganze* (Piper-Verlag, Munich, 1969).

27. Feynman's path integral technique is a widely used calculational tool in quantum field theory and equilibrium statistical mechanics. *See* R. P. Feynman and A. R. Hibbs. *Quantum Mechanics and Path Integrals* (McGraw-Hill, New York, 1965).

28. N. Herbert. *Quantum Reality*, *op. cit.*, p. 17.

29. *Ibid.*

30. J. Ourmson. *Berkeley* (Oxford University Press, 1982), p. 39.

31. Contrary to what many authors have maintained, the Copenhagen interpretation does *not* require reference to observers. It is *not* a subjective interpretation of quantum mechanics. This point is brought out later in the text.

32. But see also the comments and footnote 51 in Chapter Two.

33. M. Jammer. *The Philosophy of Quantum Mechanics* (Wiley, New York and London, 1974), Chapter Five.

34. A. Pais. *'Subtle Is the Lord'* . . . *The Science and Life of Albert Einstein*, *op. cit.*

35. *Ibid.*, p. 404.

36. Conversation between Oliver Penrose and PVC, Edinburgh, August 1989.

37. If we assume that quantum mechanics is a fundamental theory of matter, and not a statistical or ensemble theory as Einstein held to be the case.

38. Note the similarity with Aristotle's dual notions of *potentiality* and *actuality* with regard to phenomena.

39. Quantum theory has appeared to resurrect a number of Aristotle's ideas. Thus Bohr's interpretation of quantum theory is also in line with Aristotelianism concerning atomism: as we have seen, it gives short shrift to an objective independent atomic existence. Aristotle was one of the principal opponents of atomic theory in antiquity, maintaining that the idea was illogical.

40. Perhaps the probability of this would be close to zero if we were considering the wavefunction of a high-brow theatre company.

41. Roger Penrose's book *The Emperor's New Mind* (Oxford University Press, 1989), Chapter Eight, gives a simple numerical example demonstrating this.

42. An intellectual giant (and child prodigy), von Neumann did pioneering work on computers, new branches of mathematics like ergodic theory, game theory, C*-algebras and helped to invent the atomic- and H-bombs. He wrote a series of papers between 1925 and 1929 which culminated in his book, *The Mathematical Foundations of Quantum Mechanics* (trans. R. Beyer) (Princeton University Press, 1955).

43. Quoted in J. D. Barrow and F. J. Tipler. *The Anthropic Cosmological Principle* (Oxford University Press, 1986), p. 458. In a personal communication with the authors (December 1989), Barrow said it was a spoken comment by Hawking.

44. A. Leggett. 'Schrödinger's cat and its laboratory cousins', *Contemporary Physics* 25, 583 (1984). Leggett suggests certain tests of the paradox and discusses macroscopic quantum coherence. He believes that it may not be possible to extrapolate quantum laws to the macroscopic level.

45. We have not been able to establish whether Schrödinger had a particular aversion to the feline species. As regards the cat itself, the most convincing candidate seems to be Mr Mistoffelees, the 'Original Conjuring Cat', of T. S. Eliot. Eliot was at the Institute for Advanced Study in Princeton during 1948 when Einstein was there. Mistoffelees, the poem says, 'holds all the patent monopolies for performing surprising illusions'.

46. B. DeWitt and N. Graham (eds). *The Many-Worlds Interpretation of Quantum Mechanics* (Princeton University Press, 1973), p. 116.

47. W. Gratzer (ed.). *The Longman Literary Companion to Science* (Longman, Harlow, 1989), p. 156.

48. P. C. W. Davies. *Other Worlds* (Penguin, Harmondsworth, 1988), p. 192.

49. A. Rae. *Quantum Physics: Illusion or Reality?* (Cambridge University Press, 1986), p. 68.

50. B. d'Espagnat's book, *Une incertaine réalité* (Gautier-Villars, Paris, 1985, Eng. trans. *Reality and the Physicist* (Cambridge University Press, 1989)) is largely concerned with a discussion of how that notion of an independent reality, although blurred, can nevertheless be maintained within quantum theory.

51. H. Everett III. *Reviews of Modern Physics* 29, 454 (1957).

52. N. Herbert. *Quantum Reality*, *op. cit.*, p. 19. G. Benford. *Timescape* (Pocket Books, New York, 1981). J. Williamson. *The Legion of Time* (Sphere, London, 1977).

53. M. Fakhry. *A History of Islamic Philosophy* (Columbia University Press, New York, Second Edition, 1983); M. Jammer. *The Philosophy of Quantum Mechanics*, *op. cit.*, p. 517.

54. B. DeWitt. *Physics Today*, September 1970, p. 3.

55. Such cosmologists include Stephen Hawking (personal communication with the authors, November 1989).

56. P. C. W. Davies, quoted in A. Rae. *Quantum Physics: Illusion or Reality?*, *op. cit.*, p. 79. Paul Davies believes that he may have said this to Rae over the telephone which concurs with the fact that Rae 'can just about hear P. D. saying it in my mind' (personal communications of PCWD and AR with PVC, November 1989).

57. D. Deutsch. *Proc. R. Soc. London* A400, 97 (1985).

58. More correctly, the wavefunction exists in 'configuration space': if the system comprises N particles, this space is 3N-dimensional, since the wavefunction depends on the three spatial coordinates of each particle present.

59. A. Einstein, B. Podolsky and N. Rosen. *Physical Review* 47, 777 (1935).

60. These are the local hidden variable theories developed by David Bohm and others. The interested reader should consult the bibliography listed for this chapter.

61. A. Aspect, J. Dalibard and G. Roger. *Physical Review Letters* 49, 1804 (1982). This was only one of a series of experiments by various different people along similar lines over some 20 years.

62. However, this does not violate the theory of relativity since there is no *information* transferred between the two regions – the non-local correlations in the measurements are only manifest when data are compared *after* the experiment has been performed.

63. 'The Zeno's paradox in quantum theory', *Journal of Mathematical Physics* 18, 756 (1977). Earlier work along these lines was done by L. A. Khalfin. *Sov. Phys. JETP* 6, 1053 (1958); R. Winter. *Phys. Rev.* 123, 1503 (1961); W. Yourgrau, in *Problems in the Philosophy of Science*, I. Lakatos and A. Musgrave (eds.) (North-Holland, Amsterdam 1968), p. 191.

64. PVC conversation with G. Sudarshan, Brussels, 28 September 1989.

65. W. Itano, D. Heinzen, J. Bollinger and D. Wineland. 'Observation of the Quantum Zeno Effect', *Physical Review A* 41, 2295 (1990). RRH, *Daily Telegraph*, 25 November 1989.

66. One must be careful in maintaining that this experiment really does confirm the quantum Zeno effect. It can be argued that the laser causes a serious perturbation of the system and so cannot be regarded as a truly 'passive' probe.

67. W. Gratzer (ed.). *The Longman Literary Companion to Science*, *op. cit.*, p. 170.

68. cf. Chapter Three.
69. First suggested by E. Stuekelberg. *Helvetica Physica Acta* 14, 588 (1941); developed by R. Feynman. *Physical Review* 76, 749 (1949).
70. J. H. Christenson *et al. Physical Review Letters* 13, 138 (1964).
71. S. Hawking and W. Israel (eds). *General Relativity: An Einstein Centenary Survey* (Cambridge University Press, 1979), p. 583.
72. Often referred to as due to a so-called 'superweak' interaction – a sophisticated way of saying that we do not understand the nature of the forces at work.
73. By Einstein's mass–energy formula, the greater the particle-pair's mass, the greater the borrowed energy needs to be.
74. The spontaneous emission of light is a truly irreversible process, while stimulated absorption and emission are the time reverse of one another.
75. C. Llewellyn Smith, comment made during a Wolfson College lecture; the lecture was published later in J. Mulvey (ed.). *The Nature of Matter* (Oxford Univesity Press, 1981).
76. S. Hawking. *A Brief History of Time* (Bantam Books, New York, 1988), p. 157.
77. The four fundamental forces known in Nature are the strong and weak, electro-magnetic and gravitational interactions.
78. J. D. Barrow. *The World Within the World* (Oxford University Press, 1988), p. 197.
79. *See* note 30 in Chapter One.
80. A. Guth, 1982 (unknown citation), quoted in H. R. Pagels. *Perfect Symmetry* (Simon and Schuster, New York, 1985), p. 316.
81. E. Tryon. *Nature* 246, 396 (1973).
82. Robert Brout, François Englert and Edgar Günzig.
83. $E = mc^2$.
84. 'The creation of the universe as a quantum phenomenon', *Annalen der Physik* 115, 78 (1978)
85. Of Tufts University, Boston, in 1983.
86. J. Hartle and S. Hawking. *Physical Review D* 28, 2906 (1983).
87. S. Hawking. Halley Lecture, Oxford, June 1989.
88. In flat spacetime one can obtain a positive definite metric simply by writing t' = it, equivalent to the usual Wick rotation in special relativistic quantum field theory. But in curved spacetime, there is no simple way to get from a Lorentzian to a Euclidean metric. However, the path integral over all Euclidean metrics is in a sense equivalent to the path integral over all Lorentzian metrics, which is what Hawking's approach does through the introduction of imaginary time. (We are grateful to Stephen Hawking for clarification of this point in a personal communication with the authors, November 1989.)
89. S. Hawking. Halley Lecture, Oxford, June 1989.
90. *Ibid.*

CHAPTER FIVE

The introductory quotation appears in R. Penrose. *The Emperor's New Mind* (Oxford University Press, 1989), p. 371.
1. R. Graves. 'Counting the Beats', in *The Oxford Book of Twentieth Century English Verse* (Oxford University Press, 1973), p. 298.

2. C. P. Snow. *The Two Cultures and the Scientific Revolution. The Rede Lecture 1959* (Cambridge University Press), p. 14.

3. D. Hull, in *Entropy, Information, and Evolution*, B. Weber, D. Depew and J. Smith (eds) (MIT Press, 1988), p. 3.

4. M. Flanders and D. Swann. 'The First and Second Law', as sung on their record: *At the drop of another hat* (EMI Records, 1964).

5. A. Eddington. *The Nature of the Physical World* (Cambridge University Press, 1928), p. 75.

6. *Ibid*, p. 105.

7. Some people find it difficult to accept that the universe can be treated as an isolated thermodynamic system.

8. M. Flanders and D. Swann. 'The First and Second Law', *op. cit.*

9. M. Twain, (Kipling, *From Sea to Sea*, Letter 37).

10. This is because, for a classical system with gravitational attractions, the partition function diverges.

11. R. Caillois. 'Avant propos à la dissymétrie' in *Cohérences aventureuses* (Gallimard, Paris, 1976).

12. L. Boltzmann. Populäre Schriften Essay 1, *Ludwig Boltzmann, Theoretical Physics and Philosophical Problems*, B. McGuinness (ed.) (Reidel, Dordrecht, 1974), p. 15.

13. Boltzmann reasoned that it was through evolution that inanimate atoms had evolved into the human brain and laid the basis for the human spirit and emotions. *Ibid*.

14. A. Eddington. *The Nature of the Physical World, op. cit.*, p. 74.

15. In thermodynamics, remaining close to equilibrium is a necessary but not a sufficient condition for reversibility.

16. This is a subject of considerable contemporary research interest in non-equilibrium statistical mechanics.

17. L. Wheeler. *Josiah Willard Gibbs* (Yale University Press, 1962), p. 84.

18. *Ibid.*, p. 181.

19. This language is borrowed from dynamical systems theory, developed in more detail in Chapter Six.

20. This is not to say that its results are irrelevant to the real world. For example, one can calculate such time-independent features as how the boiling point of water changes with pressure and how much barometric pressure diminishes with height. These are useful pieces of information for mountaineering *aficionados* who wish to boil eggs at high altitudes.

21. P. T. Landsberg. *Thermodynamics with Quantum Statistical Illustrations* (Wiley, New York, 1961), p. vii.

22. Samuel Beckett. *Endgame* (Faber, London, 1964).

23. Awarded for his contributions to irreversible thermodynamics, specifically his derivation of the reciprocity relations.

24. The validity of this theorem is limited by the premises necessary to prove it. *See* e.g. K. G. Denbigh. *Transactions of the Faraday Society* 48, 389 (1952); R. Landauer. *Annals of the New York Academy of Sciences* 316, 433 (1975); G. Nicolis. *Reports on Progress in Physics* 42, 225 (1979).

25. I. Prigogine. *Etude thermodynamique des phénomènes irréversibles* (Desoer, Liège, 1947, and C. C. Thomas, Springfield, 1955).

26. This had already been published in *Académie Royale de Belgique, Bulletin de la Classe des Sciences* 31, 600 (1945).

27. There is at least one other 'rival' school of non-equilibrium thermodynamics that goes under the name of 'rational thermodynamics'. Pioneered by Truesdall, Coleman and others in the USA (C. Truesdall. *Rational Thermodynamics* (McGraw-Hill, New York, 1969)), it makes different but similarly restrictive assumptions, as a result of which it can be used to describe a number of quite different phenomena (notably two types of continuous media: elastic materials with linear viscosity and those with 'memory') from those to which generalised thermodynamics is applicable. Because of the great difference in their outlooks and interests, there is not much contact between these different schools of thought. However, the reader should consult the reference to D. Jou *et al.* listed in the Bibliography to this chapter for recent developments in irreversible thermodynamics by the Catalan School, which do bridge the divide in some respects.

28. The Brussels School does not use the term 'generalised thermodynamics'.

29. This is a good approximation for systems whose properties do not vary too rapidly in space.

30. P. Glansdorff and I. Prigogine. *Thermodynamic Theory of Structure, Stability and Fluctuations* (Wiley, New York, 1971).

31. Technically speaking, it provides a sufficient, rather than a necessary condition for instability of the thermodynamic branch.

32. J. Keizer and R. Fox. 'Qualms regarding the range of validity of the Glansdorff–Prigogine criterion for the stability of non-equilibrium states', *Proceedings of the National Academy of Sciences of the USA* 71, 192 (1974); *see also* the reply by G. Nicolis, P. Glansdorff and I. Prigogine. *Proceedings of the National Academy of Sciences of the USA* 71, 197 (1974).

33. The first story in *The Memoirs of Sherlock Holmes* by Arthur Conan Doyle (Penguin, Harmondsworth, 1969).

34. L. Boltzmann. *Lectures on Gas Theory*, trans. S. Brush (University of California Press, 1964), p. 9.

35. Lord Kelvin, quoted in J. D. Barrow. *The World Within the World* (Oxford University Press, 1988), p. 126. The vision of a time-reversed world has been used to great effect by Martin Amis in his short story, *Time's Arrow* (*Granta*, 33, summer 1990). Meals provide just one example of the bizarre results: 'Various materials are gulped up into my mouth, and after I've massaged them into shape with my tongue and teeth I transfer them on to the plate and sculpt them up with knife and fork.'

36. We are here pretending that the gas can be treated ideally; only two variables are needed for a fixed mass of gas. Deviations from the ideal gas laws arise due to the presence of intermolecular forces inside the gas.

37. *See* e.g. R. Balescu. *Equilibrium and Non-Equilibrium Statistical Mechanics* (Wiley-Interscience, New York, 1975); R. Baxter. *Exactly Solved Models in Statistical Mechanics* (Academic Press, New York, 1982).

38. The canonical partition function, for example, depends on the temperature, a quantity which is strictly defined at thermodynamic equilibrium.

39. These are situations in which the effects of quantum statistics, or equivalently the Pauli principle, cannot be ignored but which are not present in the classical theory.

330

40. W. Moore. *Schrödinger, Life and Thought* (Cambridge University Press, 1989), p. 37. This is an 'initial conditions' argument.

41. A. Beck and P. Havas (eds). *The Collected Papers of Albert Einstein* (Princeton University Press, 1987), Vol. 1, p. 149.

42. L. Boltzmann. Populäre Schriften, *Ludwig Boltzmann, Theoretical Physics and Philosophical Problems*, B. McGuinness (ed.) (Reidel, Dordrecht, 1974), p. 146.

43. *Ibid.*, p. 9.

44. *See* e.g. S. G. Brush. *The Kind of Motion We Call Heat* (North-Holland, Amsterdam, 1976); also L. Boltzmann. *Nature* 51, 413 (1895).

45. L. Boltzmann. *Lectures on Gas Theory, op. cit.*, p. 447; also I. Prigogine and I. Stengers. *Order out of Chaos*, (Heinemann, London, 1984), p. 254.

46. D. Flamm in *The Boltzmann Equation*, E. Cohen and W. Thirring (eds) (Springer, Vienna, 1973), p. 5; J. Blackmore. *Ernst Mach* (University of California Press, 1972), p. 205.

47. J. Blackmore. *Ernst Mach, op. cit.*, p. 207.

48. Where he took over Mach's lectures on natural philosophy.

49. E. Cohen and W. Thirring (eds). *The Boltzmann Equation, op. cit.*, p. 12.

50. J. Blackmore. *Ernst Mach, op. cit.*, p. 212.

51. Perrin received the 1926 Nobel Prize for his work on Brownian motion.

52. L. Boltzmann. Populäre Schriften 5, *Ludwig Boltzmann, Theoretical Physics and Philosophical Problems*, B. McGuinness (ed.) (Reidel, Dordrecht, 1974), p. 33. In philosophy, Boltzmann advocated the 'complementarity of contradictory hypotheses', according to which contradictory theories of the natural world may yet all be correct. Such theories should be regarded as complementary rather than antagonistic. It is remarkable that Boltzmann's thinking here adumbrated so clearly Niels Bohr's later 'principle of complementarity' based on quantum theory in which, to properly describe the world, one must be able to combine mutually exclusive concepts with varying degrees of precision.

53. E. Jaynes. *Physical Review* 108, 171 (1957), quoted in K. Denbigh and J. Denbigh. *Entropy in Relation to Incomplete Knowledge* (Cambridge University Press, 1985), p. 107.

54. One can argue that both information flow along a cable and statistical mechanics follow from a similar set of axioms. Thus in the subjective Jaynesian approach, statistical mechanics is not taken to furnish an objective description of reality but is instead regarded as a way of making statistical inferences. A refutation of this line of argument is given by J. Jauch and J. Báron in *Helvetica Physica Acta* 45, 220 (1972). It is not even clear whether the approach is capable of making predictions rather than simply furnishing retrodictive explanations. Similar approaches are sometimes advocated in quantum theory to explain wavefunction collapse as solely due to a change in our knowledge about the system's state on making a measurement.

55. K. Denbigh and J. Denbigh. *Entropy in Relation to Incomplete Knowledge, op. cit.*, p. 104.

56. I. Prigogine and I. Stengers. *Order out of Chaos, op. cit.*, p. 285.

57. It is not obvious that Heat Death could ever be attained since, at least in classical mechanics, the amount of gravitational energy available is infinite.

58. There is in addition the vexed question of the value of the cosmological constant, also mentioned in Chapter Three, which may complicate matters a great deal.

59. R. Morris. *Time's Arrows* (Touchstone, New York, 1985), p. 211. However, it is claimed that intelligent beings would have their thought processes reversed so they would not notice the difference (S. Hawking. *A Brief History of Time* (Bantam Books, New York, 1988), p. 15).

60. M. Berry. *Principles of Cosmology and Gravitation* (Adam Hilger, Bristol, 1989), p. 133.

61. This is Penrose's so-called 'Weyl Curvature Hypothesis', that the Weyl curvature vanishes at initial singularities, whereas it explodes to infinity for final ones.

62. Conversation between Roger Penrose and the authors, Oxford, June 1989.

63. See e.g. R. Penrose. *The Emperor's New Mind, op. cit.*, Chapters Seven and Eight. This much, at least, is in agreement with the Hawking 'no boundary universe' discussed in Chapter Four. However, as we pointed out there, Hawking's approach does not contain any time-asymmetric features. Hawking is a supporter of the coarse-graining route to irreversibility. He would maintain that the irreversibility inherent in all the processes of life – to be discussed in Chapters Six and Seven – is due to coarse-graining. He also does not regard thermodynamic entropy as objective. (Stephen Hawking, personal communication with the authors, November 1989).

64. Conversation between Roger Penrose and the authors, Oxford, June 1989.

65. R. Penrose. *The Emperor's New Mind, op. cit.*, p. 367.

66. *Ibid.*, p. 371.

67. Paul Ehrenfest maintained the line of thinking pioneered by Boltzmann, believing that the explanation for irreversibility must come from within mechanics. By 1933 he too had killed himself. P. Ehrenfest and T. Ehrenfest. *Enzyklopedie der Mathematischen Wissenschaften* (Leipzig, 1911), English translation, *The Conceptual Foundations of Statistical Mechanics* (Cornell University Press, Ithaca, 1959).

CHAPTER SIX

The title of this chapter is taken from Henri Bergson's *L'évolution créatrice*, in *Oeuvres*, (Editions du Centenaire, Presses Universitaires de France, Paris, 1963), English title *Creative Evolution*, trans. A. Mitchell (Macmillan, London, 1911).

The introductory quotation in I. Kant. *Universal Natural History and Theory of the Heavens*, trans. S. Jaki (Scottish Academic Press, Edinburgh, 1981), p. 87; *see also* J. D. Barrow. *The World Within the World* (Oxford University Press, 1988), p. 214.

1. R. Field and F. Schneider. *Journal of Chemical Education* 66(3) (1989).

2. D. Ruelle. *Mathematical Intelligencer*. 2, 126 (1980).

3. Recall the resolution of Zeno's Achilles paradox, mentioned in Chapter One.

4. Contained within the nucleus of every eukaryotic cell.

5. However, it is simplistic to suggest that nobody knew about non-linearity until recently. The great applied mathematicians of the nineteenth century based their analyses of shock waves and similar phenomena in fluid dynamics on the non-linear properties of the (Navier–Stokes) fluid equations of motion. Poincaré was again a major figure in this endeavour with his work on singular perturbation theory.

6. Although the underlying phenomenon of convection was recognised as long ago as 1798 by Count Rumford, who explained the transport of heat in an apple pie on this basis. *See* S. Chandrasekhar. *Hydrodynamic and Hydromagnetic Stability* (Oxford University Press, 1961), p. 9.

NOTES

7. Forces between molecules act over microscopic distances, around 10^{-8} cm, whereas the correlations in the Rayleigh–Bénard cells range over macroscopic lengths, typically around 1 cm.

8. G. Nicolis, in *The New Physics*, P. C. W. Davies (ed.) (Cambridge University Press, 1989), p. 319.

9. The precise geometries which can arise are many and varied – the hexagonal pattern is only the commonest under normal conditions. *See* M. G. Verlarde and C. Normand. *Scientific American* 243, 92 (1980).

10. Nevertheless, despite its familiarity to chemical engineers, the continuous-flow stirred-tank reactor was only introduced into experimental laboratory chemistry by the group at the Paul Pascal Research Centre in Bordeaux, France, during the 1970s.

11. Non-linearity is a necessary but not a sufficient condition. Non-linearity is actually quite common in chemical reactions yet other more subtle requirements are needed which are less easily satisfied, explaining in large measure why such phenomena are relatively uncommon in chemistry itself (*see* C. Vidal and H. Lemarchand. *La réaction créatrice*, Chapter Five, listed in the Bibliography). On the other hand, in biochemistry it is much easier to find reactions which generate self-organisation, as we shall see in Chapter Seven.

12. A. Turing. 'The Chemical Basis of Morphogenesis', *Pro. R. Soc. London* B237, 37 (1952).

13. *Ibid.*, p. 44.

14. Turing's lover, Arnold Murray, had been linked with the crime. Turing ended up telling all, even providing the police with a five-page hand-written statement on his activities.

15. A. Hodges. *Alan Turing. The Enigma* (Burnett Books, London, 1983), p. 472.

16. *Ibid.*, p. 149.

17. B. Chance, A. K. Ghosh, E. K. Pye and B. Hess. *Biological and Biochemical Oscillators* (Academic Press, New York, 1973); *also* C. Vidal and P. Hanusse. *International Review of Physical Chemistry* 5, 1 (1986).

18. I. Prigogine and R. Lefever. *Journal of Chemical Physics* 48, 1695 (1968).

19. In recent times there have been a number of mounting attacks on the integrity of the Brusselator; for instance B. Gray and T. Morley-Buchanan. *Journal of the Chemical Society, Faraday Transactions 2* 81, 77 (1985). These are refuted by R. Lefever, G. Nicolis and P. Borkmans. *Journal of the Chemical Society, Faraday Transactions 1* 84, 1013 (1988).

20. The first person to suggest a sustained oscillation in a chemical system was Alfred J. Lotka in 1920. An identical set of differential equations was proposed quite independently by Vito Volterra in 1926 to represent population dynamics in fish colonies. Their set of equations is known as the 'Lotka–Volterra model' which, however, does not exhibit limit-cycle behaviour – the long-time behaviour is dependent on the initial conditions. (*See also* Chapter Seven.)

21. *See* Chapter Two for a short discussion of how differential equations arise in these circumstances.

22. In an oscillating chemical reaction based on chlorite, iodide and malonic acid discovered by a group of scientists from the University of Bordeaux, Brandeis University and L. Eötvös University, white dots appear that grow into concentric rings and then expand to annihilate each other on collision (*Scientific American* 248, 96 (1983)).

23. J. Murray. *Mathematical Biology* (Springer-Verlag, Heidelberg, 1989), p. 450.
24. The Brusselator has found some applications in the study of instabilities in laser physics; it is, however, too contrived to describe any real chemical reaction.
25. A. Winfree. *Journal of Chemical Education* <u>61</u>, 661 (1984).
26. The Krebs cycle, also known as the citric acid or tricarboxylic acid cycle, is the final stage in the respiratory chain. It is named after Hans Adolf Krebs, awarded the Nobel Prize for Physiology and Medicine in 1953 for his work on elucidating this pathway.
27. Art Winfree believes that Belousov's observation of patterns is 'largely apocryphal'. (Personal communication with the authors, November 1989). It seems as if Heinrich Busse in Germany had the first substantial instance in 1969, but he completely misinterpreted it (as a stationary spatial pattern rather than a 'pseudo-wave').
28. A. Winfree. *When Time Breaks Down* (Princeton University Press, 1987), p. 13.
29. B. Belousov. 'Oscillation Reaction and Its Mechanism' (Russian title), *Sbornik Referatov po Radaicioni Medicine, Medgiz, Moscow* <u>1</u>, 145, 1958 Meeting (1959). Note that Belousov's first paper had been rejected before Turing committed suicide. What might have happened had they met?
30. Art Winfree has drawn our attention to the fact that the referees may just possibly have had alternative 'cogent' explanations in mind. Not much is known about what the referees really said, but the tables have now turned completely, with the making of a Russian film *Never Say Never* in which Belousov is glorified and the establishment sarcastically denounced.
31. W. Bray. *Journal of the American Chemical Society* <u>43</u>, 1262 (1921).
32. The iron catalyst was ferroin.
33. They were gathered by chain-letter: in the USA, Art Winfree rounded up scientists (A. Winfree, personal communication to the authors, November 1989).
34. One by Zhabotinsky in 1974, in Russian, the other by J. Tyson in 1976. Many Western scientists attribute principal credit for the elucidation of the mechanism of the BZ reaction to Richard Field, Endre Körös and Richard Noyes working at the University of Oregon in the USA during the early 1970s. This view is disputed by Soviet scientists, who think 'that American scientists belatedly cooked up a fanciful mechanism and sold it as gospel, while much more accurate models were already in use and have since been much improved in the USSR', Winfree told us. This is hard to sort out since Americans steadfastly refuse to read Russian or to acknowledge the Russian literature so that comparisons are not made. 'It is certainly the case that the FKN [Field, Körös, Noyes] model, especially the Oregonator version, has glaring deficiencies,' (A. Winfree, personal communication with the authors, November 1989).
35. T. Briggs and W. Rauscher. *Journal of Chemical Education* <u>50</u>, 496 (1973).
36. R. Noyes. *Journal of Chemical Education* <u>66</u>, 3 (1989).
37. See e.g. L. Kuhnert *et al. Nature* <u>337</u>, 244 (1988); or H. Busse and B. Hesse. *Nature* <u>244</u>, 203 (1973). Just as with all kinds of limit-cycle oscillators, including circadian clocks and cardiac pacemakers, the BZ reaction can be entrained by periodic stimuli, which constitute the excitation. *See* Chapter Seven and the Appendix for further discussion.
38. It has nothing to do with its ability to support spontaneous oscillations.
39. A. Winfree. *When Time Breaks Down, op. cit.*, p. 216.

40. This philosophy has even filtered down to the undergraduate level. The size of F. Cotton and G. Wilkinson's widely acclaimed treatise *Advanced Inorganic Chemistry* (Wiley-Interscience, New York, 1988) continues to increase with each edition. Yet the conceptual framework has been pared down to an absolute minimum as the authors seek to report more and more purely factual material – they themselves admit this trend in the Introduction. Exponential growth in research publications is a feature of other fields of scientific research. According to Winfree, 'starting in the mid-1960s and continuing still, the annual rate of publication on oscillating reactions has doubled every three years, five times the doubling rate for chemistry as a whole.' (A. Winfree. *When Time Breaks Down*, *op. cit.*, p. 163).

41. I. Prigogine and I. Stengers. *Order out of Chaos* (Heinemann, London, 1984), p. 153.

42. Note the distinction between this behaviour and disorder at equilibrium: deterministic chaos is due to an overloading of order.

43. D. Ruelle and F. Takens. *Communications in Mathematical Physics* 20, 167 (1971), when Ruelle was at the Princeton Institute for Advanced Study and Takens at the University of Amsterdam, although a subsequent reference (*Communications in Mathematical Physics* 23, 343 (1971)) gives credit to some Russian work in the field. *See also* S. Newhouse, D. Ruelle and F. Takens. *Communications in Mathematical Physics* 64, 35 (1978).

44. To be correct, it seems that strange attractors are useful for describing the *onset* of turbulence, but not necessarily for fully developed turbulence, which remains largely a mystery. (The same distinction holds good in cardiology, between fibrillation and its onset in tachycardia caused by rotating waves. Heart attacks are discussed in Chapter Seven.) Essentially, Ruelle and Takens showed that any system which passes through three or more successive Hopf (limit-cycle) bifurcations (regardless of the initial conditions) will necessarily end up in a chaotic state.

45. D. Ruelle. *Mathematical Intelligencer* 2, 126 (1980). To be complete, mention should be made of 'transients'. These are phenomena which arise due to the initial preparation of a system, but die out during the time evolution. Thus, they are not associated with any attractor. Often they look chaotic, but are not in the technical sense of the word. Examples include milk being poured into coffee and Ruelle's cigarette smoke. In each case the attractor is the homogeneous state of thermodynamic equilibrium.

46. *Ibid.*, p. 131.

47. E. Zeeman. *Bulletin of the London Mathematical Society*, preprint 1989. Many are indeed known today, including those of Lorenz, Hénon and Rössler, the solenoid, the logistic map (in the latter case it is strictly speaking a chaotic repellor). However, none has been rigorously demonstrated to fulfil all the mathematical conditions of the Ruelle–Takens definition. 'Chaotic attractor' is to be preferred in technical parlance, since it is possible to find strange but non-chaotic attractors. *See* note 65.

48. Mandelbrot currently divides his time between the IBM Research Laboratory at Yorktown Heights, New York State, and the Department of Mathematics at Yale University.

49. B. Mandelbrot. *The Fractal Geometry of Nature* (W. H. Freeman, New York, 1983) and *Science* 156, 636 (1967).

50. The idea of treating dimension as some kind of continuous variable was present in the work of Kenneth Wilson at Cornell University, USA, which came to fruition in the late 1960s and early 1970s on the renormalisation group approach to phase transitions and critical phenomena, for which he won the 1982 Nobel Prize for Physics. His ideas placed a heavy emphasis on the notion of self-similarity, in which the behaviour of matter looks the same on all length scales as one goes to a critical point. Wilson's work was picked up and developed by Feigenbaum.

51. H-O. Peitgen and P. Richter. *The Beauty of Fractals. Images of complex dynamical systems* (Springer-Verlag, Heidelberg, 1986).

52. B. Mandelbrot. *The Fractal Geometry of Nature, op. cit.*, p. 150. Neurons have also been shown to be fractal objects (F. Caserta, H. Stanley, W. Eldred, G. Daccord, R. Hausman and J. Nittmann. *Physical Review Letters* 64, 95 (1990)).

53. *Ibid.*, p. 113.

54. D. Ruelle. *Transactions of the New York Academy of Sciences* 35, 66 (1973).

55. The trajectories diverge at an exponential rate characterised by a positive definite Lyapounov exponent. Strange non-chaotic attractors have only non-positive Lyapounov exponents.

56. The popular books by Gleick and Stewart cited in the bibliography to this chapter are cases in point, wherein the focus is so much on chaos that self-organisation is overlooked. One could reasonably maintain that chaos is a special form of self-organisation, but certainly not the converse.

57. D. Ruelle. 'Some comments on chemical oscillators', *Transactions of the New York Academy of Sciences* 35, 66 (1973).

58. D. Ruelle. *Mathematical Intelligencer, op. cit.*, p. 136.

59. This at least was progress. Recall how the self-organising features of the Bray and BZ reactions were originally rejected as artefacts.

60. Rather complex patterns of alternate order and chaos can occur in the BZ reaction as the flow rate increases. *See* M. Markus and B. Hess, 'Isotropic Cellular Automata for Minimal Modelling of Excitable Media', submitted as a letter to *Nature* (1989), which shows a method to model the transition with cellular automata.

61. J. Roux, S. Simoyi and H. Swinney. *Physica D* 8, 257 (1983); W. Tam *et al. Physical Review Letters* 61, 2163 (1988).

62. According to B. Hess and M. Markus (*Trends in Biochemical Sciences* 12, 45 (1987)), the first example of (bio)chemical chaos was found in an enzymatic peroxidase reaction, in which the reduction of oxygen is catalysed by NADH; L. Olsen and H. Degn. *Nature* 267, 177 (1977); and L. Olsen. *Physical Letters* 94A, 454 (1983).

63. E. Lorenz. *Journal of Atmospheric Science* 20, 130 (1963).

64. Note the similarity with the Rayleigh–Bénard instability discussed earlier. In the late 1970s, Albert Libchaber studied in exquisite detail convection instabilities in liquid helium at very low temperatures and found the period-doubling cascade (described later in this chapter) to chaos as the temperature was increased. Popular accounts of his experiments can be found in the books by Gleick and Stewart mentioned in the bibliography to this chapter.

65. Y. Pomeau and P. Manneville. *Communications in Mathematical Physics* 74, 189 (1980). An interesting recent paper describes a route to chaos via strange

non-chaotic attractors: T. Kapitaniak, E. Ponce and J. Wojewoda. *Journal of Physics A: Mathematical and General* (hereinafter *J. Phys. A: Math. Gen.*) 23 L383 (1990).

66. The full subharmonic cascade is difficult to observe in the BZ reaction.

67. D. Ruelle. *Mathematical Intelligencer, op. cit.*, p. 137.

68. I. Prigogine and I. Stengers. *Order out of Chaos, op. cit.*, p. 129.

69. H. Bergson. *L'évolution créatrice*, in *Oeuvres*; Eng. title *Creative Evolution*, trans. A. Mitchell (Macmillan, London, 1911).

70. *Dictionary of Scientific Biography* (Scribner, New York, 1981). *See also* the discussion of Teilhard's ideas in J. D. Barrow and F. J. Tipler. *The Anthropic Cosmological Principle* (Oxford University Press, 1986), Chapter Three.

71. To J. D. Hooker, quoted in *Origin of Life*, Y. Wolman (ed.) (Reidel, Dordrecht, 1981), p. 1.

72. S. Ramón y Cajal. *Recollections of My Life* (Garland, New York and London, 1988), p. 294.

73. Oxygen was a 'pollutant' that appeared later through photosynthesis.

74. S. Miller. *Science* 117, 528 (1953); and S. Miller and L. Orgel. *The Origins of Life on the Earth* (Prentice-Hall, New Jersey, 1974), p. 55.

75. J. D. Bernal. *The Origin of Life* (Weidenfield & Nicolson, London, 1967), p. 8.

76. A. Cairns-Smith. *Journal of Theoretical Biology* 10, 53 (1966); *Genetic Takeover* (Cambridge University Press, 1982).

77. J. D. Bernal. *The Origin of Life, op. cit.*, p. 2.

78. M. Ho, P. Saunders and S. Fox. *New Scientist* 27 February (1986).

79. C. Avers. *Molecular Cell Biology* (Addison-Wesley, Massachusetts, 1986).

80. Nobel citation, Royal Swedish Academy of Sciences Information Department, 12 October 1989. *See also* J. Rajagopal, J. Doudna, J. Szostak. *Science* 244, 692 (1989); J. McSwiggen and T. Cech. *Science* 244, 679 (1989).

81. R. Dawkins. *The Blind Watchmaker* (Longman, Harlow, 1986), p. 116.

82. A. Wilson. *Scientific American* 253, 148. However, there is evidence that the mutation rate of mitochondrial DNA does vary. In studies of macaque monkeys, Dr Donald Melnick of Columbia University of New York found that social organisation affects the clock: in macaque groups where females always remain and males always leave, different populations diverge rapidly and the genetic clock is ten times faster than the standard clock. (Conversation with RRH, March 1990.)

83. M. Eigen, B. Lindemann, M. Tietze, R. Winkler-Oswatitsch, A. Dress and A. von Haeseler. *Science* 244, 673 (1989).

84. They have dealt with autocatalytic systems and a combination of auto- and cross-catalysis which can give rise to non-linear dynamics.

85. M. Eigen and P. Schuster. *Journal of Molecular Evolution* 19, 47 (1982); W. Fontana, W. Schnabl and P. Schuster. *Physical Review A* 40, 3301 (1989); M. Eigen, J. McCaskill and P. Schuster. *Journal of Physical Chemistry* 92, 6881 (1988).

86. Conversation between RRH and Stuart Kauffman, April 1990. *See also: Science* 247, 1543 (1990); S. Rasmussen in *Artificial Life*, C. Langton (ed.) (Addison-Wesley, Redwood City, 1989), p. 79; J. Farmer, S. Kauffman and N. Packard. *Physica D* 220, 50 (1986).

337

CHAPTER SEVEN

The introductory quotation is given in A. Peacocke. *God and the New Biology* (Dent, London, 1986), p. 160.

1. L. Glass and M. Mackey. *From Clocks to Chaos* (Princeton University Press, 1988), p. 72.

2. *Ibid.*, p. 73.

3. The fundamental difficulty is due to inherent complexity. One can never be sure, in making a mathematical model, that one has really included all the influences. One can only judge this using Occam's razor, and going for the simplest model that is in (closest) agreement with all the facts (and this may or may not produce predictions which agree with subsequent experiments – cf. falsifiability in the Popperian sense).

4. *See* e.g. D. Ruelle. 'Deterministic chaos: the science and the fiction', Claude Bernard Lecture of the Royal Society, 23 October 1989, published in *Proc. R. Soc. London* A (1990). A widely used method is the Grassberger–Procaccia algorithm (*Physica D* 9, 189 (1983)), which gives meaningless results with inadequate time-series data.

5. E. Racker. 'From Pasteur to Mitchell: a Hundred Years of Bioenergetics', *Federal Proceedings* 39, 210 (1980). Cells such as yeast can switch reversibly between fermentation and oxidative phosphorylation. While it is true that glycolysis gives stoichiometrically far fewer molecules of ATP per molecule of glucose metabolised (2 vs. 36), by virtue of the Pasteur effect, when the cell does switch from aerobic to anaerobic existence, it makes up for the drop in stoichiometry by increasing the glycolytic flux.

6. B. Hess. 'Oscillating Reactions', *Trends in Biochemical Sciences* 2, 193 (1977).

7. A. Babloyantz. *Molecules, Dynamics and Life* (Wiley, New York, 1986), p. 255. The Egyptians were also the first to make beer.

8. An excellent quantitative model was later given by M. Markus and B. Hess. *Proceedings of the National Academy of Sciences of the USA* 81, 4394 (1984). This predicted limit-cycles, hysteresis and chaos. Dissipative structures have recently been discovered in solutions of microtubules, believed to be the principal organisers of the cell interior (J. Tabony, private communication with RRH).

9. B. Hess. *Trends in Biochemical Sciences* 2, 37 (1985).

10. B. Hess and M. Markus. *Trends in Biochemical Sciences* 12, 45 (1987). Analysis of experimental data for glycolysis has indicated a strange attractor of dimension 2.2, implying that any three independent variables are enough to describe the complex dynamics, in spite of the far larger number of metabolites involved.

11. B. Alberts *et al.* *The Molecular Biology of the Cell* (Garland Publishing, New York, Second Edition, 1989), pp. 13–25.

12. The mechanism of cell ageing is still something of a puzzle. One common explanation is that ageing occurs as errors accumulate in the cell's biosynthetic machinery. For instance, telomeres have been found to decrease as a function of cell ageing (C. Harley *et al.* *Nature* 345, 458, (1990)).

13. A. Picard, J. Labbé and M. Dorée. *La Recherche* 20, 800 (1989).

14. M. Lohka *et al.* *Proceedings of the National Academy of Science of the USA* 85, 3009 (1988); and J. Gautier *et al.* *Cell* 54, 433 (1988).

338

15. M. Lee and P. Nurse. *Nature* 327, 31 (1987). Subsequent quotations in the text attributed to Paul Nurse arose in an interview with the authors, Trinity 1989.

16. She took yeast cells, unable to divide because they contained a defective *cdc 2* kinase gene, and inserted a range of human genes. One was found to restore the ability of the yeast to divide. While the yeast protein was made up of 297 protein building blocks – amino acids – the human one had one more. Two-thirds of the amino acids in the human protein were identical to the yeast protein. It was indeed a human version of *cdc 2*.

17. *See* A. Winfree. *The Geometry of Biological Time* (Springer-Verlag, Heidelberg, 1980), Chapter 22, Section D. The quotation is from A. Winfree (personal communication with the authors, November 1989).

18. A. Winfree. *The Geometry of Biological Time, op. cit.*, p. 337.

19. A. Winfree. *The Timing of Biological Clocks* (W. H. Freeman, New York, 1987), p. 175.

20. Located on the interior of the cell wall.

21. Via a coupling through the cell membrane.

22. A. Goldbeter. *Nature* 253, 540 (1975); for more recent developments *see* J. L. Martiel and A. Goldbeter. *Biophysical Journal* 52, 807 (1987); J. J. Tyson and J. D. Murray. *Development* 106, 421 (1989).

23. B. Hess and M. Markus. *Trends in Biochemical Science* 12, 45 (1987); also M. B. Coukell and F. K. Chan. *FEBS Lett.* 110, 39 (1980).

24. J. Seger. *Nature* 337, 305 (1989).

25. *Brewer's Dictionary of Phrase and Fable* (Cassell, London, 1981), p. 580.

26. H. Meinhardt. *Models of Biological Pattern Formation* (Academic Press, London, 1982), p. 12. Reaction-diffusion models account for the symmetry-breaking of homogeneous steady-states, and so 'potentiate' evolutionary transitions. Other things must happen subsequently to stabilise the resulting forms (BZ reaction structures are relatively fragile). Evolutionary theorists subdivide biological evolution into 'anagenesis', 'stasigenesis' and 'cladogenesis'. Thus it would seem that the Brussels School notion of dissipative structures applies primarily to the anagenesis part.

27. B. Alberts *et al. The Molecular Biology of the Cell, op. cit.*, p. 911.

28. *See* J. Murray. *Mathematical Biology* (Springer-Verlag, Heidelberg, 1989).

29. S. Turing. *Alan M. Turing* (Heffer, Cambridge, 1959), p. 105.

30. Interview of J. D. Murray with RRH, *Daily Telegraph*, 30 November 1987.

31. C. Thaller and G. Eichele. *Nature* 327, 625 (1987).

32. W. Driever and C. Nüsslein-Volhard. *Cell* 54, 83 and 95 (1988).

33. J. Murray and P. Maini. *Science Progress* (Oxford) 70, 539 (1986).

34. Actually, the figure may be 800 million for all other species but Man and shrew, because both are off the universal regression line by a factor of around three, the former on account of anomalous lifespan, the latter due to anomalous heart rate. *See* T. McMahon and J. Bonner. *On Size and Life* (W. H. Freeman, New York, 1983).

35. B. van der Pol. *Philosophical Magazine* 2, 978 (1926); and J. van der Mark. *Philosophical Magazine* 6, (suppl.), 763 (1928); the van der Pol oscillator is also well known to laser physicists, where it again gives different stability regimes arising from bifurcation phenomena.

339

36. L. Glass and M. Mackey. *From Clocks to Chaos*, *op. cit.*, p. 21.

37. Conversation between Denis Noble and the authors, June 1989.

38. L. Glass and M. Mackey. *From Clocks to Chaos*, *op. cit.*, p. 160.

39. S. Muller, T. Plesser and B. Hess. *Leonardo* 22, 5 (1989).

40. It is through triggering the release of adenylate cyclase that adrenaline quickens the heart beat.

41. L. Glass and M. Mackey. *From Clocks to Chaos*, *op. cit.*, p. 141. The latest evidence to emerge as the book went into production came from D. Chialvo, R. Gilmour and J. Jalife. *Nature* 343, 653 (1990).

42. M. Markus, D. Kuschmitz and B. Hess. *Biophysical Chemistry* 22, 95 (1985).

43. R. Pool. *Science* 243, 604 (1989).

44. A. Winfree. *The Timing of Biological Clocks*, *op. cit.*, p. 178.

45. N. Shibata, P-S. Chen, Dixon, P. Wolf, N. Daniely, W. Smith and R. Ideker. *American Journal of Physiology* 255, H891 (1988).

46. A. Winfree, personal communication with the authors, November 1989.

47. A. Winfree. *When Time Breaks Down* (Princeton University Press, 1987), p. 288. *See also* note 44 to Chapter Six.

48. J. Horne. *Why We Sleep* (Oxford University Press, 1988), p. 8.

49. If the time-series data used were inadequate, such results become meaningless. David Ruelle has studied EEG data, but he concluded that the data had been pre-filtered and was hence useless for such analyses; Claude-Bernard Lecture, Royal Society of London, 23 November 1989.

50. L. Glass and M. Mackey. *From Clocks to Chaos*, *op. cit.*, p. 178.

51. *Ibid.*, p. 176.

52. I. Petersen and I. Stener. *Electromyography* 10, 23 (1970); L. Glass and G. Mackey. *From Clocks to Chaos*, *op. cit.*, p. 93.

53. L. Glass and M. Mackey. *From Clocks to Chaos*, *op. cit.*, p. 97.

54. *Ibid.*, p. 95.

55. I. Stewart. *Does God Play Dice? The Mathematics of Chaos* (Blackwell, Oxford, 1989), p. 267.

56. The best known model of the 'living planet' where life influences its environment is the 'Gaia' hypothesis of the British scientist James Lovelock. *See* J. Lovelock. *The Ages of Gaia* (Oxford University Press, 1988).

57. Technically, these are often actually discrete *difference* equations rather than continuum differential equations, but this distinction is not significant in the analysis.

58. The Lotka–Volterra equations do not lead to a limit-cycle as it happens, the oscillations being sensitive to the initial predator-prey populations. This was mentioned in note 20 to Chapter Six.

59. J. Murray. *Mathematical Biology* (Springer-Verlag, Heidelberg, 1989), p. 68.

60. R. May, in *The Fragile Environment*, L. Friday and R. Lasky (eds) (Cambridge University Press, 1989), p. 65.

61. R. May. *Science* 186, 645 (1974).

62. *Ibid.*

63. Actually, chaos in the logistic map is due to a repellor, not an attractor.

64. We have seen these features repeatedly throughout recent parts of the book; this is an example of so-called period doubling. The term bifurcating cascade was introduced by Robert May in 1976.

65. Properties of such generic maps are universal, independent of the particular model chosen.

66. R. May and G. Oster. *American Naturalist* 110, 573 (1976).

67. The fractal properties of the bifurcation diagram are not directly related to those of the strange attractor.

68. Conversation between Robert May and the authors, June 1989.

69. R. May. 'Simple Mathematical Models with very complicated dynamics', *Nature* 261, 459 (1976).

70. *Ibid.*, p. 467.

71. R. Pool. *Science* 243, 310 (1989), p. 312.

72. L. Olsen, G. Truty and W. Schaffer. 'Oscillations and chaos in epidemics: a non-linear dynamics study of six childhood diseases in Copenhagen, Denmark', *Theoretical Population Biology* 33, 344 (1988); R. Pool. *Science* 243, 25 (1989).

73. L. Croft. *The Life and Death of Charles Darwin* (Elmwood, Chorley, 1989).

74. Formerly Sir William Thomson.

75. C. Albritton. *The Abyss of Time* (W. H. Freeman Cooper, San Francisco, 1980), p. 182.

76. See J. D. Barrow and F. J. Tipler. *The Anthropic Cosmological Principle* (Oxford University Press, 1986), Chapter Three, for an interesting discussion and more details of these calculations. Kelvin's work followed on from the publication of Darwin's *Origin of Species*, and was designed to contradict the notion of biological evolution. Kelvin wanted to show that there had not been enough time for the process of evolution to have taken place on Earth. It certainly led to some worries amongst the biologists and geologists.

77. Thermonuclear fusion within the Sun, converting hydrogen into helium, provides an enormous additional source of heat which was unknown until some 30 years later.

78. S. J. Gould. *Time's Arrow, Time's Cycle* (Penguin, Harmondsworth, 1988), p. 3.

79. L. Croft. *The Life and Death of Charles Darwin, op. cit.*, p. 11.

80. *Ibid.*, p. 12.

81. P. Barrett and R. Freeman (eds). *The Works of Charles Darwin* (William Pickering, London, 1986) Vol. 1, p. 19.

82. L. Croft. *The Life and Death of Charles Darwin, op. cit.*, p. 14.

83. T. Malthus. *Essay on the Principle of Population*.

84. L. Croft. *The Life and Death of Charles Darwin, op. cit.*, p. 65.

85. R. May, in *The Fragile Environment*, Darwin College Lectures, L. Friday and R. Laskey (eds), *op. cit.*, p. 61.

86. D. Brooks and E. Wiley. *Evolution as Entropy* (University of Chicago Press, Second Edition, 1986), p. 33.

87. J. Maynard Smith. 'Time in the evolutionary process', *Studium Generale* 23, 266 (1970).

88. R. Dawkins. *The Blind Watchmaker* (Longman, Harlow, 1986), p. 94.

89. Interview of C. Zeeman with the authors, June 1989.

90. We are indebted to Robert May for this quotation.

91. S. Luria. *Life, the Unfinished Experiment* (Scribner, New York, 1973), pp. 22–3.

92. S. J. Gould, N. Gilinsky and R. German. *Science* 236, 1437 (1987).

93. S. J. Gould. *The Flamingo's Smile* (Penguin, Harmondsworth, 1988), p. 257.

341

94. *Ibid.*, p. 255. *See also* S. J. Gould. *Wonderful Life* (Hutchinson Radius, London, 1990), p. 302.

95. A. Peacocke. *God and the New Biology*, *op. cit.*, p. 160.

96. *Ibid.*, p. 64.

97. *Ibid.*, p. 159.

98. B. Weber *et al.* (eds). *Entropy, Information and Evolution* (MIT Press, Cambridge, Massachusetts, 1988); D. Brooks and E. Wiley, *Evolution as Entropy, op. cit.*; M. Denton. *Evolution: a Theory in Crisis* (Burnett Books, London, 1985).

99. In the language of dynamical systems theory, it would be better to use the term 'punctuated steady-states'.

100. Even Darwin himself suggests in the *Origin* 'that the duration of each formation is, perhaps, short compared with the average duration of specific forms'. (Summary of Ch. X).

101. Interview of E. C. Zeeman with the authors, June 1989.

102. From E. C. Zeeman. *Dynamics of Darwinian Evolution, Colloque des Systèmes Dynamiques* (Fondation Louis de Broglie, September 1984).

103. J. Maynard Smith. *The Evolution of Sex* (Cambridge University Press, 1978) and conversation with RRH, May 1990.

104. This is the long prevalent Weismann–Muller–Fisher view in which the principal achievement of sexual reproduction is in bringing 'good mutations' together into the evolving stock as fast as possible.

105. W. Hamilton, R. Axelrod and R. Tanese. 'Sexual reproduction as an adaptation to resist parasites', preprint (1989).

106. Hamilton believes that sex works best when there are competing hosts for the parasites rather than non-competing ones. The quotations are based on a conversation with the authors, June 1989.

107. *See* e.g. A. Kondrashov. 'Deleterious mutations and the evolution of sexual reproduction', *Nature* <u>336</u>, 435 (1988).

108. B. d'Espagnat. *Une incertaine réalité* (Gauthier-Villars, Paris, 1985), English title *Reality and the Physicist* (Cambridge University Press, 1989), p. 123.

CHAPTER EIGHT

The introductory quotation is given in A. Whitehead. *Science and the Modern World* (The Free Press, New York, 1967), p. 186.

1. Poincaré's recurrence theorem holds provided that the system is isolated, finite and not expanding. It should also probably need to be non-gravitating as well. The situation in relativistic cosmology where there is expansion and gravitation is discussed by F. J. Tipler. *Nature* <u>280</u>, 203 (1979); *see also* F. J. Tipler in *Essays in General Relativity*, F. J. Tipler (ed.) (Academic Press, New York, 1980). Of course, for any macroscopic system a Poincaré recurrence time would be fantastically large; some typical values are estimated in K. Denbigh. *Three Concepts of Time* (Springer-Verlag, Berlin, 1981), pp. 106 and 120.

2. This is a supposition very similar to the final anthropic principle, as discussed by J. D. Barrow and F. J. Tipler in their book, *The Anthropic Cosmological Principle* (Oxford University Press, 1986). But see also the discussion of Roger Penrose's view in Chapter Five.

3. We should in any case remember that in quantum theory, Heisenberg's uncertainty principle rules out the exact specification of such initial conditions.

4. Along lines implicit within kaon decay, which does not alter the deterministic character of the description.

5. Letter to Thomas Manning, 2 January 1810. However, Lamb also wrote: 'In everything that relates to science, I am a whole encyclopaedia behind the rest of the world' in Essays of Elia.

6. P. Coveney. D. Phil. Thesis, University of Oxford (1985), p. 1.

7. R. Thom. *Structural Stability and Morphogenesis*, trans. D. Fowler (Benjamin-Addison-Wesley, New York, 1975).

8. I. Prigogine. *From Being to Becoming* (W. H. Freeman, San Francisco, 1980), p. xii.

9. The microscopic world is, of course, believed to be quantum mechanical, but this would only serve to complicate matters further at the beginning of the discussion. We shall consider quantum mechanics later on.

10. The French mathematician Jacques Hadamard and the thermodynamicist Pierre Duhem were also acutely aware of this limitation. Heinrich Bruns was in fact the first to prove the non-integrability of three-body systems under restricted conditions.

11. 'The mathematical heritage of Henri Poincaré', *Proceedings of Symposia in Pure Mathematics* 39 (American Mathematical Society, 1983). Jules Henri Poincaré was born on 29 April 1854 in Nancy, France. By the age of 25 he was Professor of Mathematical Analysis at Caen and subsequently became Professor of Mathematics and Science at the University of Paris. In mathematics, he initiated the study of automorphic functions, did pioneering work in topology and made major contributions to probability theory and astronomy. He was also a philosopher and member of the Académie Française and became President of the French National Academy of Science. What of Poincaré the man? Ian Stewart wrote in his book, *Does God Play Dice? The Mathematics of Chaos* (Blackwell, Oxford, 1989), p. 61: 'The traditional stereotype of the mathematician is the absent-minded dreamer – bearded, bespectacled, forever searching for those spectacles, unaware that they are perched on his nose. Few of the great (or ordinary) mathematicians actually fit this stereotype; but Poincaré was one who did. More than once he forgetfully took hotel linen with him on departure.'

12. H. Eves. *An Introduction to the History of Mathematics* (CBS College Publishing, 1983), p. 432.

13. H. Poincaré. *The Value of Science* (Dover, New York, 1958), p. 8.

14. H. Eves. *An Introduction to the History of Mathematics*, op. cit., p. 432.

15. I. Newton. *Principia, Book III*.

16. H. Poincaré. *Science and Hypothesis* (Walter Scott Publishing, New York, 1905), p. 130.

17. Quoted in A. Winfree. *The Geometry of Biological Time* (Springer-Verlag, Heidelberg, 1980), p. 345.

18. *Ibid.*, p. 358.

19. J. C. Maxwell, cited in I. Prigogine. *From Being to Becoming*, op. cit., p. 33. This is the so-called 'ergodic hypothesis'.

20. Now also used in other branches of mathematics, such as number theory.

21. Poincaré believed that dynamics and thermodynamics were incompatible, as he pointed out in his *Leçon de Thermodynamique*.
22. M. Walsh. Letter to the Editor, *The Independent* (10 June 1989).
23. M. Berry, in *A Passion for Science*, L. Wolpert and A. Richards (eds) (Oxford University Press, 1988). Berry's anecdote is based on calculations originally performed by B. Chirikov; a rather similar computation can be found in the work of E. Borel. *Le Hasard* (Alcan, Paris, 1928). *See* M. Berry, *American Institute of Physics Conference Proceedings* 46, 16 (1978), who also discusses the sensitivity of quantum systems containing many particles.
24. Technically, they are said to form a 'set of zero measure'.
25. J. Lighthill. 'The recently recognised failure of predictability in Newtonian dynamics', *Proc. R. Soc. London A* 407, 38 (1986).
26. The blob is nothing other than the probability distribution function described in Chapter Five.
27. This result is known in the trade as Liouville's theorem, a direct consequence of the measure-preserving nature of the time evolution.
28. Also by Koopman and von Neumann.
29. Gibbs envisioned a *Gedankenexperiment* in which a drop of ink is placed in a layer of glycerol wedged between two concentric jars of slightly different radii – as you twist the outer jar, holding the inner one fixed, the drop is smeared into a uniform state.
30. The distribution function (the evolving blob) becomes the basic entity: only in the case of stable motion can we conceive of a limiting procedure whereby we may focus the blob down to a single point in a physically meaningful manner. When chaos is present, such a procedure represents an idealisation beyond the bounds of physical realisation. Technically speaking, it is a singular limit.
31. There are even entities called Bernoulli flows, which are at the very top of the instability tree.
32. Assumed here to have a classical as opposed to a quantum undercoat.
33. However, unusual properties may appear in dynamics because of the singular nature of the interaction potential. *See* the discussion of the KAM theorem below.
34. Such a function is called a Lyapounov functional or variable in the theory of dynamical systems. The idea of constructing such a functional was first mooted by I. Prigogine, C. George, F. Henin and L. Rosenfeld. *Chemica Scripta* 4, 5 (1973).
35. B. Misra. *Proceedings of the National Academy of Science USA* 75, 1627 (1978).
36. Technically, this is because the entropy operator is 'non-factorisable'. B. Misra *et al. Proceedings of the National Academy of Science USA* 76, 3607 (1979); *also Physica* 98A, 1 (1979).
37. Both the internal time and the microscopic entropy are actually operators which act on the distribution function. Some readers might think that operators only exist in quantum theory. However, when one works with distribution functions and the Liouville equation, one can construct a Hilbert space formalism ('Koopman formalism') entirely analogous to that of quantum mechanics. An 'uncertainty relationship' appears between the thermodynamic internal time operator and the dynamical Liouville operator which is very similar to that between, for example, position and momentum operators in quantum mechanics. *See* e.g. P. Coveney. *Nature* 333, 409 (1988).
38. I. Prigogine. *From Being to Becoming, op. cit.*, p. 51.

39. *Science* <u>246</u>, 998 (1989).

40. Numerical studies by J. Wisdom, G. Sussman *et al.* in *Dynamical Chaos*, M. Berry, I. Percival and R. May (eds) (Royal Society Publications, London, 1989).

41. Some rigorous results are known which suggest that regular motions of finite measure can persist for larger systems than generally thought possible. *See* e.g. E. Wayne. *Communications in Mathematical Physics* <u>96</u>, 311 (1984); <u>103</u>, 351 (1986); <u>104</u>, 1 (1986).

42. There is an interesting comparison here. Equilibrium statistical mechanics is an accepted description of the macroscopic world. It predicts the absence of any phase transition unless the system contains an infinite number of particles. This is why the thermodynamic limit must be used. Yet we observe phase transitions with a mere 10^{23} particles. Thus there is a scale problem here too: how large must a system be to be indistinguishable from an infinite system?

43. Actually a compact manifold. The dynamics is now called an Anosov flow. J. Hadamard. *Journal de mathématiques pures et appliquées* <u>4</u>, 27 (1898). *See also* P. Duhem. *The Aim and Structure of Physical Theory* (Princeton University Press, 1954).

44. B. Misra. *Journal of Statistical Physics* <u>48</u>, 1295 (1987) demonstrates this for the classical Klein–Gordon equation.

45. *See* e.g. the article by J. Ford in *The New Physics*, P. C. W. Davies (ed.) (Cambridge University Press, 1989).

46. In particular, even the density matrix is destined to repeat its behaviour, in contradistinction from the classical distribution function for mixing flows shown in the earlier phase space portraits.

47. J. Bell. *Speakable and Unspeakable in Quantum Mechanics* (Cambridge University Press, 1987), p. 170.

48. *Ibid.*, p. 171.

49. *Ibid.*, p. 117.

50. The basic description must then be made in terms of a density matrix: the conversion of pure quantum states (described by wavefunctions) into mixtures under irreversible time evolution corresponds precisely to 'the collapse of the wavefunction'. It is interesting to note that Leon Rosenfeld, one of the greatest advocates of the Copenhagen interpretation of quantum mechanics, nevertheless felt the need for further work to make sense of the measurement process. In his later years, he collaborated enthusiastically with the Brussels Group on some of these questions. *See* e.g. C. George, I. Prigogine and L. Rosenfeld. *Kon. Danske Vidensk. Selsk. Mat.-fys. Med.* <u>38</u> (12), 1 (1972).

51. J. Bell. *Speakable and Unspeakable in Quantum Mechanics*, *op. cit.*, p. 27. However, we should point out that Bell seems to have little sympathy with the concept of irreversibility being of primary importance in physical theory (PVC discussion with JSB, Schrödinger Centenary Conference, Imperial College, London, 1987). He apparently feels that this is too great a leap, although he endorses the ideas of G. Ghirardhi, A. Rimini and T. Weber. *Physical Review D* <u>34</u>, 470 (1986) concerning wavefunction collapse. Yet the model these authors propose is attractive to Roger Penrose (whose general approach to wavefunction collapse as a feature of a correct quantum gravity was mentioned in Chapter Four). His ideas have certain points in common with those which we have been describing in the text.

52. J. Bell. *Speakable and Unspeakable in Quantum Mechanics*, *op. cit.*, p. 172.

53. This division holds provided a system fulfils certain criteria, called 'dissipativity conditions'.

54. The formalism is algebraically exact but non-rigorous. The problem is to determine the necessary and sufficient conditions guaranteeing its existence, currently under investigation by PVC and Oliver Penrose. Additionally, the connection between 'subdynamics' and the work of Misra *et al.* still remains to be clarified.

55. V. Škarka and P. V. Coveney. 'Solution of the linearised Vlasov equation for collisionless plasmas evolving in external fields of arbitrary spatial and time dependence' 'I' and 'II', *J. Phys A: Math. Gen.* 23, 2439 and 2463 (1990).

56. O. Lanford III. *Lecture Notes in Physics* 38, 1 (1975).

57. H. Spohn. *Reviews of Modern Physics* 52, 569 (1980).

58. Strictly these processes are adiabatic and hence isentropic.

59. E. Günzig, J. Géhéniau and I. Prigogine. *Nature* 330, 621 (1987).

60. There is some dispute over whether the Minkowskian vacuum is actually unstable in this sense. In later work, these authors avoid this problem by applying phenomenological thermodynamics of open systems to describe the situation.

61. Described by the Robertson–Walker metric.

62. Recall, for comparison, that in the scenario favoured by Roger Penrose, which we discussed in Chapter Five, an extremely precisely chosen low-entropy initial state is preferred for the Big Bang. Compared with the entropy residing in black holes in the present universe (computed from the Bekenstein–Hawking formula), which are believed to reside at the heart of every galaxy, the entropy in the black-body background radiation is then 'utter chicken feed', as Penrose puts it. If the universe is closed, the Big Crunch corresponds to the conglomeration of all these black holes, which produces an unimaginably large entropy. It is this extraordinary difference between the Big Bang and Big Crunch entropies which leads Penrose to believe in a fundamentally time-asymmetric theory of quantum gravity.

CHAPTER NINE

The introductory quotation is given in I. Prigogine. *From Being to Becoming* (W. H. Freeman, San Francisco, 1980), p. 14.

1. See e.g. *The Economy as an Evolving Complex System*, P. Anderson, K. Arrow and D. Pines (eds) (Santa Fé Institute, Addison-Wesley, Redwood City, 1989).

2. Any simple-minded statement about how the economy is causally linked to variations in one or more parameters, for example interest and exchange rates, is guaranteed baloney in the long term. The only statements which can be made for chaotic systems are probabilistic ones, based on the inherent uncertainties in the initial conditions.

3. The interactions between families, villages, cities, and even nations may be modelled in this way, according to C. Dyke, in *Entropy, Information and Evolution*, B. Weber, D. Depew and J. Smith (eds) (MIT Press, Cambridge, Massachusetts, 1988), p. 355.

4. The recently established Santa Fé Institute in New Mexico is doing just this, bringing together physicists, economists, biologists and social scientists to search for new ways of formulating old problems.

5. M. Gibbons and J. Metcalfe, in *The Laws of Nature and Human Conduct*, I. Prigogine and M. Sanglier (eds) (Brussels Task Force of Research Information and Study on Science, Brussels, 1985), p. 253. *See also* M. Markus, S. Müller and G. Nicolis (eds). *From Chemical to Biological Organisation* (Springer-Verlag, Berlin, 1988), p. 348.

6. E. Laszlo, in *The Laws of Nature and Human Conduct, op. cit.*, p. 298.

7. B. West and J. Salk, in *The Laws of Nature and Human Conduct, op. cit.*, p. 324.

8. I. Prigogine. *From Being to Becoming, op. cit.*, p. xiii.

9. A. Toffler, in the foreword to I. Prigogine and I. Stengers. *Order Out of Chaos* (Heinemann, London, 1984), p. xi.

NOTES TO APPENDIX

The introductory quotation is given in M. Young. *The Metronomic Society* (Harvard University Press, 1988), p. 20.

1. R. Coleman. *Wide Awake at 3.00 am* (W. H. Freeman, New York, 1986), p. 3.

2. R. Ward. *The Living Clocks* (Mentor, New York, 1971), p. 35.

3. A. Winfree. *The Geometry of Biological Time* (Springer-Verlag, New York, 1980), p. 375.

4. The day length 600 million years ago was only 21 hours and there were more than 420 days in each year. *See* D. Saunders. *An Introduction to Biological Rhythms* (Blackie, Glasgow, 1977), p. 9.

5. M. Moore-Ede, F. Sulzman and C. Fuller. *The Clocks that Time Us* (Harvard University Press, 1982), p. 9.

6. *Ibid.*

7. J. Cloudsley-Thompson. *Biological Clocks: Their Functions in Nature* (Weidenfield & Nicolson, London, 1980), p. 8.

8. E. Pengelley (ed.). *Circannual Clocks* (Academic Press, New York, 1974), p. 394.

9. J. Murray. *Mathematical Biology* (Springer-Verlag, Heidelberg, 1989), p. 100.

10. A. Winfree, personal communication with the authors, November 1989.

11. The species are *Neurospora crassa* and *Drosophila pseudoobscura*. A. Winfree. *The Geometry of Biological Time, op. cit.*, p. 172; these observations have not been repeated or confirmed by anyone else.

12. D. Mergenhagen and H. Schweiger. *Experimental Cell Research* 92, 127 (1975); *Experimental Cell Research* 94, 321 (1975).

13. A. Winfree. *The Timing of Biological Clocks* (W. H. Freeman, Scientific American Library, San Francisco, 1987), p. 113.

14. R. Coleman. *Wide Awake at 3.00 am, op. cit.*, p. 7.

15. Attempts to develop mathematical models of these rhythms have been made since 1960. M. Moore-Ede and C. Czeisler (eds). *Mathematical Models of the Sleep-Wake Cycle* (Raven Press, New York, 1984), p. v.

16. We tend to overestimate the passing of a minute in the morning and under-estimate it during the day. *See* J. Palmer. *An Introduction to Biological Rhythms* (Academic Press, New York, 1976), p. 136.

17. A. Aveni. *Empires of Time* (Basic Books, New York, 1989), p. 100.

18. A. Winfree. *The Timing of Biological Clocks*, *op. cit.*, p. 84.

19. *Ibid.*, p. 89.

20. R. Konopka and S. Benzer. *Proceedings of the National Academy of Science USA* 68, 2112 (1971). In recent work on the *periodic* gene (P. Hardin, J. Hall and M. Rosbash. *Nature* 343, 536 (1990)) it is suggested that there is a feedback loop through which the activity of *per*-encoded protein causes cycling of its own RNA. This feedback mechanism appears to be an important component of the circadian pacemaker.

21. C. Kyriacou, personal communication with the authors, November 1989.

22. M. Heisenberg. *Nature* 333, 19 (1988).

23. D. Saunders. *Journal of Comparative Physiology* 124, 75 (1978).

24. F. Cummings. *Journal of Theoretical Biology* 55, 455 (1975).

25. A. Winfree. *The Geometry of Biological Time*, *op. cit.*, p. 379. This view is one of many on the mechanisms of circadian clocks.

26. B. Rusak and I. Zucker. 'Neural regulation of circadian rhythms', *Physiology Review* 59, 449 (1979).

27. M. Ralph *et al. Neuroscientific Abstracts* 14, 462 (1988).

28. N. Mrosovsky. *Nature* 337, 213 (1989).

29. J. Campbell. *Winston Churchill's Afternoon Nap: a Wide Awake Inquiry into the Human Nature of Time* (Aurum, London, 1988), p. 145.

30. By, respectively, photo-inhibition and photo-activation non-linear feedback loops.

31. Serotonin-N-acetyltransferase.

32. Serotonin.

33. This is due to the movements of melanin in specialised dermic cells, caused by secreted melatonin, as was discovered only recently.

34. Other documented effects of melatonin are: it lowers the optimum preferential body temperature in lizards and reduces the internal body temperature in sparrows.

35. J. Arendt. *ISI Atlas of Science Pharmacology* (1987), p. 257.

36. *Agricultural and Food Research Council News* (Swindon, July 1989), p. 15.

37. J. Arendt, M. Aldhous and V. Marks. *British Medical Journal* 292, 1170 (1986).

38. J. Arendt, personal communication with the authors, October 1989.

39. Cited in T. Wehr and N. Rosenthal. *North American Journal of Psychiatry* 146, 829 (1989).

40. Erythemal response to histamine injection is greatest at 11 p.m. and smallest between 7 a.m. and 11 a.m. Reactions to penicillin and house dust also show a circadian rhythm. *See* M. Moore-Ede, F. Sulzman and C. Fuller. *The Clocks that Time Us*, *op. cit.*, p. 360.

41. Cited in T. Wehr and N. Rosenthal. *North American Journal of Psychiatry* 146, 829 (1989).

42. J. Arendt, personal communication with the authors, October 1989.

43. C. Czeisler, R. Kronauer, J. Allan, J. Duffy, M. Jewett, E. Brown and J. Ronda. *Science* 244, 1328 (1989).

44. G. Richardson, interview with RRH, March 1990. The key to their success was the use of a constant routine so that the circadian rhythm of their subjects could be accurately measured (J. Mills, D. Minors and J. Waterhouse. *J. Physiol.* (London) 285, 455 (1978)).

45. G. Richardson, interview with RRH, March 1990.
46. C. Czeisler, M. Johnson, J. Duffy, E. Brown, J. Ronda and R. Kronauer. *New England Journal of Medicine* 322, 1253 (1990).
47. M. Moore-Ede, F. Sulzman and C. Fuller. *The Clocks that Time Us*, *op. cit.*, p. 352.
48. A. Winfree. *The Timing of Biological Clocks*, *op. cit.*, p. 7. M. Gardner *Science, Good, Bad and Bogus* (Prometheus Books, Loughton, Essex, 1990) gives a historical account of biorhythms, 'one of the most extraordinary and absurd episodes in the history of numerological pseudoscience.' But see also F. J. Sulloway, *Freud: Biologist of the Mind*, Chapter 5 (Fontana, 1980).

Bibliography

CHAPTER ONE

Aveni, A. *Empires of Time: Calendars, Clocks, and Cultures* (Basic Books, New York, 1989).

Bergson, H. *Oeuvres* (Presses Universitaires de France, Paris, 1959).

Blackmore, J. T. *Ernst Mach. His Work, Life and Influence* (University of California Press, 1972).

Eliade, M. *The Myth of the Eternal Return* (Routledge & Kegan Paul, London, 1955).

Fitzgerald, E. *Rubáiyát of Omar Khayyám and Other Writings by Edward Fitzgerald* (Collins, London and Glasgow, 1953).

Fraser, J. *Time, the Familiar Stranger* (Tempus, Washington, 1987).

Gould, S. J. *Time's Arrow, Time's Cycle* (Penguin, Harmondsworth, 1988).

Hoffmann, B. *Albert Einstein, Creator and Rebel* (Viking Press, New York, 1972).

Horwich, P. *Asymmetries in Time* (MIT Press, Cambridge, Mass., 1987).

Jaffé, G. 'Recollections of three great laboratories', *J. Chem. Ed.* 29, 230 (1952).

Kant, I. *Critique of Pure Reason*, trans. N. Kemp-Smith (St Martin's, New York, 1961).

Kolakowski, L. *Bergson* (Oxford University Press, 1985).

Landsberg, P. T., in *The Study of Time III*. J. Fraser, N. Lawrence, D. Park (eds) (Springer-Verlag, Heidelberg, 1978).

Plato. *Timæus and Critias*, trans. D. Lee (Penguin, Harmondsworth, 1971).

Schrödinger, E. *What is Life?* (Cambridge University Press, 1944).

Shakespeare. *The Sonnets of William Shakespeare* (Shepheard-Walwyn, London, 1975).

Thirring, H. 'Ludwig Boltzmann', *J. Chem. Ed.* 29, 298 (1952).

Whitehead, A. *Process and Reality: An Essay in Cosmology*, D. R. Griffin and D. W. Sherburne (eds) (Free Press, New York, Corrected Edition, 1979).

Whitrow, G. J. *The Natural Philosophy of Time* (Oxford University Press, Second Edition, 1980).

———*Time in History* (Oxford University Press, 1989).

CHAPTER TWO

Bergson, H. 'L'évolution créatrice', in *Oeuvres* (Editions du Centenaire, Presses Universitaires de France, Paris, 1963).

Bohm, D. *Causality and Chance in Modern Physics* (Routledge & Kegan Paul, London, 1984).

Boyer, C. B. *A History of Mathematics* (Wiley, New York, 1968).

Coveney, P. V. *La Recherche* 20, 190 (1989).

Dawkins, R. *The Blind Watchmaker* (Longman, Harlow, 1986).

Fakhry, M. *History of Islamic Philosophy* (Columbia University Press, New York, Second Edition, 1983).

BIBLIOGRAPHY

Fauvel, J. K., Flood, R., Shortland, M. and Wilson, R. (eds). *Let Newton Be!* (Oxford University Press, 1988).

Gjertsen, D. *The Newton Handbook* (Routledge & Kegan Paul, London and New York, 1986).

Goldstein, H. *Classical Mechanics* (Addison-Wesley, Massachusetts, Second Edition, 1980).

Gratzer, W. (ed.), *The Longman Literary Companion to Science* (Longman, Harlow, 1989).

Hammond, N. and Scullard, H. (eds). *The Oxford Classical Encyclopedia* (Oxford University Press, Second Edition, 1970).

Hayes, J. (ed.). *The Genius of Arab Civilisation* (Eurabia, London, Second Edition, 1983).

Jackson, J. *Classical Electrodynamics* (Wiley, New York, Second Edition, 1975).

Kline, M. *Mathematical Thought from Ancient to Modern Times* (Oxford University Press, 1972).

————*Mathematics in Western Culture* (Penguin, Harmondsworth, 1972).

Koyré, A. *Etudes Newtoniennes* (Gallimard, Paris, 1968).

Manuel, F. *A Portrait of Isaac Newton* (Harvard University Press, 1968).

Nasr, S. H. *Islamic Science – An Illustrated Study* (World of Islam Festival Publishing Co., Westerham, 1976).

Pais, A. *'Subtle is the Lord' . . . The Science and Life of Albert Einstein* (Oxford University Press, 1982).

Paley, W. *Natural Theology* (J. Vincent, Oxford, Second Edition, 1828).

Prigogine, I. *From Being to Becoming* (W. H. Freeman, San Francisco, 1980).

Ronan, C. *The Cambridge Illustrated History of the World's Science* (Cambridge University Press, 1983).

Russell, B. *History of Western Philosophy* (Allen & Unwin, London, 1979).

Westfall, R. S. *Never at Rest: a Biography of Isaac Newton* (Cambridge University Press, 1980).

Whitrow, G. J. *Time in History* (Oxford University Press, 1989).

CHAPTER THREE

Barrow, J. D. *The World Within the World* (Oxford University Press, 1988).

Barrow, J. D. and Tipler, F. J. *The Anthropic Cosmological Principle* (Oxford University Press, 1986).

Beck, A. and Havas, P. *The Collected Papers of Albert Einstein.* Vol. 1. (Princeton University Press, 1987).

Bertola, F. and Curi, V. (eds). *The Anthropic Principle* (Cambridge University Press, 1989).

Calder, N. *Einstein's Universe* (BBC, London, 1979).

Davies, P. C. W. *The Accidental Universe* (Cambridge University Press, 1981).

————*The Physics of Time Asymmetry* (Surrey University Press and the University of California Press, 1974).

Douglas, A. V. *Arthur Stanley Eddington* (Nelson, 1956).

Einstein, A. *Annalen der Physik* 17, 891 (1905).

Flood, R. and Lockwood, M. (eds). *The Nature of Time* (Blackwell, Oxford, 1988).

Forward, R. L. *Journal of the British Interplanetary Society* 42, 533 (1989).

French, A. P. (ed.). *Einstein, a Centenary Volume* (Heinemann, Oxford, 1979).

Gribbin, J. *In Search of the Big Bang* (Corgi, Ealing, 1987).

Hawking, S. *A Brief History of Time* (Bantam Books, New York, 1988).

Hawking, S. and Israel, W. (eds). *300 Years of Gravitation* (Cambridge University Press, 1989).

Hoffmann, B. *Albert Einstein, Creator and Rebel* (Viking Press, New York, 1972).

Krauss, L. M. *The Fifth Essence: The Search for Dark Matter in the Universe* (Hutchinson Radius, London, 1989).

Landsberg, P. T. (ed.). *The Enigma of Time* (Adam Hilger, Bristol, 1982).

Mach, E. *The Science of Mechanics* (Open Court, Chicago, Fourth Edition, 1919).

Marder, L. *Time and the Space-traveller* (Allen & Unwin, London, 1971).

Mehra, J. *Einstein, Hilbert and the Theory of Gravitation* (Reidel, Dordrecht, 1974).

Misner, C., Thorne, K. and Wheeler, J. *Gravitation* (W. H. Freeman, San Francisco, 1973).

Pais, A. *'Subtle is the Lord' . . . The Science and Life of Albert Einstein* (Oxford University Press, 1982).

Parker, B. *Invisible Matter and the Fate of the Universe* (Plenum, New York, 1989).

Penrose, R. *The Emperor's New Mind* (Oxford University Press, 1989).

Regis, E. *Who Got Einstein's Office? Eccentricity and Genius at the Institute for Advanced Study* (Addison-Wesley, Reading, Mass., 1987).

Rindler, W. *Essential Relativity* (Springer-Verlag, Heidelberg, Second Edition, 1977).

Weinberg, S. *The First Three Minutes* (André Deutsch, London, 1977).

———*Gravitation and Cosmology* (Wiley, New York, 1972).

Whitrow, G. J. *The Structure and Evolution of the Universe* (Hutchinson, London, 1959).

Zukav, G. *The Dancing Wu Li Masters: an Overview of the New Physics* (Rider and Hutchinson, London, 1979).

CHAPTER FOUR

Barrow, J. D. *The World Within The World* (Oxford University Press, 1988).

Barrow, J. D. and Tipler, F. J. *The Anthropic Cosmological Principle* (Oxford University Press, 1986).

Bell, J. *Speakable and Unspeakable in Quantum Mechanics* (Cambridge University Press, 1987).

Bohr, N., in *Albert Einstein: Philosopher-Scientist*, P. Schilpp (ed.), (Tudor, New York, 1949).

Brush, S. *The Kind of Motion We Call Heat* (North-Holland, Amsterdam, 1976).

Carus, Titus Lucretius. *De Rerum Natura*, trans. R. E. Latham (Penguin, Harmondsworth, 1951).

Davies, P. C. W. *The Physics of Time Asymmetry* (Surrey University Press, 1974).

———(ed.). *The New Physics* (Cambridge University Press, 1989).

d'Espagnat, B. *The Conceptual Foundations of Quantum Mechanics* (Benjamin, Massachusetts, 1976).

———*Une incertaine réalité* (Gauthier-Villars, Paris, 1985); English title *Reality and the Physicist* (Cambridge University Press, 1989).

deWitt, B. S. and Graham, N. (eds). *The Many Worlds Interpretation of Quantum Mechanics* (Princeton University Press, 1973).

BIBLIOGRAPHY

Dirac, P. *Directions in Physics* (Wiley-Interscience, New York, 1978).

French, A. and Kennedy, P. *Niels Bohr. A Centenary Volume* (Harvard University Press, 1985).

Gamow, G. *Biography of Physics* (Hutchinson, London, 1962).

Gardner, M. *The Ambidexterous Universe* (Scribner, New York, 1979).

Greenaway, F. *John Dalton and the Atom* (Heinemann, London, 1966).

Hammond, N. and Scullard, H. (eds). *The Oxford Classical Encyclopedia* (Oxford University Press, Second Edition, 1970).

Herbert, N. *Quantum Reality. Beyond the New Physics* (Anchor Press, New York, 1985).

Hey, T. and Walters, P. *The Quantum Universe* (Cambridge University Press, 1987).

Jammer, M. *The Philosophy of Quantum Mechanics* (Wiley, New York, 1974).

Kabir, P. *The CP Puzzle* (Academic Press, New York, 1968).

Levi-Leblond, J. M. and Cini, M. (eds). *Quantum Theory Without Reduction* (Adam Hilger, Bristol, 1990).

Moore, W. *Physical Chemistry* (Longman, London, 1972).

———*Schrödinger, Life and Thought* (Cambridge University Press, 1989).

Mulvey, J. (ed.). *The Nature of Matter* (Oxford University Press, 1981).

Pagels, H. *Perfect Symmetry: The Search for the Beginning of Time* (Simon and Schuster, New York, 1985).

Pais, A. *'Subtle is the Lord' . . . The Science and Life of Albert Einstein* (Oxford University Press, 1982).

Penrose, R. *The Emperor's New Mind* (Oxford University Press, 1989).

Popper, K. *The Open Universe. An Argument for Indeterminism* (Hutchinson, London, 1982).

———*Quantum Theory and the Schism in Physics* (Hutchinson, London, 1982).

Popper, K. and Eccles, J. *The Self and its Brain* (Springer-Verlag, Berlin, 1977).

Prigogine, I. and Stengers, I. *Entre le temps et l'éternité* (Fayard, Paris, 1988).

Rae, A. *Quantum Physics: Illusion or Reality?* (Cambridge University Press, 1986).

Reichenbach, H. *The Direction of Time* (University of California Press, 1956).

Sachs, R. *The Physics of Time Reversal* (University of Chicago Press, 1987).

Schrödinger, E. *Collected Papers on Wave Mechanics* (Chelsea Publishing Co., New York, Second Edition, 1978).

Wheeler, J. and Zurek, W. *Quantum Theory and Measurement* (Princeton University Press, 1983).

CHAPTER FIVE

Babloyantz, A. *Molecules, Dynamics and Life* (Wiley-Interscience, New York, 1986).

Balsecu, R. *Equilibrium and Non-Equilibrium Statistical Mechanics* (Wiley, New York, 1975).

Boltzmann, L. *Vorlesungen über Gastheorie* (Leipzig, 1896); English title *Lectures on Gas Theory* (University of California Press, 1964).

Barrow, J. D. *The World Within the World* (Oxford University Press, 1988).

Barrow, J. D. and Tipler, F. J. *The Anthropic Cosmological Principle* (Oxford University Press, 1986).

Beck, A. and Havas, P. *The Collected Papers of Albert Einstein* Vol. 1. (Princeton University Press, 1987).

Blackmore, J. T. *Ernst Mach. His Work, Life and Influence* (University of California Press, 1972).

Born, M. *Natural Philosophy of Cause and Chance* (Oxford University Press, 1949).

Brillouin, L. *Science and Information Theory* (Academic Press, New York, 1962).

Brush, S. G. *The Kind of Motion We Call Heat* (North-Holland, Amsterdam, 1976).

Cohen, E. and Thirring, W. (eds). *The Boltzmann Equation* (Springer-Verlag, Vienna, 1973).

Coveney, P. 'The Second Law of Thermodynamics: entropy, irreversibility and dynamics', *Nature* 333, 409 (1988).

———'L'irréversibilité du temps', *La Recherche* 20, 190 (1989).

Denbigh, K. *Three Concepts of Time* (Springer-Verlag, Berlin, 1981).

Denbigh, K. and Denbigh J. *Entropy in Relation to Incomplete Knowledge* (Cambridge University Press, 1985).

Eddington, A. *The Nature of the Physical World* (Cambridge University Press, 1928).

Ehrenfest, P. and Ehrenfest, T. *The Conceptual Foundations of the Statistical Approach in Mechanics*, trans. M. J. Moravcsik (Cornell University Press, Ithaca NY, 1959).

Glansdorff, P. and Prigogine, I. *Thermodynamic Theory of Structure, Stability and Fluctuations* (Wiley, New York, 1971).

de Groot, S. and Mazur, P. *Non-Equilibrium Thermodynamics* (Dover, New York, 1984).

Hollinger, H. and Zenzen, M. *The Nature of Irreversibility* (Reidel, Dordrecht, 1985).

Johnston, W. *The Austrian Mind: An Intellectual and Social History 1848–1938* (University of California Press, 1972).

Jou, D., Casas-Vázquez, J. and Lebon, G. 'Extended Irreversible Thermodynamics', *Reports on Progress in Physics* 51, 1105 (1988).

Landsberg, P. T. *Thermodynamics with Quantum Statistical Illustrations* (Wiley-Interscience, New York, 1961).

Lavenda, B. *Thermodynamics of Irreversible Processes* (Macmillan Press, London, 1978).

Layzer, D. *Scientific American* 236, 56 (1975).

Nicolis, G. and Prigogine, I. *Self-Organisation in Non-Equilibrium Systems* (Wiley-Interscience, New York, 1977).

Penrose, O. *Foundations of Statistical Mechanics* (Pergamon Press, Oxford, 1970).

———'Foundations of Statistical Mechanics', *Reports on Progress in Physics* 42, 1937 (1979).

Penrose, R. *The Emperor's New Mind* (Oxford University Press, 1989).

Perrin, J. *Les atomes* (Librairie Alcan, Paris, Fourth Edition, 1914); English title *Atoms*, trans. D. L. Hammick (Van Nostrand, New York, 1916).

Prigogine, I. *Etude thermodynamique des phénomènes irréversibles* (Desoer, Liège, 1947).

———*From Being to Becoming* (W. H. Freeman, San Francisco, 1980).

———*Thermodynamics of Irreversible Processes* (C. C. Thomas, Springfield, 1955).

Prigogine, I. and Stengers, I. *Entre le temps et l'éternité* (Fayard, Paris, 1988).

Shannon, C. and Weaver, W. *The Mathematical Theory of Communication* (University of Illinois Press, Urbana, 1949).

Tolman, R. *Principles of Statistical Mechanics* (Dover, New York, 1979).

Wheeler, L. *Josiah Willard Gibbs* (Yale University Press, New Haven, 1952).

354

BIBLIOGRAPHY

CHAPTER SIX

Alberts, B., Bray, D., Lewis, J., Raff, M., Roberts, K. and Watson, J. *The Molecular Biology of the Cell* (Garland, New York, Second Edition, 1989).

Babloyantz, A. *Molecules, Dynamics and Life* (Wiley-Interscience, New York, 1986).

Barrow, J. D. *The World Within the World* (Oxford University Press, 1988).

Barrow, J. D. and Tipler, F. J. *The Anthropic Cosmological Principle* (Oxford University Press, 1986).

Bendall, D. (ed.). *Evolution from Molecules to Men* (Cambridge University Press, 1982).

Bergé, P., Pomeau, Y. and Vidal, C. *Order Within Chaos* (Wiley-Interscience, New York, 1986).

Cairns-Smith, A. G. *Seven Clues to the Origin of Life* (Cambridge University Press, 1985).

Christiansen, P. and Parmentier, R. (eds). *Structure, Coherence and Chaos in Dynamical Systems* (Manchester University Press, 1989).

Coveney, P. 'L'irréversibilité du temps', *La Recherche* 20, 190 (1989).

Cramer, F. *Chaos und Ordnung. Die komplexe Struktur des Lebendingen* (Deutsche Verlags-Anstalt, Stuttgart, 1988).

Cuénot, C. *Teilhard de Chardin* (Helicon, Baltimore, 1965).

Davies, P. C. W. (ed.). *The New Physics* (Cambridge University Press, 1989).

Dyson, F. *Origins of Life* (Cambridge University Press, 1985).

Eigen, M. and Schuster, P. *The Hypercycle. A Principle of Natural Self-Organisation* (Springer-Verlag, New York, 1979).

Fages, J-B. *Teilhard de Chardin et le nouvel âge scientifique* (Privat, Toulouse, 1985).

Glass, L. and Mackey, M. *From Clocks to Chaos, The Rhythms of Life* (Princeton University Press, 1988).

Gleick, J. *Chaos* (Sphere, London, 1988).

Gray, P. and Scott, S. *Chemical Oscillations and Instabilities* (Oxford University Press, 1990).

Hodges, A. *Alan Turing, The Enigma* (Burnett Books, London, 1983).

Langton, C. *Artificial Life. Proceedings of an Interdisciplinary Workshop on the Synthesis and Simulation of Living Systems* (Addison-Wesley, Redwood City, 1989).

Mandelbrot, B. *The Fractal Geometry of Nature* (W. H. Freeman, New York, 1977).

Markus, M., Müller, S. and Nicolis, G. (eds). *From Chemical to Biological Organisation* (Springer-Verlag, Heidelberg, 1988).

Müller, S., Markus, M., Plesser, T., and Hess, B. *Dynamic Pattern Formation in Chemistry and Biology* (Catalogue of an Exhibition) (Max-Planck-Institut, Dortmund, 1988).

Nicolis, G. 'Physics of far-from-equilibrium systems and self-organisation', in *The New Physics*, P. C. W. Davies (ed.) (Cambridge University Press, 1989).

Nicolis, G. and Prigogine, I. *Exploring Complexity* (W. H. Freeman, New York, 1989).

————*Self-Organisation in Non-Equilibrium Systems* (Wiley-Interscience, New York, 1977).

Orgel, L. *The Origins of Life* (Wiley, New York, 1973).

Peacocke, A. *An Introduction to the Physical Chemistry of Biological Organisation* (Oxford University Press, 1983).

Peitgen, H-O. and Richter, P. *The Beauty of Fractals. Images of Complex Dynamical Systems* (Springer-Verlag, Berlin, 1986).

Prigogine, I. *From Being to Becoming* (W. H. Freeman, San Francisco, 1980).

Prigogine, I. and Stengers, I. *Order out of Chaos* (Heinemann, London, 1984).

Ruelle, D. 'Strange Attractors', *Mathematical Intelligencer* 2, 127 (1980).

Schrödinger, E. *What is Life?* (Cambridge University Press, 1944).

Schuster, H. *Deterministic Chaos* (Physik-Verlag, Weinheim, 1984).

Stewart, I. *Does God Play Dice? The Mathematics of Chaos* (Blackwell, Oxford, 1989).

Tyson, J. *The Belousov–Zhabotinski Reaction* (Springer-Verlag, Heidelberg, 1976).

Vidal, C. and Lemarchand, H. *La réaction créatrice* (Hermann, Paris, 1988).

Winfree, A. *The Geometry of Biological Time* (Springer-Verlag, Heidelberg, 1980).

Zeeman, E. 'On the classification of dynamical systems', *Bulletin of the London Mathematical Society*, November (1988).

———'Stability of dynamical systems', *Non-linearity* 1, 115 (1988).

CHAPTER SEVEN

Alberts, B., Bray, D., Lewis, J., Raff, M., Roberts, K. and Watson, J. *The Molecular Biology of the Cell* (Garland, New York, Second Edition, 1989).

Albritton, C. *The Abyss of Time* (Freeman Cooper, San Francisco, 1980).

Barrow, J. D. and Tipler, F. J. *The Anthropic Cosmological Principle* (Oxford University Press, 1986).

Bell, G. *The Masterpiece of Nature: The Evolution and Genetics of Sexuality* (University of California Press, 1982).

Brooks, D. and Wiley, E. *Evolution as Entropy: Towards a Unified Theory of Biology* (University of Chicago Press, Second Edition, 1988).

Cramer, F. *Chaos und Ordnung. Die komplexe Struktur des Lebendigen* (Deutsche Verlags-Anstalt, Stuttgart, 1988).

Crick, F. *Life Itself* (Macdonald, London, 1981).

Darwin, C. *The Origin of Species by Means of Natural Selection* (Murray, London, First Edition, 1859); *also* J. W. Burrow (ed.) (Penguin Harmondsworth, 1968).

Dawkins, R. *The Blind Watchmaker* (Longman, Harlow, 1986).

———*The Selfish Gene* (Oxford University Press, 1976).

DiFrancesco, D. and Noble, D. and DiFrancesco, D., Noble, D. and Denyer, J., in I. W. Jacklet (ed.). *Neuronal and Cellular Oscillators* (Dekker, New York, 1989), pp. 31–57 and 59–85.

Eckert, R. and Randall, D. *Animal Physiology: Mechanisms and Adaptations* (W. H. Freeman, San Francisco, 1983).

Glass, L., Goldberger, A., Courtemanche, M. and Shrier, A. *Proceedings of the Royal Society of London A* 413, 9 (1987).

Glass, L. and Mackey, M. *From Clocks to Chaos, The Rhythms of Life* (Princeton University Press, 1988).

Gleick, J. *Chaos* (Sphere, London, 1987).

Goldbeter, A. (ed.). *Cell to Cell Signalling* (Academic Press, London, 1989).

Gould, S. J. *Ever Since Darwin* (Penguin, Harmondsworth, 1980).

———*The Panda's Thumb* (Penguin, Harmondsworth, 1983).

———*Wonderful Life* (Hutchinson Radius, London, 1990).

Grene, M. (ed.). *Dimensions of Darwinism* (Cambridge University Press, 1983).

BIBLIOGRAPHY

Hille, B. *Ionic Channels of Excitable Membranes* (Sinauer, Massachusetts, 1984).

May, R. (ed.). *Stability and Complexity in Model Ecosystems* (Princeton University Press, 1973).

————*Theoretical Ecology: Principles and Applications* (Blackwell, Oxford, 1976).

Maynard Smith, J. *The Problems of Biology* (Oxford University Press, 1986).

Meinhardt, H. *Models of Biological Pattern Formation* (Academic Press, London, 1982).

Michod, R. and Levins, B. (eds). *The Evolution of Sex* (Sinauer, Massachusetts, 1988).

Milkman, R. (ed.). *Perspectives on Evolution* (Sinauer, Massachusetts, 1982).

Murray, J. *Mathematical Biology* (Springer-Verlag, Berlin, 1989).

Nicolis, G. and Prigogine, I. *Exploring Complexity* (W. H. Freeman, New York, 1989).

————*Self-Organisation in Non-Equilibrium Systems* (Wiley-Interscience, New York, 1977).

Noble, D. *The Initiation of the Heart Beat* (Oxford University Press, 1979).

Peacocke, A. *God and the New Biology* (Dent, London, 1986).

————*An Introduction to the Physical Chemistry of Biological Organisation* (Oxford University Press, 1983).

Prigogine, I. and Sanglier, M. *The Laws of Nature and Human Conduct* (Task Force of Research Information and the Study of Science, Brussels, 1985).

Stewart, I. *Does God Play Dice? The Mathematics of Chaos* (Blackwell, Oxford, 1989).

Thom, R. *Structural Stability and Morphogenesis*, trans. D. H. Fowler (Benjamin, Massachusetts, 1975).

Thomson, D. *On Growth and Form* (Cambridge University Press, 1942).

Waddington, C. *Towards a Theoretical Biology*, 4 vols. (Edinburgh University Press, 1968–72).

Weber, B., Depew, D. and Smith, J. *Entropy, Information and Evolution* (MIT Press, Massachusetts, 1988).

Winfree, A. *The Geometry of Biological Time* (Springer-Verlag, Heidelberg, 1980).

————*The Timing of Biological Clocks* (Scientific American Library, W. H. Freeman, New York, 1987).

————*When Time Breaks Down* (Princeton University Press, 1987).

Zeeman, E. *Dynamics of Darwinian Evolution* (preprint, 1988).

————'Population Dynamics from Game Theory', in *Global Theory of Dynamical Systems*, Proceedings, Northwestern 1979, Z. Nitecki and R. C. Robinson (eds) Springer Lecture Notes in Mathematics <u>819</u>, 471 (1980).

CHAPTER EIGHT

Arnold, V. and Avez, A. *Ergodic Problems of Classical Mechanics* (Benjamin, New York, 1968).

Balescu, R. *Equilibrium and Non-Equilibrium Statistical Mechanics* (Wiley-Interscience, New York, 1975).

Barrow, J. D. *The World Within the World* (Oxford University Press, 1988).

Bell, J. *Speakable and Unspeakable in Quantum Mechanics* (Cambridge University Press, 1987).

Coveney, P. 'The Second Law of Thermodynamics: entropy, irreversibility and dynamics', *Nature* <u>333</u>, 409 (1988).

————'Chaos, entropy and the arrow of time', *New Scientist* (<u>1736</u>, 49, 1990).

————'L'irréversibilité du temps', *La Recherche* 20, 190 (1989).

Davies, P. C. W. *The Cosmic Blueprint* (Unwin Hyman, London, 1989).

d'Espagnat, B. *The Conceptual Foundations of Quantum Mechanics* (Benjamin, Massachusetts, 1976).

————*Une incertaine réalité* (Gauthier-Villars, Paris, 1985), English title *Reality and the Physicist* (Cambridge University Press, 1989).

Ford, J. 'What is chaos, that we should be mindful of it?', in P. C. W. Davies (ed.). *The New Physics* (Cambridge University Press, 1989).

Lebowitz, J. and Penrose, O. 'Modern Ergodic Theory', *Physics Today* 23, 2 (1973).

Lighthill, J. 'The recently recognised failure of predictability in Newtonian dynamics', *Proceedings of the Royal Society of London A* 407, 35 (1986).

Mehra, J. (ed.). *The Physicist's Conception of Nature* (Reidel, Dordrecht, 1973).

Penrose, O. 'Foundations of statistical mechanics', *Reports on Progress in Physics* 42, 1937 (1979).

Penrose, O. and Percival, I. C. 'The Direction of Time', *Proceedings of the Physical Society* 79, 605 (1962).

Penrose, R. *The Emperor's New Mind* (Oxford University Press, 1989).

Polkinghorne, J. *The Way the World Is* (Triangle, London, 1983).

Popper, K. *The Open Universe: An Argument for Indeterminism* (Hutchinson, London, 1982).

————*Quantum Theory and the Schism in Physics* (Hutchinson, London, 1982).

————*Realism and the Aim of Science* (Hutchinson, London, 1983).

Prigogine, I. *From Being to Becoming* (W. H. Freeman, San Francisco, 1980).

Prigogine, I. and Stengers, I. *Order out of Chaos* (Heinemann, London, 1984).

Rae, A. *Quantum Physics: Illusion or Reality?* (Cambridge University Press, 1986).

Reichenbach, H. *The Direction of Time* (University of California Press, Berkeley, 1956).

Schuster, H. *Deterministic Chaos* (Physik-Verlag, Weinheim, 1984).

Whitehead, A. *Process and Reality: An Essay in Cosmology*, D. R. Griffin and D. W. Sherburne (eds) (Free Press, New York, Corrected Edition, 1979).

————*Science and the Modern World* (Free Press, New York, 1967).

CHAPTER NINE

Anderson, P., Arrow, K. and Pines, D. (eds). *The Economy as an Evolving Complex System* (Santa Fé Institute Studies in the Sciences of Complexity Vol. 5, Addison-Wesley, Redwood City, 1989).

Hell, V. *L'idée de culture* (Presses Universitaires de France, Paris, 1981).

Langton, C. *Artificial Life. Proceedings of an Interdisciplinary Workshop on the Synthesis and Simulation of Living Systems* (Santa Fé Institute Studies in the Sciences of Complexity Vol. 6, Addison-Wesley, Redwood City, 1989).

Nicolis, G. and Prigogine, I. *Exploring Complexity* (W. H. Freeman, New York, 1989).

Prigogine, I. *From Being to Becoming* (W. H. Freeman, San Francisco, 1980).

Prigogine, I. and Sanglier, M. (eds). *The Laws of Nature and Human Conduct* (Brussels Task Force of Research Information and Study on Science, Brussels, 1985).

BIBLIOGRAPHY

APPENDIX

Binkley, S. *The Pineal: Endocrine and Nonendocrine Function* (Prentice-Hall, New Jersey, 1988).

Campbell, J. *Winston Churchill's Afternoon Nap: a Wide Awake Inquiry into the Human Nature of Time* (Aurum, London, 1988).

Cloudsley Thompson, J. *Biological Clocks. Their Functions in Nature* (Weidenfeld & Nicolson, London, 1980).

Collin, J-P., Arendt, J. and Gem, W. A. 'Le troisième œil', *La Recherche* <u>19</u>, 1154, (1988).

Evered, D. and Clark, S. (eds). *Photoperiodism, Melatonin and the Pineal* (Ciba Foundation Symposium, 117) (Pitman Press, Bath, 1985).

Horne, J. *Why We Sleep: The Functions of Sleep in Humans and Other Mammals* (Oxford University Press, 1988).

Pribram, K. *The Language of the Brain* (Brandon House, New York, 1983).

Saunders, D. *An Introduction to Biological Rhythms* (Blackie, Glasgow, 1977).

Winfree, A. *The Geometry of Biological Time* (Springer-Verlag, Heidelberg, 1980).

——*The Timing of Biological Clocks* (W. H. Freeman, San Francisco, 1987).

359

Glossary of terms

Absolute time Newton's view of time, in which time flows at the same rate throughout the universe and people at different locations experience the same 'now'.

Acceleration The rate at which the speed of an object changes.

Æther A hypothetical medium formerly believed to fill all space and support the passage of light and other electromagnetic waves.

Amino acids The molecular building blocks of proteins (qv).

Anthropic principle The set of ideas which maintains that the fact that we are present in the universe puts constraints on its properties. More extreme versions of the principle border on the claim that the universe was designed for our benefit.

Atom A unit of matter consisting of a positively charged nucleus orbited by negatively charged electrons. The nucleus is made of protons and neutrons.

Atomism A philosophy of science based on belief in the existence of atoms.

Attractor A way to describe the long-term behaviour of a dissipative system in phase space (qv). Equilibrium and steady-states correspond to fixed-point attractors, periodic states to limit-cycle attractors (qv) and chaotic states to strange attractors (qv).

Autocatalysis The ability of a chemical species to catalyse its own production.

Belousov–Zhabotinsky reaction A chemical clock reaction named after two Russian scientists which displays a remarkable wealth of self-organising features.

Bifurcation A point at which there are two distinct choices open to a system; similar to a fork at which a path divides into two. Beyond this critical point the properties of a system can change abruptly. For instance, in the case of a chemical clock reaction (qv), the bifurcation can mark the concentration of a reactant beyond which the chemical clock goes into action.

Big Bang A widely espoused cosmological theory according to which some 15,000 million years ago all the matter and energy in the universe was born in a cataclysmic explosion. Since then the universe has expanded and cooled to its present state.

Big Crunch Similar to the Big Bang (qv), but marking the end of the universe (presupposing that it contains sufficient matter).

Black hole An object that is so dense that nothing, not even light, can escape from it except by quantum mechanical means.

Brownian motion Erratic motions of minute particles, such as smoke particles in air, due to irregular bombardment by surrounding molecules.

Brusselator A simplified theoretical model of a chemical reaction showing self-organising features – like regular colour changes – consistent with the Second Law of Thermodynamics (qv).

Catalyst A substance able to accelerate a chemical reaction yet left chemically unchanged in the process.

Causality The doctrine that everything has a cause which is antecedent in time.

Cell The ultimate component of all living organisms, the smallest that can function independently.

Chaos Term used to describe unpredictable and apparently random behaviour in dynamical systems (qv).

Chemical clock A chemical reaction which oscillates regularly, for instance changing colour periodically from red to blue and back again.

Circadian rhythm A biological rhythm that oscillates on a roughly daily basis.

Classical mechanics The mechanics (qv) formulated by Isaac Newton.

Closed system One that exchanges energy but not matter with its surroundings.

Closed universe A universe with sufficient matter to recollapse to a Big Crunch (qv).

Coarse-graining An *ad hoc* averaging procedure designed to explain irreversibility on the basis of reversible mechanics.

Conservation of energy An alternative statement of the First Law of Thermodynamics which says that energy can neither be created nor destroyed during any process.

Copenhagen interpretation The orthodox interpretation of quantum mechanics due to Bohr, Heisenberg and their followers in which no reality can be ascribed to the microscopic world.

Cosmology The study of the origin and nature of the universe.

CPT theorem A theorem maintaining that the laws of physics should be left unchanged if a specific series of operations is carried out on a process: that its coordinates are reflected in a mirror, matter is swapped with antimatter and *vice versa*, and the direction of time is reversed.

Density matrix The quantum analogue of the probability distribution function (qv) in classical mechanics.

Determinism The doctrine that events are completely determined by previous causes rather than being affected by free will or random factors.

Differential equations Equations involving the instantaneous rate of change of certain quantities with respect to others. Newton's equations of motion, for example, are differential equations linking the force experienced by a body to the instantaneous rate of change of its velocity with time.

Dissipative structure An organised state of matter arising beyond the first bifurcation point (qv) when a system is maintained far from thermodynamic equilibrium (qv).

DNA (deoxyribonucleic acid) A very large nucleic acid molecule carrying the genetic blueprint for the design and assembly of proteins, the basic building blocks of life.

Dynamical systems General term for systems whose properties change with time. Dynamical systems can be divided into two kinds, conservative and dissipative. In the former the time evolution is reversible, in the latter it is irreversible.

Dynamics The science of matter in motion.

Electromagnetic radiation General term for visible light, ultraviolet light, infrared radiation, X-rays, gamma rays and radio waves. All are disturbances of the electromagnetic field which propagate at the same speed (300 million metres per second) and differ only in wavelength (qv) and frequency (qv).

Electron An elementary particle of negative electric charge which orbits the nucleus of an atom and carries electrical current.

Entropy In thermodynamics, a measure of the capacity of an isolated macroscopic system for change.

Enzyme A biological catalyst (qv), usually comprised of a large protein molecule which accelerates essential biochemical reactions in living cells.

Equilibrium In thermodynamics, the final state of time evolution at which all capacity for change is spent. Equilibrium thermodynamics is concerned exclusively with the properties of such static states.

Equivalence principle The equivalence between acceleration and the force of gravity.

Ergodicity In a dynamical system (qv), the property of passing through all the available points in phase space (qv) compatible with the system's energy.

Event horizon The boundary of a black hole (qv).

Evolution General term for the unfolding of behaviour with the passage of time. In biology, the Darwinian theory according to which higher forms of life have arisen out of lower forms with the passage of time.

First Law of Thermodynamics *See* 'Conservation of energy'.

Fractal geometry The geometry used to describe irregular patterns. Fractals display the characteristic feature of self-similarity – an unending series of motifs within motifs repeated at all length scales.

Frame of reference Technical term for the point of view of an observer in relativity, according to which all such frames are of equal status.

Frequency The number of cycles of a wave passing a point in one second.

Gene A unit of heredity comprised of DNA (qv), responsible for passing on specific characteristics from parents to offspring.

Genome The entire genetic material of an organism.

Geodesic The shortest path between two neighbouring points on a surface or manifold (qv).

Gravitation The universal force of attraction between all forms of matter.

Heat The sensation of hotness caused by the random motions of molecules. The hotter an object, the more frantic the motion of its constituent atoms and molecules.

Heat Death The presumed implication of the Second Law of Thermodynamics that the universe will eventually reach thermodynamic equilibrium, a thin uniform cosmic gruel containing no hot spots and no capacity for further change. In fact, if the effects of gravity are included, after long times the universe must go further and further away from such a uniform state.

Initial conditions The quantities (such as position and velocity) which must be specified at an initial moment in time in order to predict subsequent behaviour.

Interference pattern The pattern produced when two waves overlap. The relationship between the overlapping waves dictates the pattern: a crest will add to another to make a larger crest, while a crest and a trough will cancel.

Irreversibility The one-way time evolution of a system, giving rise to an arrow of time.

Irreversibility paradox The paradox arising from the fact that macroscopic systems (qv) are irreversible, whereas microscopic ones (qv) are reversible.

Isolated system One that can exchange neither energy nor matter with its surroundings.

Kaons Sub-atomic particles. The long-lived kaon undergoes a decay process which has no time-reversed equivalent.

GLOSSARY

Laser Acronym for Light Amplification by Stimulated Emission of Radiation. Laser light is highly coherent and has extremely large photon (qv) densities.

Light year The distance that light travels in one year, moving at a rate of some 300 million metres per second.

Limit-cycle An attractor (qv) describing regular (periodic or quasi-periodic) temporal behaviour, for instance in a chemical clock undergoing periodic colour changes.

Lorentz transformation In special relativity, the connection between descriptions recorded in different frames of reference (qv) which ensures that the laws of physics remain unchanged (invariant).

Macroscopic systems Systems of everyday dimensions such as billiard balls, tables, people, and so on, for which thermodynamics and Newtonian mechanics are usually applicable.

Manifold In mathematics, the general term for a space described by the discipline of differential topology, of which spacetime is an example.

Mass Either the property of matter through which the force of gravity acts ('gravitational mass') or the resistance of a body to acceleration ('inertial mass'). The two are identical by Einstein's equivalence principle.

Maxwell's equations The four equations which together completely describe all the phenomena of electromagnetism.

Measurement problem In quantum theory, the problem of accounting for the outcome of experimental measurements on microscopic systems (qv).

Mechanics The branch of physics dealing with all aspects of the dynamic and static properties of bodies.

Microscopic systems Systems of atomic and molecular dimensions ruled by quantum mechanics.

Minimum entropy production theorem A theorem according to which, for thermodynamic systems close to equilibrium, the steady-state (qv) corresponds to that in which the system's entropy production is at its minimum.

Molecule The smallest unit of a chemical compound which still possesses the properties of the original substance.

Morphogenesis The evolution of form in animals and plants.

Muon Also known as a mu meson. Particles that are heavy relatives of electrons.

Mutation A change in the genetic material of an organism.

Neutrino An uncharged elementary particle generally believed to have no mass. Its anti-particle is an anti-neutrino.

Non-equilibrium The state of a macroscopic system which has not attained thermodynamic equilibrium and thus still has the capacity to change in time.

Non-linear systems Behaviour typical of many real systems, meaning in a qualitative sense 'getting more than you bargained for', unlike linear systems, which produce no surprises. Dissipative non-linear dynamical systems (qv) are capable of exhibiting self-organisation (qv) and chaos (qv).

Open system One that can exchange energy and matter with its surroundings.

Open universe A universe with insufficient matter to collapse to a Big Crunch (qv).

Partition function In statistical mechanics (qv), a theoretical quantity which can be used to calculate the equilibrium thermodynamic properties of a system from a knowledge of the structure of the atoms and molecules which comprise it.

Period-doubling A well-studied pathway to chaos in dissipative dynamical systems.

Phase space An abstract space in which a single point completely defines the instantaneous state of a dynamical system (qv). The dimension of the space depends on the number of variables needed to describe the system. As the system evolves in time it maps out a trajectory in the phase space.

Photon A quantum 'particle' carrying the energy in electromagnetic radiation.

Pion Also known as a pi meson. An unstable particle, responsible for binding protons and neutrons within the nucleus.

Poincaré's recurrence The phenomenon of eternal repetition of behaviour in an isolated mechanical system. For macroscopic systems (qv), the recurrence time (the time taken before repetition begins) is astronomical.

Positivism The doctrine that we can have no knowledge other than that provided directly by our senses; according to this philosophy, it is pointless to talk about atoms and molecules.

Positron The positively charged anti-particle of an electron.

Precession The act of encircling a fixed axis. For example, a gyroscope precesses about a vertical axis.

Probability distribution function A mathematical function used in classical mechanics to work out how probable it is that a system occupies a given state in phase space (qv).

Proteins A class of large molecules found in living organisms, consisting of strings of amino acids folded into complex but well defined three-dimensional structures.

Pulsar A celestial object that emits very regular pulses of radiation. It is thought to be a rapidly rotating neutron star.

Quantum (pl. quanta) A discrete lump of energy in quantum theory.

Quantum mechanics The mechanics (qv) that rules the microscopic world, where energy changes occur in abrupt quantum jumps.

Quasar Quasi-stellar radio sources – heavenly bodies first detected by their radio emissions. Believed to be the most distant objects in the universe.

Red shift Displacement of spectral lines towards the red end of the spectrum. It can be caused by objects receding at high speeds, or by gravity, when it is known as the gravitational red shift.

Reductionism A doctrine according to which complex phenomena can be explained in terms of something simpler. In particular, atomistic reductionism contends that macroscopic phenomena can be explained in terms of the properties of atoms and molecules.

Relativity Einstein's theories of relativity, an extension of Newtonian physics, deal with the concepts of space, time and matter. Special relativity starts from the premise that the laws of physics are the same for observers moving at constant speeds relative to one another. General relativity is based on the idea that the laws of physics should be the same for all observers, regardless of how they are moving relative to one another. In the general theory, gravity is explained as the curvature of spacetime (qv).

Renormalisation A mathematically dubious method used to remove absurd infinities from quantum field theory.

RNA (ribonucleic acid) The genetic material used to translate DNA into proteins (qv). In some organisms, it can also be the principal genetic material.

Schrödinger's equation An equation for the wavefunction (qv) which describes the propagation of matter-waves in time.

364

Second Law of Thermodynamics The law stating that, during an irreversible process, entropy (qv) always increases. The future state of any isolated system (qv) has higher entropy than its present or past states.

Self-organisation The emergence of structural organisation. It occurs within dissipative non-linear dynamical systems (qv).

Singularity A region where the mathematics underpinning a theory becomes ill defined.

Spacetime Four-dimensional synthesis of space and time occurring in relativity (qv).

Standing wave A stationary wave produced by the overlap of two waves travelling in opposite directions.

Statistical mechanics The discipline which attempts to express the properties of macroscopic systems (qv) in terms of their atomic and molecular constituents.

Steady-state A non-equilibrium state which does not change with time.

Strange attractor An attractor (qv) which has a fractal (fractional) dimension; describes chaotic dynamics in dissipative dynamical systems (qv).

Teleology The study (or doctrine) of final causes, particularly in relation to design or purpose in nature.

Thermodynamics The science of heat and work.

Time dilation The effect of time slowing down, caused by increasing speed or the pull of gravity. In the latter case it is known as gravitational time dilation.

Time-symmetry The property of Newtonian, Einsteinian and quantum mechanics that both directions of time are equally permissible.

Uncertainty principle A quantum mechanical principle stating that there is a fundamental limitation on the simultaneous measurement of pairs of quantities such as position and momentum.

Velocity The rate at which the position of an object changes.

Wavefunction The central quantity in quantum theory which is used to calculate the probability of an event occurring (for instance, an atom emitting a photon) when a measurement is made.

Wavelength The distance between two adjacent troughs or two adjacent crests of a wave.

White hole The time reverse of a black hole (qv).

Wormholes Passages from one region of spacetime to another.

Zero point energy The energy possessed by atoms, molecules or fields at absolute zero, the lowest achievable temperature.

Index

Note: entries in *italic* indicate terms defined in the glossary

dissipation 150, 155
dissipative structures 168, 187, 195, 259, 293
DNA (deoxyribonucleic acid) 36, 184, 216–18, 252, 307
Dollo, Louis 253
Dollo's law 253
Donder, Théophile de 161
Dorée, Marcel 227
Duhem, Pierre 22
Dukas, Helen 75
Dumas, Jean Baptiste 109
Duysens, L. N. 224
dynamical chaos 271
dynamical systems 222, 247, 273–4, 281, 291
Dyson, Freeman 142

earth: age of 250
economics 293–4
Eddington, Arthur 24–5, 33, 70, 83, 92–4, 151, 154
Egyptians, ancient 42
Ehrenfest, Paul 22
Eigen, Manfred 218, 255
Einstein, Albert: relativity theory 23, 30–2, 66, 68, 71, 73–4, 84–6, 96, 296; on death 30, 64; and Boltzmann's principles 35, 173; on Galileo 49; and aether 62, 73, 78, 142; and absolute rest 63; and molecular theory 67; demolishes notion of absolute time 68, 70–1, 76–82; appearance 71; background and career 74–5, 85; and position of observer 84; and problem of gravitation 84–6, 89–92; and geometry 88–9; and age of universe 98; and cosmological constant 99; and invariance of causality 105–6, 121, 127–8, 135–6, 261; on Brownian motion 110, 113, 176; contribution and attitude to quantum theory 113, 121, 126–7, 135; praises de Broglie 115; dispute with Bohr 126–7; on Schrödinger's cat

paradox 131; and structure of universe 291; 'Cosmological Considerations on the General Theory of Relativity' 98; 'The Field Equations of Gravitation' 89; 'On the Electrodynamics of Moving Bodies' 76
Eldredge, Niles 255–6
electroencephalograph (EEG) 241–2
electromagnetic radiation 59–60, 98, 112–14
electromagnetism: Maxwell and 59–63, 66, 72; Einstein and 76; Dirac and 142; and lasers 290
electrons 111–13, 117–20, 123–4, 135, 138, 271–2
electrostatics 58–9
elements: creation of 102
Eliade, Mircea: *The Myth of the Eternal Return* 26
embryos: self-organisation 189
energy: creation 141; and thermodynamics 149–51; free 156; and biological fuel 223–6
energy–time uncertainty 126, 128, 141
Englert, François 291
entropy: Boltzmann and 21, 34, 176; link with time 33–4; and probability 34–5; and thermodynamics 147, 151–2, 156, 279–80; and 'disorder' 160, 164; minimum production 162, 164; and partition function 171; and information theory 177–8; and Big Bang 179–81; and living systems 222; and irreversibility 281; and creation 290–2
enzymes 184, 197, 220–1
EPR paradox 135–6
equilibrium (thermodynamic) 153–9, 161–8, 171, 182–3, 193, 280, 289; prevention of 159–62; and cyclic time 261
equivalence principle 85
ergodicity 269–70, 274–8, 281
Espagnat, Bernard d' 259
Euclid 86–8